Fundamentals of Number Theory

Fundamentals of Number Theory

Edited by
Emanuel Patterson

Larsen & Keller
www.larsen-keller.com

Fundamentals of Number Theory
Edited by Emanuel Patterson
ISBN: 978-1-63549-199-9 (Hardback)

Larsen & Keller

Published by Larsen and Keller Education,
5 Penn Plaza,
19th Floor,
New York, NY 10001, USA

Cataloging-in-Publication Data

Fundamentals of number theory / edited by Emanuel Patterson
 p. cm.
Includes bibliographical references and index.
ISBN 978-1-63549-199-9
1. Number theory. 2. Algebra. I. Patterson, Emanuel.
QA241 .F86 2017
512.7--dc23

The publisher's policy is to use permanent paper from mills that operate a sustainable forestry policy. Furthermore, the publisher ensures that the text paper and cover boards used have met acceptable environmental accreditation standards.

Printed and bound in the United States of America.

For more information regarding Larsen and Keller Education and its products, please visit the publisher's website www.larsen-keller.com

Table of Contents

Preface

The branch of pure mathematics that is dedicated to study of integers is called number theory or arithmetic. Number theory studies the properties of prime numbers, rational numbers, and algebraic integers. This book elucidates the concepts and innovative models around prospective developments with respect to number theory. Such selected concepts that redefine this subject have been presented in it. It will provide comprehensive knowledge to the readers. Those in search of information to further their knowledge will be greatly assisted by this textbook. Coherent flow of topics, student-friendly language and extensive use of examples make this book an invaluable source of information.

To facilitate a deeper understanding of the contents of this book a short introduction of every chapter is written below:

Chapter 1- Number theory is a branch of mathematics that concerns itself with the study of integers. They study prime numbers and also the properties of objects that are made of integers. The chapter on number theory offers an insightful focus, keeping in mind the subject matter.

Chapter 2- The two branches of number theory are analytic number theory and algebraic number theory. Analytic number theory focuses on the methods of mathematical analysis that is used to solve problems that are related to integers. This section is a compilation of the various branch of number theory that forms an integral part of the broader subject matter.

Chapter 3- A number is used to count and measure objects. Apart from counting and measuring, numbers are also used for codes and mathematical abstraction. This chapter also explains theories such as natural numbers, rational numbers, integers, prime numbers, real numbers and complex numbers. This section is an overview of the subject matter incorporating all the major aspects of numbers.

Chapter 4- Fraction represents a part of an entire object or a number. An example of a fraction would be 5/25 and ¾. Unit fraction, dyadic rational, repeating decimal, cyclic number and Egyptian fraction are the aspects elucidated in the following chapter.

Chapter 5- Some of the arithmetic operations explained in the section are algebraic operation, addition, subtraction, method of complements, multiplication and division. These operations are performed on numbers and variables. The topics elucidated in this chapter are of vital importance and provide a better understanding of arithmetic operations.

Chapter 6- Division algorithm calculates the quotient of two given integers N and D. Division algorithm falls under two major categories which are slow division and fast division. Multiplication algorithm, Euclidean algorithm, greatest common divisor, least common multiple and fundamental theorem of arithmetic are some of the aspects elucidated in the section. The chapter discusses the methods of division and multiplication algorithm in a critical manner providing key analysis to the subject matter.

I owe the completion of this book to the never-ending support of my family, who supported me throughout the project.

Editor

Introduction to Number Theory

Number theory is a branch of mathematics that concerns itself with the study of integers. They study prime numbers and also the properties of objects that are made of integers. The chapter on number theory offers an insightful focus, keeping in mind the subject matter.

Number theory or, in older usage, arithmetic is a branch of pure mathematics devoted primarily to the study of the integers. It is sometimes called "The Queen of Mathematics" because of its foundational place in the discipline. Number theorists study prime numbers as well as the properties of objects made out of integers (e.g., rational numbers) or defined as generalizations of the integers (e.g., algebraic integers).

A Lehmer sieve, which is a primitive digital computer once used for finding primes and solving simple Diophantine equations.

Integers can be considered either in themselves or as solutions to equations (Diophantine geometry). Questions in number theory are often best understood through the study of analytical objects (e.g., the Riemann zeta function) that encode properties of the integers, primes or other number-theoretic objects in some fashion (analytic number theory). One may also study real numbers in relation to rational numbers, e.g., as approximated by the latter (Diophantine approximation).

The older term for number theory is *arithmetic*. By the early twentieth century, it had been superseded by "number theory". (The word "arithmetic" is used by the general public to mean "elementary calculations"; it has also acquired other meanings in mathematical logic, as in *Peano arithmetic*, and computer science, as in *floating point arithmetic*.) The use of the term *arithmetic* for *number theory* regained some ground in the second half of the 20th century, arguably in part due to French influence. In particular, *arithmetical* is preferred as an adjective to *number-theoretic*.

History

Origins

Dawn of Arithmetic

The first historical find of an arithmetical nature is a fragment of a table: the broken clay tablet Plimpton 322 (Larsa, Mesopotamia, ca. 1800 BCE) contains a list of "Pythagorean triples", i.e., integers (a,b,c) such that $a^2 + b^2 = c^2$. The triples are too many and too large to have been obtained by brute force. The heading over the first column reads: "The *takiltum* of the diagonal which has been subtracted such that the width..."

The Plimpton 322 tablet

The table's layout suggests that it was constructed by means of what amounts, in modern language, to the identity

$$\left(\frac{1}{2}\left(x-\frac{1}{x}\right)\right)^2 + 1 = \left(\frac{1}{2}\left(x+\frac{1}{x}\right)\right)^2,$$

which is implicit in routine Old Babylonian exercises. If some other method was used, the triples were first constructed and then reordered by c/a, presumably for actual use as a "table", i.e., with a view to applications.

It is not known what these applications may have been, or whether there could have been any; Babylonian astronomy, for example, truly flowered only later. It has been suggested instead that the table was a source of numerical examples for school problems.

While Babylonian number theory—or what survives of Babylonian mathematics that can be called thus—consists of this single, striking fragment, Babylonian algebra (in the secondary-school sense of "algebra") was exceptionally well developed. Late Neoplatonic sources state that Pythagoras learned mathematics from the Babylonians. Much earlier sources state that Thales and Pythagoras traveled and studied in Egypt.

Euclid IX 21—34 is very probably Pythagorean; it is very simple material ("odd times even is even", "if an odd number measures [= divides] an even number, then it also measures [= divides] half of it"), but it is all that is needed to prove that $\sqrt{2}$ is irrational. Pythagorean mystics gave great importance to the odd and the even. The discovery that $\sqrt{2}$ is irrational is credited to the early Pythagoreans (pre-Theodorus). By revealing (in modern terms) that numbers could be irrational, this discovery seems to have provoked the first foundational crisis in mathematical history; its proof or its divulgation are sometimes credited to Hippasus, who was expelled or split from the Pythagorean sect. This forced a distinction between *numbers* (integers and the rationals—the subjects of arithmetic), on the one hand, and *lengths* and *proportions* (which we would identify with real numbers, whether rational or not), on the other hand.

The Pythagorean tradition spoke also of so-called polygonal or figurate numbers. While square numbers, cubic numbers, etc., are seen now as more natural than triangular numbers, pentagonal numbers, etc., the study of the sums of triangular and pentagonal numbers would prove fruitful in the early modern period (17th to early 19th century).

We know of no clearly arithmetical material in ancient Egyptian or Vedic sources, though there is some algebra in both. The Chinese remainder theorem appears as an exercise in Sun Zi's *Suan Ching*, also known as *The Mathematical Classic of Sun Zi* (3rd, 4th or 5th century CE.) (There is one important step glossed over in Sun Zi's solution: it is the problem that was later solved by Āryabhaṭa's *kuṭṭaka*)

There is also some numerical mysticism in Chinese mathematics, but, unlike that of the Pythagoreans, it seems to have led nowhere. Like the Pythagoreans' perfect numbers, magic squares have passed from superstition into recreation.

Classical Greece and The Early Hellenistic Period

Aside from a few fragments, the mathematics of Classical Greece is known to us either through the reports of contemporary non-mathematicians or through mathematical works from the early Hellenistic period. In the case of number theory, this means, by and large, *Plato* and *Euclid*, respectively.

Plato had a keen interest in mathematics, and distinguished clearly between arithmetic and calculation. (By *arithmetic* he meant, in part, theorising on number, rather than what *arithmetic* or *number theory* have come to mean.) It is through one of Plato's dialogues—namely, *Theaetetus*—that we know that Theodorus had proven that $\sqrt{3}, \sqrt{5}, \ldots, \sqrt{17}$ are irrational. Theaetetus was, like Plato, a disciple of Theodorus's; he worked on distinguishing different kinds of incommensurables, and was thus arguably a pioneer in the study of number systems. (Book X of Euclid's Elements is described by Pappus as being largely based on Theaetetus's work.)

Euclid devoted part of his *Elements* to prime numbers and divisibility, topics that belong unambiguously to number theory and are basic to it (Books VII to IX of Euclid's Elements). In particular, he gave an algorithm for computing the greatest common divisor of two numbers (the Euclidean algorithm; *Elements*, Prop. VII.2) and the first known proof of the infinitude of primes (*Elements*, Prop. IX.20).

In 1773, Lessing published an epigram he had found in a manuscript during his work as a librarian; it claimed to be a letter sent by Archimedes to Eratosthenes. The epigram proposed what has become known as Archimedes' cattle problem; its solution (absent from the manuscript) requires solving an indeterminate quadratic equation (which reduces to what would later be misnamed Pell's equation). As far as we know, such equations were first successfully treated by the Indian school. It is not known whether Archimedes himself had a method of solution.

Diophantus

Very little is known about Diophantus of Alexandria; he probably lived in the third century CE, that is, about five hundred years after Euclid. Six out of the thirteen books of Diophantus's *Arithmetica* survive in the original Greek; four more books survive in an Arabic translation. The *Arithmetica* is a collection of worked-out problems where the task is invariably to find rational solutions to a system of polynomial equations, usually of the form $f(x, y) = z^2$ or $f(x, y, z) = w^2$ $f(x_1, x_2, x_3) = 0$. Thus, nowadays, we speak of *Diophantine equations* when we speak of polynomial equations to which rational or integer solutions must be found.

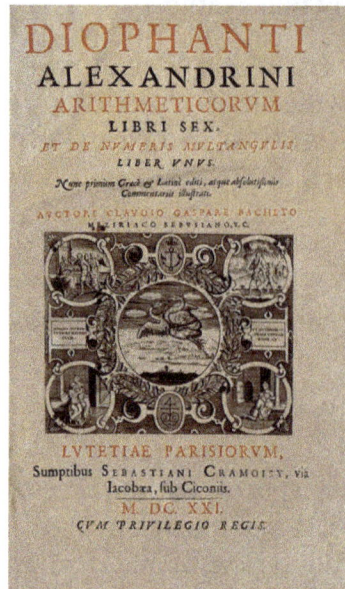

Title page of the 1621 edition of Diophantus' *Arithmetica*, translated into Latin by Claude Gaspard Bachet de Méziriac.

One may say that Diophantus was studying rational points — i.e., points whose coordinates are rational — on curves and algebraic varieties; however, unlike the Greeks of the Classical period, who did what we would now call basic algebra in geometrical terms, Diophantus did what we would now call basic algebraic geometry in purely algebraic terms. In modern language, what Diophantus did was to find rational parametrizations of varieties; that is, given an equation of the form (say) $f(x_1, x_2, x_3) = 0$, his aim was to find (in essence) three rational functions g_1, g_2, g_3 such that, for all values of r and s, , setting $x_i = g_i(r, s)$ for $i = 1, 2, 3$ gives a solution to $f(x_1, x_2, x_3) = 0$.

Diophantus also studied the equations of some non-rational curves, for which no rational parametrisation is possible. He managed to find some rational points on these curves (elliptic curves, as it happens, in what seems to be their first known occurrence) by means of what amounts to a tangent construction: translated into coordinate geometry (which did not exist in Diophantus's time), his method would be visualised as drawing a tangent to a curve at a known rational point, and then finding the other point of intersection of the tangent with the curve; that other point is a new rational point. (Diophantus also resorted to what could be called a special case of a secant construction.)

While Diophantus was concerned largely with rational solutions, he assumed some results on integer numbers, in particular that every integer is the sum of four squares (though he never stated as much explicitly).

Āryabhata, Brahmagupta, Bhāskara

While Greek astronomy probably influenced Indian learning, to the point of introducing trigonometry, it seems to be the case that Indian mathematics is otherwise an indigenous tradition; in particular, there is no evidence that Euclid's Elements reached India before the 18th century.

Āryabhata (476–550 CE) showed that pairs of simultaneous congruences $n \equiv a_1 \bmod m_1$, $n \equiv a_2 \bmod m_2$ could be solved by a method he called *kuṭṭaka*, or *pulveriser*; this is a procedure close to (a generalisation of) the Euclidean algorithm, which was probably discovered independently in India. Āryabhata seems to have had in mind applications to astronomical calculations.

Brahmagupta (628 CE) started the systematic study of indefinite quadratic equations—in particular, the misnamed Pell equation, in which Archimedes may have first been interested, and which did not start to be solved in the West until the time of Fermat and Euler. Later Sanskrit authors would follow, using Brahmagupta's technical terminology. A general procedure (the chakravala, or "cyclic method") for solving Pell's equation was finally found by Jayadeva (cited in the eleventh century; his work is otherwise lost); the earliest surviving exposition appears in Bhāskara II's Bīja-gaṇita (twelfth century).

Indian mathematics remained largely unknown in Europe until the late eighteenth century; Brahmagupta and Bhāskara's work was translated into English in 1817 by Henry Colebrooke.

Arithmetic in The Islamic Golden Age

Al-Haytham seen by the West: frontispice of *Selenographia*, showing Alhasen [*sic*] representing knowledge through reason, and Galileo representing knowledge through the senses.

In the early ninth century, the caliph Al-Ma'mun ordered translations of many Greek mathematical works and at least one Sanskrit work (the *Sindhind*, which may or may not be Brahmagupta's Brāhmasphuṭasiddhānta). Diophantus's main work, the *Arithmetica*, was translated into Arabic by Qusta ibn Luqa (820–912). Part of the treatise *al-Fakhri* (by al-Karajī, 953 – ca. 1029) builds on it to some extent. According to Rashed Roshdi, Al-Karajī's contemporary Ibn al-Haytham knew what would later be called Wilson's theorem.

Western Europe in The Middle Ages

Other than a treatise on squares in arithmetic progression by Fibonacci — who lived and studied in north Africa and Constantinople during his formative years, ca. 1175–1200 — no number theory to speak of was done in western Europe during the Middle Ages. Matters started to change in Europe in the late Renaissance, thanks to a renewed study of the works of Greek antiquity. A catalyst was the textual emendation and translation into Latin of Diophantus's *Arithmetica* (Bachet, 1621, following a first attempt by Xylander, 1575).

Early Modern Number Theory

Fermat

Pierre de Fermat (1601–1665) never published his writings; in particular, his work on number theory is contained almost entirely in letters to mathematicians and in private marginal notes. He wrote down nearly no proofs in number theory; he had no models in the area. He did make repeated use of mathematical induction, introducing the method of infinite descent.

Pierre de Fermat

One of Fermat's first interests was perfect numbers (which appear in Euclid, *Elements* IX) and amicable numbers; this led him to work on integer divisors, which were from the beginning among the subjects of the correspondence (1636 onwards) that put him in touch with the mathematical community of the day. He had already studied Bachet's edition of Diophantus carefully; by 1643, his interests had shifted largely to Diophantine problems and sums of squares (also treated by Diophantus).

Fermat's achievements in arithmetic include:

- Fermat's little theorem (1640), stating that, if a is not divisible by a prime p, then $a^{p-1} \equiv 1 \bmod p$.

- If a and b are coprime, then $a^2 + b^2$ is not divisible by any prime congruent to −1 modulo 4; and every prime congruent to 1 modulo 4 can be written in the form $a^2 + b^2$. These two statements also date from 1640; in 1659, Fermat stated to Huygens that he had proven the latter statement by the method of infinite descent. Fermat and Frenicle also did some work (some of it erroneous) on other quadratic forms.

- Fermat posed the problem of solving $x^2 - Ny^2 = 1$ as a challenge to English mathematicians (1657). The problem was solved in a few months by Wallis and Brouncker. Fermat considered their solution valid, but pointed out they had provided an algorithm without a proof (as had Jayadeva and Bhaskara, though Fermat would never know this.) He states that a proof can be found by descent.

- Fermat developed methods for (doing what in our terms amounts to) finding points on curves of genus 0 and 1. As in Diophantus, there are many special procedures and what amounts to a tangent construction, but no use of a secant construction.

- Fermat states and proves (by descent) in the appendix to *Observations on Diophantus* (Obs. XLV) that $x^4 + y^4 = z^4$ has no non-trivial solutions in the integers. Fermat also mentioned to his correspondents that $x^3 + y^3 = z^3$ has no non-trivial solutions, and that this could be proven by descent. The first known proof is due to Euler (1753; indeed by descent).

Fermat's claim ("Fermat's last theorem") to have shown there are no solutions to $x^n + y^n = z^n$ for all $n \geq 3$ (the only known proof of which is beyond his methods) appears only in his annotations on the margin of his copy of Diophantus; he never claimed this to others and thus would have had no need to retract it if he found any mistake in his supposed proof.

Euler

Leonhard Euler

The interest of Leonhard Euler (1707–1783) in number theory was first spurred in 1729, when a friend of his, the amateur Goldbach, pointed him towards some of Fermat's work on the subject.

This has been called the "rebirth" of modern number theory, after Fermat's relative lack of success in getting his contemporaries' attention for the subject. Euler's work on number theory includes the following:

- *Proofs for Fermat's statements.* This includes Fermat's little theorem (generalised by Euler to non-prime moduli); the fact that $p = x^2 + y^2$ if and only if $p \equiv 1 \bmod 4$; initial work towards a proof that every integer is the sum of four squares (the first complete proof is by Joseph-Louis Lagrange (1770), soon improved by Euler himself); the lack of non-zero integer solutions to $x^4 + y^4 = z^2$ (implying the case *n=4* of Fermat's last theorem, the case *n=3* of which Euler also proved by a related method).

- *Pell's equation*, first misnamed by Euler. He wrote on the link between continued fractions and Pell's equation.

- *First steps towards analytic number theory.* In his work of sums of four squares, partitions, pentagonal numbers, and the distribution of prime numbers, Euler pioneered the use of what can be seen as analysis (in particular, infinite series) in number theory. Since he lived before the development of complex analysis, most of his work is restricted to the formal manipulation of power series. He did, however, do some very notable (though not fully rigorous) early work on what would later be called the Riemann zeta function.

- *Quadratic forms.* Following Fermat's lead, Euler did further research on the question of which primes can be expressed in the form $x^2 + Ny^2$, some of it prefiguring quadratic reciprocity.

- *Diophantine equations.* Euler worked on some Diophantine equations of genus 0 and 1. In particular, he studied Diophantus's work; he tried to systematise it, but the time was not yet ripe for such an endeavour – algebraic geometry was still in its infancy. He did notice there was a connection between Diophantine problems and elliptic integrals, whose study he had himself initiated.

Lagrange, Legendre and Gauss

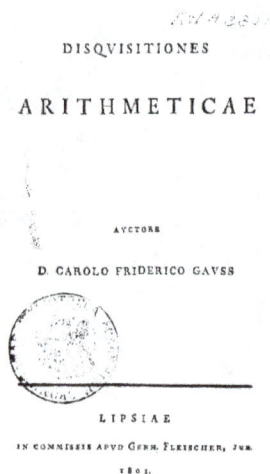

DISQVISITIONES

ARITHMETICAE

AVCTORE

D. CAROLO FRIDERICO GAVSS

LIPSIAE

IN COMMISSIS APVD GERR. FLEISCHER, JVN.

1801.

Carl Friedrich Gauss's Disquisitiones Arithmeticae, first edition

Joseph-Louis Lagrange (1736–1813) was the first to give full proofs of some of Fermat's and Euler's work and observations – for instance, the four-square theorem and the basic theory of the misnamed "Pell's equation" (for which an algorithmic solution was found by Fermat and his contemporaries, and also by Jayadeva and Bhaskara II before them.) He also studied quadratic forms in full generality (as opposed to $mX^2 + nY^2$) — defining their equivalence relation, showing how to put them in reduced form, etc.

Adrien-Marie Legendre (1752–1833) was the first to state the law of quadratic reciprocity. He also conjectured what amounts to the prime number theorem and Dirichlet's theorem on arithmetic progressions. He gave a full treatment of the equation $ax^2 + by^2 + cz^2 = 0$ and worked on quadratic forms along the lines later developed fully by Gauss. In his old age, he was the first to prove "Fermat's last theorem" for $n = 5$ (completing work by Peter Gustav Lejeune Dirichlet, and crediting both him and Sophie Germain).

Carl Friedrich Gauss

In his *Disquisitiones Arithmeticae* (1798), Carl Friedrich Gauss (1777–1855) proved the law of quadratic reciprocity and developed the theory of quadratic forms (in particular, defining their composition). He also introduced some basic notation (congruences) and devoted a section to computational matters, including primality tests. The last section of the *Disquisitiones* established a link between roots of unity and number theory:

The theory of the division of the circle…which is treated in sec. 7 does not belong by itself to arithmetic, but its principles can only be drawn from higher arithmetic.

In this way, Gauss arguably made a first foray towards both Évariste Galois's work and algebraic number theory.

Maturity and Division Into Subfields

Starting early in the nineteenth century, the following developments gradually took place:

- The rise to self-consciousness of number theory (or *higher arithmetic*) as a field of study.

- The development of much of modern mathematics necessary for basic modern number theory: complex analysis, group theory, Galois theory—accompanied by greater rigor in analysis and abstraction in algebra.

- The rough subdivision of number theory into its modern subfields—in particular, analytic and algebraic number theory.

Peter Gustav Lejeune Dirichlet   Ernst Kummer

Algebraic number theory may be said to start with the study of reciprocity and cyclotomy, but truly came into its own with the development of abstract algebra and early ideal theory and valuation theory. A conventional starting point for analytic number theory is Dirichlet's theorem on arithmetic progressions (1837), whose proof introduced L-functions and involved some asymptotic analysis and a limiting process on a real variable. The first use of analytic ideas in number theory actually goes back to Euler (1730s), who used formal power series and non-rigorous (or implicit) limiting arguments. The use of *complex* analysis in number theory comes later: the work of Bernhard Riemann (1859) on the zeta function is the canonical starting point; Jacobi's four-square theorem (1839), which predates it, belongs to an initially different strand that has by now taken a leading role in analytic number theory (modular forms).

Main Subdivisions

Elementary Tools

The term *elementary* generally denotes a method that does not use complex analysis. For example, the prime number theorem was first proven using complex analysis in 1896, but an elementary proof was found only in 1949 by Erdős and Selberg. The term is somewhat ambiguous: for example, proofs based on complex Tauberian theorems (e.g. Wiener–Ikehara) are often seen as quite enlightening but not elementary, in spite of using Fourier analysis, rather than complex analysis as such. Here as elsewhere, an *elementary* proof may be longer and more difficult for most readers than a non-elementary one.

Number theory has the reputation of being a field many of whose results can be stated to the layperson. At the same time, the proofs of these results are not particularly accessible, in part because the range of tools they use is, if anything, unusually broad within mathematics.

Analytic Number Theory

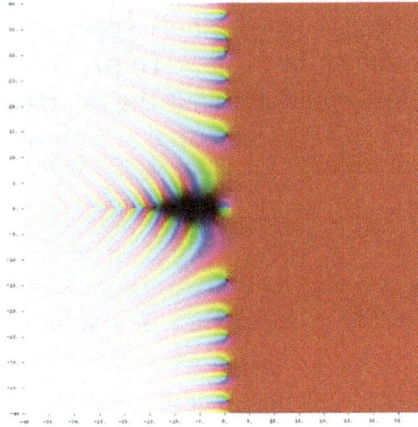

Riemann zeta function $\zeta(s)$ in the complex plane. The color of a point s gives the value of $\zeta(s)$: dark colors denote values close to zero and hue gives the value's argument.

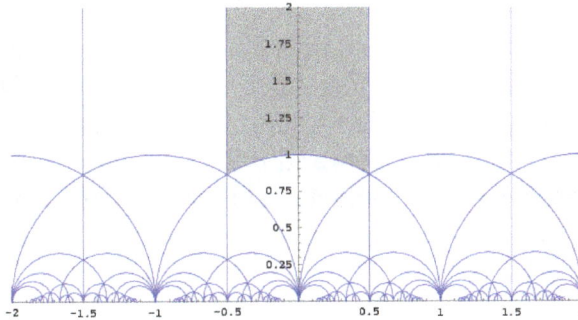

The action of the modular group on the upper half plane. The region in grey is the standard fundamental domain.

Analytic number theory may be defined

- in terms of its tools, as the study of the integers by means of tools from real and complex analysis; or

- in terms of its concerns, as the study within number theory of estimates on size and density, as opposed to identities.

Some subjects generally considered to be part of analytic number theory, e.g., sieve theory, are better covered by the second rather than the first definition: some of sieve theory, for instance, uses little analysis, yet it does belong to analytic number theory.

The following are examples of problems in analytic number theory: the prime number theorem, the Goldbach conjecture (or the twin prime conjecture, or the Hardy–Littlewood conjectures), the Waring problem and the Riemann hypothesis. Some of the most important tools of analytic number theory are the circle method, sieve methods and L-functions (or, rather, the study of their properties). The theory of modular forms (and, more generally, automorphic forms) also occupies an increasingly central place in the toolbox of analytic number theory.

One may ask analytic questions about algebraic numbers, and use analytic means to answer such questions; it is thus that algebraic and analytic number theory intersect. For example, one may define prime ideals (generalizations of prime numbers in the field of algebraic numbers) and ask how many prime ideals there are up to a certain size. This question can be answered by means of an examination of Dedekind zeta functions, which are generalizations of the Riemann zeta function, a key analytic object at the roots of the subject. This is an example of a general procedure in analytic number theory: deriving information about the distribution of a sequence (here, prime ideals or prime numbers) from the analytic behavior of an appropriately constructed complex-valued function.

Algebraic Number Theory

An *algebraic number* is any complex number that is a solution to some polynomial equation $f(x) = 0$ with rational coefficients; for example, every solution x of $x^5 + (11/2)x^3 - 7x^2 + 9 = 0$ (say) is an algebraic number. Fields of algebraic numbers are also called *algebraic number fields*, or shortly *number fields*. Algebraic number theory studies algebraic number fields. Thus, analytic and algebraic number theory can and do overlap: the former is defined by its methods, the latter by its objects of study.

It could be argued that the simplest kind of number fields (viz., quadratic fields) were already studied by Gauss, as the discussion of quadratic forms in *Disquisitiones arithmeticae* can be restated in terms of ideals and norms in quadratic fields. (A *quadratic field* consists of all numbers of the form $a + b\sqrt{d}$, where a and b are rational numbers and d is a fixed rational number whose square root is not rational.) For that matter, the 11th-century chakravala method amounts—in modern terms—to an algorithm for finding the units of a real quadratic number field. However, neither Bhāskara nor Gauss knew of number fields as such.

The grounds of the subject as we know it were set in the late nineteenth century, when *ideal numbers*, the *theory of ideals* and *valuation theory* were developed; these are three complementary ways of dealing with the lack of unique factorisation in algebraic number fields. (For example, in the field generated by the rationals and $\sqrt{-5}$, the number 6 can be factorised both as $6 = 2 \cdot 3$ and $6 = (1 + \sqrt{-5})(1 - \sqrt{-5})$; all of , 3, $1 + \sqrt{-5}$ and $1 - \sqrt{-5}$ are irreducible, and thus, in a naïve sense, analogous to primes among the integers.) The initial impetus for the development of ideal numbers (by Kummer) seems to have come from the study of higher reciprocity laws,i.e., generalisations of quadratic reciprocity.

Number fields are often studied as extensions of smaller number fields: a field L is said to be an *extension* of a field K if L contains K. (For example, the complex numbers C are an extension of the reals R, and the reals R are an extension of the rationals Q.) Classifying the possible extensions of a given number field is a difficult and partially open problem. Abelian extensions—that is, extensions L of K such that the Galois group Gal(L/K) of L over K is an abelian group—are relatively well understood. Their classification was the object of the programme of class field theory, which was initiated in the late 19th century (partly by Kronecker and Eisenstein) and carried out largely in 1900—1950.

An example of an active area of research in algebraic number theory is Iwasawa theory. The Langlands program, one of the main current large-scale research plans in mathematics, is sometimes described as an attempt to generalise class field theory to non-abelian extensions of number fields.

Diophantine Geometry

The central problem of *Diophantine geometry* is to determine when a Diophantine equation has solutions, and if it does, how many. The approach taken is to think of the solutions of an equation as a geometric object.

For example, an equation in two variables defines a curve in the plane. More generally, an equation, or system of equations, in two or more variables defines a curve, a surface or some other such object in n-dimensional space. In Diophantine geometry, one asks whether there are any *rational points* (points all of whose coordinates are rationals) or *integral points* (points all of whose coordinates are integers) on the curve or surface. If there are any such points, the next step is to ask how many there are and how they are distributed. A basic question in this direction is: are there finitely or infinitely many rational points on a given curve (or surface)? What about integer points?

An example here may be helpful. Consider the Pythagorean equation $x^2 + y^2 = 1$; we would like to study its rational solutions, i.e., its solutions (x, y) such that x and y are both rational. This is the same as asking for all integer solutions to $a^2 + b^2 = c^2$; any solution to the latter equation gives us a solution $x = a/c$, $y = b/c$ to the former. It is also the same as asking for all points with rational coordinates on the curve described by $x^2 + y^2 = 1$. (This curve happens to be a circle of radius 1 around the origin.)

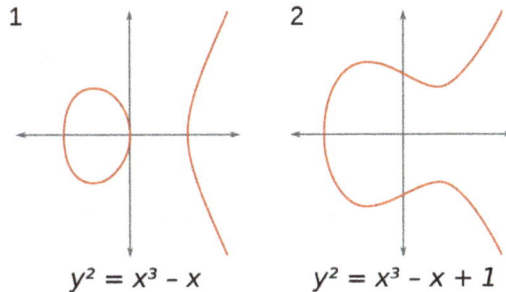

$$y^2 = x^3 - x \qquad\qquad y^2 = x^3 - x + 1$$

Two examples of an elliptic curve, i.e., a curve of genus 1 having at least one rational point. (Either graph can be seen as a slice of a torus in four-dimensional space.)

The rephrasing of questions on equations in terms of points on curves turns out to be felicitous. The finiteness or not of the number of rational or integer points on an algebraic curve—that is, rational or integer solutions to an equation $f(x, y) = 0$, where f is a polynomial in two variables—turns out to depend crucially on the *genus* of the curve. The *genus* can be defined as follows: allow the variables in $f(x, y) = 0$ to be complex numbers; then $f(x, y) = 0$ defines a 2-dimensional surface in (projective) 4-dimensional space (since two complex variables can be decomposed into four real variables, i.e., four dimensions). Count the number of (doughnut) holes in the surface; call this number the *genus* of $f(x, y) = 0$. Other geometrical notions turn out to be just as crucial.

There is also the closely linked area of Diophantine approximations: given a number x, how well can it be approximated by rationals? (We are looking for approximations that are good relative to the amount of space that it takes to write the rational: call $a >$ (with $\gcd(a, q) = 1$) a good ap-

proximation to x if $|x - a/q| < \dfrac{1}{q^c}$,, where c is large.) This question is of special interest if x is an algebraic number. If x cannot be well approximated, then some equations do not have integer or rational solutions. Moreover, several concepts (especially that of height) turn out to be crucial both in Diophantine geometry and in the study of Diophantine approximations. This question is also of special interest in transcendental number theory: if a number can be better approximated than any algebraic number, then it is a transcendental number. It is by this argument that π and e have been shown to be transcendental.

Diophantine geometry should not be confused with the geometry of numbers, which is a collection of graphical methods for answering certain questions in algebraic number theory. *Arithmetic geometry*, on the other hand, is a contemporary term for much the same domain as that covered by the term *Diophantine geometry*. The term *arithmetic geometry* is arguably used most often when one wishes to emphasise the connections to modern algebraic geometry (as in, for instance, Faltings' theorem) rather than to techniques in Diophantine approximations.

Recent Approaches and Subfields

The areas below date as such from no earlier than the mid-twentieth century, even if they are based on older material. For example, as is explained below, the matter of algorithms in number theory is very old, in some sense older than the concept of proof; at the same time, the modern study of computability dates only from the 1930s and 1940s, and computational complexity theory from the 1970s.

Probabilistic Number Theory

Take a number at random between one and a million. How likely is it to be prime? This is just another way of asking how many primes there are between one and a million. Further: how many prime divisors will it have, on average? How many divisors will it have altogether, and with what likelihood? What is the probability that it will have many more or many fewer divisors or prime divisors than the average?

Much of probabilistic number theory can be seen as an important special case of the study of variables that are almost, but not quite, mutually independent. For example, the event that a random integer between one and a million be divisible by two and the event that it be divisible by three are almost independent, but not quite.

It is sometimes said that probabilistic combinatorics uses the fact that whatever happens with probability greater than 0 must happen sometimes; one may say with equal justice that many applications of probabilistic number theory hinge on the fact that whatever is unusual must be rare. If certain algebraic objects (say, rational or integer solutions to certain equations) can be shown to be in the tail of certain sensibly defined distributions, it follows that there must be few of them; this is a very concrete non-probabilistic statement following from a probabilistic one.

At times, a non-rigorous, probabilistic approach leads to a number of heuristic algorithms and open problems, notably Cramér's conjecture.

Arithmetic Combinatorics

Let A be a set of N integers. Consider the set $A + A = \{ m + n \mid m, n \in A \}$ consisting of all sums of two elements of A. Is $A + A$ much larger than A? Barely larger? If $A + A$ is barely larger than A, must

A have plenty of arithmetic structure, for example, does A resemble an arithmetic progression?

If we begin from a fairly "thick" infinite set A, does it contain many elements in arithmetic progression: a, $a+b, a+2b, a+3b, \ldots, a+10b$, say? Should it be possible to write large integers as sums of elements of A?

These questions are characteristic of *arithmetic combinatorics*. This is a presently coalescing field; it subsumes *additive number theory* (which concerns itself with certain very specific sets A of arithmetic significance, such as the primes or the squares) and, arguably, some of the *geometry of numbers*, together with some rapidly developing new material. Its focus on issues of growth and distribution accounts in part for its developing links with ergodic theory, finite group theory, model theory, and other fields. The term *additive combinatorics* is also used; however, the sets A being studied need not be sets of integers, but rather subsets of non-commutative groups, for which the multiplication symbol, not the addition symbol, is traditionally used; they can also be subsets of rings, in which case the growth of $A+A$ and $A \cdot A$ may be compared.

Computations in Number Theory

While the word *algorithm* goes back only to certain readers of al-Khwārizmī, careful descriptions of methods of solution are older than proofs: such methods (that is, algorithms) are as old as any recognisable mathematics—ancient Egyptian, Babylonian, Vedic, Chinese—whereas proofs appeared only with the Greeks of the classical period. An interesting early case is that of what we now call the Euclidean algorithm. In its basic form (namely, as an algorithm for computing the greatest common divisor) it appears as Proposition 2 of Book VII in *Elements*, together with a proof of correctness. However, in the form that is often used in number theory (namely, as an algorithm for finding integer solutions to an equation $ax + by = c$, or, what is the same, for finding the quantities whose existence is assured by the Chinese remainder theorem) it first appears in the works of Āryabhaṭa (5th–6th century CE) as an algorithm called *kuṭṭaka* ("pulveriser"), without a proof of correctness.

There are two main questions: "can we compute this?" and "can we compute it rapidly?". Anybody can test whether a number is prime or, if it is not, split it into prime factors; doing so rapidly is another matter. We now know fast algorithms for testing primality, but, in spite of much work (both theoretical and practical), no truly fast algorithm for factoring.

The difficulty of a computation can be useful: modern protocols for encrypting messages (e.g., RSA) depend on functions that are known to all, but whose inverses (a) are known only to a chosen few, and (b) would take one too long a time to figure out on one's own. For example, these functions can be such that their inverses can be computed only if certain large integers are factorized. While many difficult computational problems outside number theory are known, most working encryption protocols nowadays are based on the difficulty of a few number-theoretical problems.

On a different note — some things may not be computable at all; in fact, this can be proven in some instances. For instance, in 1970, it was proven, as a solution to Hilbert's 10th problem, that there is no Turing machine which can solve all Diophantine equations. In particular, this means that, given a computably enumerable set of axioms, there are Diophantine equations for which there is no proof, starting from the axioms, of whether the set of equations has or does not have integer solutions. (We would necessarily be speaking of Diophantine equations for which there are no integer solutions, since, given a Diophantine equation with at least one solution, the solution itself provides a proof of the fact that a solution exists. We cannot prove, of course, that a particular Diophantine equation is of this kind, since this would imply that it has no solutions.)

Applications

The number-theorist Leonard Dickson (1874–1954) said "Thank God that number theory is unsullied by any application". Such a view is no longer applicable to number theory. In 1974, Donald Knuth said "...virtually every theorem in elementary number theory arises in a natural, motivated way in connection with the problem of making computers do high-speed numerical calculations". Elementary number theory is taught in discrete mathematics courses for computer scientists; and, on the other hand, number theory also has applications to the continuous in numerical analysis. As well as the well-known applications to cryptography, there are also applications to many other areas of mathematics.

Literature

Two of the most popular introductions to the subject are:

- *G. H. Hardy; E. M. Wright (2008) [1938]. An introduction to the theory of numbers (rev. by D. R. Heath-Brown and J. H. Silverman, 6th ed.). Oxford University Press. ISBN 978-0-19-921986-5. Retrieved 2016-03-02.*

- *Vinogradov, I. M. (2003) [1954]. Elements of Number Theory (reprint of the 1954 ed.). Mineola.*

Hardy and Wright's book is a comprehensive classic, though its clarity sometimes suffers due to the authors' insistence on elementary methods. Vinogradov's main attraction consists in its set of problems, which quickly lead to Vinogradov's own research interests; the text itself is very basic and close to minimal. Other popular first introductions are:

- *Ivan M. Niven; Herbert S. Zuckerman; Hugh L. Montgomery (2008) [1960]. An introduction to the theory of numbers (reprint of the 5th edition 1991 ed.). John Wiley & Sons. ISBN 978-81-265-1811-1. Retrieved 2016-02-28.*

- *Kenneth H. Rosen (2010). Elementary Number Theory (6th ed.). Pearson Education. ISBN 978-0-321-71775-7. Retrieved 2016-02-28.*

Popular choices for a second textbook include:

- *Borevich, A. I.; Shafarevich, Igor R. (1966). Number theory. Pure and Applied Mathematics. 20. Boston, MA: Academic Press. ISBN 978-0-12-117850-5. MR 0195803.*

- *Serre, Jean-Pierre (1996) [1973]. A course in arithmetic. Graduate texts in mathematics.* 0 Neugebauer & Sachs 1945, p. 40. The term *takiltum* is problematic. Robson prefers the rendering "The holding-square of the diagonal from which 1 is torn out, so that the short side comes up...".Robson 2001, p. 192

Prizes

The American Mathematical Society awards the Cole Prize in Number Theory. Moreover number theory is one of the three mathematical subdisciplines rewarded by the Fermat Prize.

References

- "The Unreasonable Effectiveness of Number Theory", Stefan Andrus Burr, George E. Andrews, American Mathematical Soc., 1992, ISBN 978-0-8218-5501-0

- "Applications of number theory to numerical analysis", Lo-keng Hua, Luogeng Hua, Yuan Wang, Springer-Verlag, 1981, ISBN 978-3-540-10382-0

Branches of Number Theory

The two branches of number theory are analytic number theory and algebraic number theory. Analytic number theory focuses on the methods of mathematical analysis that is used to solve problems that are related to integers. This section is a compilation of the various branch of number theory that forms an integral part of the broader subject matter.

Analytic Number Theory

In mathematics, analytic number theory is a branch of number theory that uses methods from mathematical analysis to solve problems about the integers. It is often said to have begun with Peter Gustav Lejeune Dirichlet's 1837 introduction of Dirichlet L-functions to give the first proof of Dirichlet's theorem on arithmetic progressions. It is well known for its results on prime numbers (involving the Prime Number Theorem and Riemann zeta function) and additive number theory (such as the Goldbach conjecture and Waring's problem).

Branches of Analytic Number Theory

Analytic number theory can be split up into two major parts, divided more by the type of problems they attempt to solve than fundamental differences in technique.

- Multiplicative number theory deals with the distribution of the prime numbers, such as estimating the number of primes in an interval, and includes the prime number theorem and Dirichlet's theorem on primes in arithmetic progressions.

- Additive number theory is concerned with the additive structure of the integers, such as Goldbach's conjecture that every even number greater than 2 is the sum of two primes. One of the main results in additive number theory is the solution to Waring's problem.

History

Precursors

Much of analytic number theory was inspired by the prime number theorem. Let $\varpi(x)$ be the prime-counting function that gives the number of primes less than or equal to x, for any real number x. For example, $\varpi(10) = 4$ because there are four prime numbers (2, 3, 5 and 7) less than or equal to 10. The prime number theorem then states that $x / \ln(x)$ is a good approximation to $\varpi(x)$, in the sense that the limit of the *quotient* of the two functions $\varpi(x)$ and $x / \ln(x)$ as x approaches infinity is 1:

$$\lim_{x \to \infty} \pi(x)x$$

known as the asymptotic law of distribution of prime numbers.

Adrien-Marie Legendre conjectured in 1797 or 1798 that $\varpi(a)$ is approximated by the function $a/(A\ln(a) + B)$, where A and B are unspecified constants. In the second edition of his book on number theory (1808) he then made a more precise conjecture, with $A = 1$ and $B = -1.08366$. Carl Friedrich Gauss considered the same question: "Im Jahr 1792 oder 1793", according to his own recollection nearly sixty years later in a letter to Encke (1849), he wrote in his logarithm table (he was then 15 or 16) the short note "Primzahlen unter $a(=\infty)\dfrac{a}{\ln a}$". But Gauss never published this conjecture. In 1838 Peter Gustav Lejeune Dirichlet came up with his own approximating function, the logarithmic integral li(x) (under the slightly different form of a series, which he communicated to Gauss). Both Legendre's and Dirichlet's formulas imply the same conjectured asymptotic equivalence of $\varpi(x)$ and $x / \ln(x)$ stated above, although it turned out that Dirichlet's approximation is considerably better if one considers the differences instead of quotients.

Dirichlet

Johann Peter Gustav Lejeune Dirichlet is credited with the creation of analytic number theory, a field in which he found several deep results and in proving them introduced some fundamental tools, many of which were later named after him. In 1837 he published Dirichlet's theorem on arithmetic progressions, using mathematical analysis concepts to tackle an algebraic problem and thus creating the branch of analytic number theory. In proving the theorem, he introduced the Dirichlet characters and L-functions. In 1841 he generalized his arithmetic progressions theorem from integers to the ring of Gaussian integers $\mathbb{Z}[i]$..

Chebyshev

In two papers from 1848 and 1850, the Russian mathematician Pafnuty L'vovich Chebyshev attempted to prove the asymptotic law of distribution of prime numbers. His work is notable for the use of the zeta function $\zeta(s)$ (for real values of the argument "s", as are works of Leonhard Euler, as early as 1737) predating Riemann's celebrated memoir of 1859, and he succeeded in proving a slightly weaker form of the asymptotic law, namely, that if the limit of $\varpi(x)/(x/\ln(x))$ as x goes to infinity exists at all, then it is necessarily equal to one. He was able to prove unconditionally that this ratio is bounded above and below by two explicitly given constants near to 1 for all x. Although Chebyshev's paper did not prove the Prime Number Theorem, his estimates for $\varpi(x)$ were strong enough for him to prove Bertrand's postulate that there exists a prime number between n and $2n$ for any integer $n \geq 2$.

Riemann

"…esistsehrwahrscheinlich,dassalleWurzelnreellsind.HiervonwäreallerdingseinstrengerBeweiszu wünschen;ichhabeindessdieAufsuchungdesselbennacheinigenflüchtigenvergeblichenVersuchen vorläufig bei Seite gelassen, da er für den nächsten Zweck meiner Untersuchung entbehrlich schien."

"…it is very probable that all roots are real. Of course one would wish for a rigorous proof here; I have for the time being, after some fleeting vain attempts, provisionally put aside the search for this, as it appears dispensable for the next objective of my investigation."

Riemann's statement of the Riemann hypothesis, from his 1859 paper. (He was discussing a version of the zeta function, modified so that its roots are real rather than on the critical line.)

Bernhard Riemann made some famous contributions to modern analytic number theory. In a single short paper (the only one he published on the subject of number theory), he investigated the Riemann zeta function and established its importance for understanding the distribution of prime numbers. He made a series of conjectures about properties of the zeta function, one of which is the well-known Riemann hypothesis.

Hadamard and De La Vallée-Poussin

Extending the ideas of Riemann, two proofs of the prime number theorem were obtained independently by Jacques Hadamard and Charles Jean de la Vallée-Poussin and appeared in the same year (1896). Both proofs used methods from complex analysis, establishing as a main step of the proof that the Riemann zeta function $\zeta(s)$ is non-zero for all complex values of the variable s that have the form $s = 1 + it$ with $t > 0$.

Modern Times

The biggest technical change after 1950 has been the development of *sieve methods*, particularly in multiplicative problems. These are combinatorial in nature, and quite varied. The extremal branch of combinatorial theory has in return been greatly influenced by the value placed in analytic number theory on quantitative upper and lower bounds. Another recent development is *probabilistic number theory*, which uses methods from probability theory to estimate the distribution of number theoretic functions, such as how many prime divisors a number has.

Developments within analytic number theory are often refinements of earlier techniques, which reduce the error terms and widen their applicability. For example, the *circle method* of Hardy and Littlewood was conceived as applying to power series near the unit circle in the complex plane; it is now thought of in terms of finite exponential sums (that is, on the unit circle, but with the power series truncated). The needs of diophantine approximation are for auxiliary functions that are not generating functions—their coefficients are constructed by use of a pigeonhole principle—and involve several complex variables. The fields of diophantine approximation and transcendence theory have expanded, to the point that the techniques have been applied to the Mordell conjecture.

Problems and Results

Theorems and results within analytic number theory tend not to be exact structural results about the integers, for which algebraic and geometrical tools are more appropriate. Instead, they give approximate bounds and estimates for various number theoretical functions, as the following examples illustrate.

Multiplicative Number Theory

Euclid showed that there are an infinite number of primes but it is very difficult to find an efficient method for determining whether or not a number is prime, especially a large number. A related but easier problem is to determine the asymptotic distribution of the prime numbers; that is, a rough description of how many primes are smaller than a given number. Gauss, amongst others, after computing a large list of primes, conjectured that the number of primes less than or equal to a large number N is close to the value of the integral

$$\int_2^N \frac{1}{\log t} dt.$$

In 1859 Bernhard Riemann used complex analysis and a special meromorphic function now known as the Riemann zeta function to derive an analytic expression for the number of primes less than or equal to a real number x. Remarkably, the main term in Riemann's formula was exactly the above integral, lending substantial weight to Gauss's conjecture. Riemann found that the error terms in this expression, and hence the manner in which the primes are distributed, are closely related to the complex zeros of the zeta function. Using Riemann's ideas and by getting more information on the zeros of the zeta function, Jacques Hadamard and Charles Jean de la Vallée-Poussin managed to complete the proof of Gauss's conjecture. In particular, they proved that if

$$\pi(x) = (\text{number of primes } \leq x),$$

then

$$\lim_{x \to \infty} \frac{\pi(x)}{x / \log x} = 1.$$

This remarkable result is what is now known as the *Prime Number Theorem*. It is a central result in analytic number theory. Loosely speaking, it states that given a large number N, the number of primes less than or equal to N is about $N/\log(N)$.

More generally, the same question can be asked about the number of primes in any arithmetic progression $a+nq$ for any integer n. In one of the first applications of analytic techniques to number theory, Dirichlet proved that any arithmetic progression with a and q coprime contains infinitely many primes. The prime number theorem can be generalised to this problem; letting

$$\pi(x, a, q) = (\text{number of primes} \leq x \text{ such that } p \text{ is in the arithmetic progression } a + nq, n \in \mathbf{Z}),$$

then if a and q are coprime,

$$\lim_{x \to \infty} \frac{\pi(x, a, q)\phi(q)}{x / \log x} = 1.$$

There are also many deep and wide ranging conjectures in number theory whose proofs seem too difficult for current techniques, such as the Twin prime conjecture which asks whether there are infinitely many primes p such that $p + 2$ is prime. On the assumption of the Elliott–Halberstam conjecture it has been proven recently that there are infinitely many primes p such that $p + k$ is prime for some positive even k at most 12. Also, it has been proven unconditionally (i.e. not depending on unproven conjectures) that there are infinitely many primes p such that $p + k$ is prime for some positive even k at most 246.

Additive Number Theory

One of the most important problems in additive number theory is Waring's problem, which asks whether it is possible, for any $k \geq 2$, to write any positive integer as the sum of a bounded number of kth powers,

$$n = x_1^k + \cdots + x_\ell^k.$$

The case for squares, $k = 2$, was answered by Lagrange in 1770, who proved that every positive integer is the sum of at most four squares. The general case was proved by Hilbert in 1909, using algebraic techniques which gave no explicit bounds. An important breakthrough was the application of analytic tools to the problem by Hardy and Littlewood. These techniques are known as the circle method, and give explicit upper bounds for the function $G(k)$, the smallest number of kth powers needed, such as Vinogradov's bound

$$G(k) \leq k(3 \log k + 11).$$

Diophantine Problems

Diophantine problems are concerned with integer solutions to polynomial equations: one may study the distribution of solutions, that is, counting solutions according to some measure of "size" or *height*.

An important example is the Gauss circle problem, which asks for integers points $(x\,y)$ which satisfy

$$x^2 + y^2 \leq r^2.$$

In geometrical terms, given a circle centered about the origin in the plane with radius r, the problem asks how many integer lattice points lie on or inside the circle. It is not hard to prove that the answer is $\pi r^2 + E(r)$, where $E(r)/r^2 \to 0$ as $r \to \infty$. Again, the difficult part and a great achievement of analytic number theory is obtaining specific upper bounds on the error term $E(r)$.

It was shown by Gauss that $E(r) = O(r)$. In general, an $O(r)$ error term would be possible with the unit circle (or, more properly, the closed unit disk) replaced by the dilates of any bounded planar region with piecewise smooth boundary. Furthermore, replacing the unit circle by the unit square, the error term for the general problem can be as large as a linear function of r. Therefore, getting an error bound of the form $O(r^\delta)$ for some $\delta < 1$ in the case of the circle is a significant improvement. The first to attain this was Sierpiński in 1906, who showed $E(r) = O(r^{2/3})$. In 1915, Hardy and Landau each showed that one does *not* have $E(r) = O(r^{1/2})$. Since then the goal has been to show that for each fixed $\epsilon > 0$ there exists a real number $C(\epsilon)$ such that $E(r) \leq C(\epsilon) r^{1/2+\epsilon}$.

In 2000 Huxley showed that $E(r) = O(r^{131/208})$, which is the best published result.

Methods of Analytic Number Theory

Dirichlet Series

One of the most useful tools in multiplicative number theory are Dirichlet series, which are functions of a complex variable defined by an infinite series of the form

$$f(s) = \sum_{n=1}^{\infty} a_n n^{-s}.$$

Depending on the choice of coefficients a_n, this series may converge everywhere, nowhere, or on some half plane. In many cases, even where the series does not converge everywhere, the holomorphic function it defines may be analytically continued to a meromorphic function on the entire complex plane. The utility of functions like this in multiplicative problems can be seen in the formal identity

$$\left(\sum_{n=1}^{\infty} a_n n^{-s}\right)\left(\sum_{n=1}^{\infty} b_n n^{-s}\right) = \sum_{n=1}^{\infty} \left(\sum_{k\ell=n} a_k b_\ell\right) n^{-s};$$

hence the coefficients of the product of two Dirichlet series are the multiplicative convolutions of the original coefficients. Furthermore, techniques such as partial summation and Tauberian theorems can be used to get information about the coefficients from analytic information about the Dirichlet series. Thus a common method for estimating a multiplicative function is to express it as a Dirichlet series (or a product of simpler Dirichlet series using convolution identities), examine this series as a complex function and then convert this analytic information back into information about the original function.

Riemann Zeta Function

Euler showed that the fundamental theorem of arithmetic implies (at least formally) the *Euler product*

$$\sum_{n=1}^{\infty} \frac{1}{n^s} = \prod_{p} \frac{1}{1 - p^{-s}} \text{ for } s > 1 \ (p \text{ is prime number})$$

Euler's proof of the infinity of prime numbers makes use of the divergence of the term at the left hand side for $s = 1$ (the so-called harmonic series), a purely analytic result. Euler was also the first to use analytical arguments for the purpose of studying properties of integers, specifically by constructing generating power series. This was the beginning of analytic number theory.

Later, Riemann considered this function for complex values of s and showed that this function can be extended to a meromorphic function on the entire plane with a simple pole at $s = 1$. This function is now known as the Riemann Zeta function and is denoted by $\zeta(s)$. There is a plethora of literature on this function and the function is a special case of the more general Dirichlet L-functions.

Analytic number theorists are often interested in the error of approximations such as the prime number theorem. In this case, the error is smaller than $x/\log x$. Riemann's formula for $\varpi(x)$ shows that the error term in this approximation can be expressed in terms of the zeros of the zeta function. In his 1859 paper, Riemann conjectured that all the "non-trivial" zeros of ζ lie on the line $\mathfrak{R}(s) = 1/2$ but never provided a proof of this statement. This famous and long-standing conjecture is known as the *Riemann Hypothesis* and has many deep implications in number theory; in fact, many important theorems have been proved under the assumption that the hypothesis is true. For example, under the assumption of the Riemann Hypothesis, the error term in the prime number theorem is $O(x^{1/2+\varepsilon})$.

In the early 20th century G. H. Hardy and Littlewood proved many results about the zeta function in an attempt to prove the Riemann Hypothesis. In fact, in 1914, Hardy proved that there were infinitely many zeros of the zeta function on the critical line

$$\Re(z) = 1/2.$$

This led to several theorems describing the density of the zeros on the critical line.

Multiplicative Number Theory

Multiplicative number theory is a subfield of analytic number theory that deals with prime numbers and with factorization and divisors. The focus is usually on developing approximate formulas for counting these objects in various contexts. The prime number theorem is a key result in this subject. The Mathematics Subject Classification for multiplicative number theory is 11Nxx.

Scope

Multiplicative number theory deals primarily in asymptotic estimates for arithmetic functions. Historically the subject has been dominated by the prime number theorem, first by attempts to prove it and then by improvements in the error term. The Dirichlet divisor problem that estimates the average order of the divisor function $d(n)$ and Gauss's circle problem that estimates the average order of the number of representations of a number as a sum of two squares are also classical problems, and again the focus is on improving the error estimates.

The distribution of primes numbers among residue classes modulo an integer is an area of active research. Dirichlet's theorem on primes in arithmetic progressions shows that there are an infinity of primes in each co-prime residue class, and the prime number theorem for arithmetic progressions shows that the primes are asymptotically equidistributed among the residue classes. The Bombieri–Vinogradov theorem gives a more precise measure of how evenly they are distributed. There is also much interest in the size of the smallest prime in an arithmetic progression; Linnik's theorem gives an estimate.

The twin prime conjecture, namely that there are an infinity of primes p such that $p+2$ is also prime, is the subject of active research. Chen's theorem shows that there are an infinity of primes p such that $p+2$ is either prime or the product of two primes.

Methods

The methods belong primarily to analytic number theory, but elementary methods, especially sieve methods, are also very important. The large sieve and exponential sums are usually considered part of multiplicative number theory.

The distribution of prime numbers is closely tied to the behavior of the Riemann zeta function and the Riemann hypothesis, and these subjects are studied both from a number theory viewpoint and a complex analysis viewpoint.

Additive Number Theory

In number theory, the specialty additive number theory studies subsets of integers and their behavior under addition. More abstractly, the field of "additive number theory" includes the study of abelian groups and commutative semigroups with an operation of addition. Additive number theory has close ties to combinatorial number theory and the geometry of numbers. Two principal objects of study are the sumset of two subsets A and B of elements from an abelian group G,

$$A + B = \{a + b : a \in A, b \in B\},$$

and the h-fold sumset of A,

$$hA = \underbrace{A + \cdots + A}_{h}.$$

There are two main subdivisions listed below.

Additive Number Theory

The first is principally devoted to consideration of *direct problems* over (typically) the integers, that is, determining the structure of hA from the structure of A: for example, determining which elements can be represented as a sum from hA, where A is a fixed subset. Two classical problems of this type are the Goldbach conjecture (which is the conjecture that $2P$ contains all even numbers greater than two, where P is the set of primes) and Waring's problem (which asks how large must h be to guarantee that hA_k contains all positive integers, where

$$A_k = \{0^k, 1^k, 2^k, 3^k, \ldots\}$$

is the set of k-th powers). Many of these problems are studied using the tools from the Hardy-Littlewood circle method and from sieve methods. For example, Vinogradov proved that every sufficiently large odd number is the sum of three primes, and so every sufficiently large even integer is the sum of four primes. Hilbert proved that, for every integer $k > 1$, every nonnegative integer is the sum of a bounded number of k-th powers. In general, a set A of nonnegative integers is called a basis of order h if hA contains all positive integers, and it is called an asymptotic basis if hA contains all sufficiently large integers. Much current research in this area concerns properties of general asymptotic bases of finite order. For example, a set A is called a minimal asymptotic basis of order h if A is an asymptotic basis of order h but no proper subset of A is an asymptotic basis of order h. It has been proved that minimal asymptotic bases of order h exist for all h, and that there also exist asymptotic bases of order h that contain no minimal asymptotic bases of order h. Another question to be considered is how small can the number of representations of n as a sum of h elements in an asymptotic basis can be. This is the content of the Erdős–Turán conjecture on additive bases.

Additive Combinatorics

The second is principally devoted to consideration of *inverse problems*, often over more general groups than just the integers, that is, given some information about the sumset $A+B$, the aim is find information about the structure of the individual sets A and B. (A more recent name sometimes associated to this sub-division is additive combinatorics.) Unlike problems related to classical bases, as described above, this sub-area often deals with finite subsets rather than infinite ones. A typical question is what is the structure of a pair of subsets whose sumset has small cardinality (in relation to $|A|$ and $|B|$). In the case of the integers, the classical Freiman's theorem provides a potent partial answer to this question in terms of multi-dimensional arithmetic progressions. Another typical problem is simply to find a lower bound for $|A+B|$ in terms of $|A|$ and $|B|$ (this can be view as an inverse problem with the given information for $A+B$ being that $|A+B|$ is sufficiently small and the structural conclusion then being that either A or B is the empty set; such problems are often considered direct problems as well). Examples of this type include the Erdős–Heilbronn Conjecture (for a restricted sumset) and the Cauchy–Davenport Theorem. The methods used for tackling such questions draw from across the spectrum of mathematics, including combinatorics, ergodic theory, analysis, graph theory, group theory, and linear algebraic and polynomial methods.

Maier's Matrix Method

Maier's matrix method is a technique in analytic number theory due to Helmut Maier that is used to demonstrate the existence of intervals of natural numbers within which the prime numbers are distributed with a certain property. In particular it has been used to prove Maier's theorem (Maier 1985) and also the existence of chains of large gaps between consecutive primes (Maier 1981). The method uses estimates for the distribution of prime numbers in arithmetic progressions to prove the existence of a large set of intervals where the number of primes in the set is well understood and hence that at least one of the intervals contains primes in the required distribution.

The Method

The method first selects a primorial and then constructs an interval in which the distribution of integers coprime to the primorial is well understood. By looking at copies of the interval translated by multiples of the primorial an array (or matrix) of integers is formed where the rows are the translated intervals and the columns are arithmetic progressions where the difference is the primordial. By Dirichlet's theorem on arithmetic progressions the columns will contain many primes if and only if the integer in the original interval was coprime to the primordial. Good estimates for the number of small primes in these progressions due to (Gallagher 1971) allows the estimation of the primes in the matrix which guarantees the existence of at least one row or interval with at least a certain number of primes.

Algebraic Number Theory

Algebraic number theory is a branch of number theory that uses the techniques of abstract algebra to study the integers, rational numbers, and their generalizations. Number-theoretic questions are expressed in terms of properties of algebraic objects such as algebraic number fields and their rings of integers, finite fields, and function fields. These properties, such as whether a ring admits unique factorization, the behavior of ideals, and the Galois groups of fields, can resolve questions of primary importance in number theory, like the existence of solutions to Diophantine equations.

History of Algebraic Number Theory

Diophantus

The beginnings of algebraic number theory can be traced to Diophantine equations, named after the 3rd-century Alexandrian mathematician, Diophantus, who studied them and developed methods for the solution of some kinds of Diophantine equations. A typical Diophantine problem is to find two integers x and y such that their sum, and the sum of their squares, equal two given numbers A and B, respectively:

$$A = x + y$$

$$B = x^2 + y^2.$$

Diophantine equations have been studied for thousands of years. For example, the solutions to the quadratic Diophantine equation $x^2 + y^2 = z^2$ are given by the Pythagorean triples, originally solved by the Babylonians (c. 1800 BC). Solutions to linear Diophantine equations, such as $26x + 65y = 13$, may be found using the Euclidean algorithm (c. 5th century BC).

Diophantus's major work was the *Arithmetica*, of which only a portion has survived.

Fermat

Fermat's last theorem was first conjectured by Pierre de Fermat in 1637, famously in the margin of a copy of *Arithmetica* where he claimed he had a proof that was too large to fit in the margin. No successful proof was published until 1995 despite the efforts of countless mathematicians during the 358 intervening years. The unsolved problem stimulated the development of algebraic number theory in the 19th century and the proof of the modularity theorem in the 20th century.

Gauss

One of the founding works of algebraic number theory, the *Disquisitiones Arithmeticae* (Latin: *Arithmetical Investigations*) is a textbook of number theory written in Latin by Carl Friedrich Gauss in 1798 when Gauss was 21 and first published in 1801 when he was 24. In this book Gauss brings together results in number theory obtained by mathematicians such as Fermat, Euler, Lagrange and Legendre and adds important new results of his own. Before the *Disquisitiones* was published, number theory consisted of a collection of isolated theorems and conjectures. Gauss brought the work of his predecessors together with his own original work into a systematic framework, filled in gaps, corrected unsound proofs, and extended the subject in numerous ways.

The *Disquisitiones* was the starting point for the work of other nineteenth century European mathematicians including Ernst Kummer, Peter Gustav Lejeune Dirichlet and Richard Dedekind. Many of the annotations given by Gauss are in effect announcements of further research of his own, some of which remained unpublished. They must have appeared particularly cryptic to his contemporaries; we can now read them as containing the germs of the theories of L-functions and complex multiplication, in particular.

Dirichlet

In a couple of papers in 1838 and 1839 Peter Gustav Lejeune Dirichlet proved the first class number formula, for quadratic forms (later refined by his student Kronecker). The formula, which Jacobi called a result "touching the utmost of human acumen", opened the way for similar results regarding more general number fields. Based on his research of the structure of the unit group of quadratic fields, he proved the Dirichlet unit theorem, a fundamental result in algebraic number theory.

He first used the pigeonhole principle, a basic counting argument, in the proof of a theorem in diophantine approximation, later named after him Dirichlet's approximation theorem. He published important contributions to Fermat's last theorem, for which he proved the cases $n = 5$ and $n = 14$, and to the biquadratic reciprocity law. The Dirichlet divisor problem, for which he found the first results, is still an unsolved problem in number theory despite later contributions by other researchers.

Dedekind

Richard Dedekind's study of Lejeune Dirichlet's work was what led him to his later study of algebraic number fields and ideals. In 1863, he published Lejeune Dirichlet's lectures on number theory as *Vorlesungen über Zahlentheorie* ("Lectures on Number Theory") about which it has been written that:

"Although the book is assuredly based on Dirichlet's lectures, and although Dedekind himself referred to the book throughout his life as Dirichlet's, the book itself was entirely written by Dedekind, for the most part after Dirichlet's death." (Edwards 1983)

1879 and 1894 editions of the *Vorlesungen* included supplements introducing the notion of an ideal, fundamental to ring theory. (The word "Ring", introduced later by Hilbert, does not appear in Dedekind's work.) Dedekind defined an ideal as a subset of a set of numbers, composed of algebraic integers that satisfy polynomial equations with integer coefficients. The concept underwent further development in the hands of Hilbert and, especially, of Emmy Noether. Ideals generalize Ernst Eduard Kummer's ideal numbers, devised as part of Kummer's 1843 attempt to prove Fermat's Last Theorem.

Hilbert

David Hilbert unified the field of algebraic number theory with his 1897 treatise *Zahlbericht* (literally "report on numbers"). He also resolved a significant number-theory problem formulated by Waring in 1770. As with the finiteness theorem, he used an existence proof that shows there must be solutions for the problem rather than providing a mechanism to produce the answers. He then had little more to publish on the subject; but the emergence of Hilbert modular forms in the dissertation of a student means his name is further attached to a major area.

He made a series of conjectures on class field theory. The concepts were highly influential, and his own contribution lives on in the names of the Hilbert class field and of the Hilbert symbol of local class field theory. Results were mostly proved by 1930, after work by Teiji Takagi.

Artin

Emil Artin established the Artin reciprocity law in a series of papers (1924; 1927; 1930). This law is a general theorem in number theory that forms a central part of global class field theory. The term "reciprocity law" refers to a long line of more concrete number theoretic statements which it generalized, from the quadratic reciprocity law and the reciprocity laws of Eisenstein and Kummer to Hilbert's product formula for the norm symbol. Artin's result provided a partial solution to Hilbert's ninth problem.

Modern Theory

Around 1955, Japanese mathematicians Goro Shimura and Yutaka Taniyama observed a possible link between two apparently completely distinct, branches of mathematics, elliptic curves and modular forms. The resulting modularity theorem (at the time known as the Taniyama–Shimura conjecture) states that every elliptic curve is modular, meaning that it can be associated with a unique modular form.

It was initially dismissed as unlikely or highly speculative, and was taken more seriously when number theorist André Weil found evidence supporting it, but no proof; as a result the "astounding" conjecture was often known as the Taniyama–Shimura-Weil conjecture. It became a part of the Langlands programme, a list of important conjectures needing proof or disproof.

From 1993 to 1994, Andrew Wiles provided a proof of the modularity theorem for semistable elliptic curves, which, together with Ribet's theorem, provides a proof for Fermat's Last Theorem. Both Fermat's Last Theorem and the Modularity Theorem were almost universally considered inaccessible to proof by contemporaneous mathematicians (meaning, impossible or virtually impossible to prove using current knowledge). Wiles first announced his proof in June 1993 in a version that was soon recognized as having a serious gap in a key point. The proof was corrected by Wiles, in part via collaboration with Richard Taylor, and the final, widely accepted, version was released in September 1994, and formally published in 1995. The proof uses many techniques from algebraic geometry and number theory, and has many ramifications in these branches of mathematics. It also uses standard constructions of modern algebraic geometry, such as the category of schemes and Iwasawa theory, and other 20th-century techniques not available to Fermat.

Basic Notions

Failure of Unique Factorization

An important property of the ring of integers is that it satisfies the fundamental theorem of arithmetic, that every integer has a factorization into a product of prime numbers, and this factorization is unique up to the ordering of the factors. This may no longer be true in the ring of integers O of an algebraic number field K.

A *prime element* is an element p of O such that if p divides a product ab, then it divides one of the factors a or b. This property is closely related to primality in the integers, because any positive integer satisfying this property is either 1 or a prime number. However, it is strictly weaker. For example, -2 is not a prime number because it is negative, but it is a prime element. If factorizations into prime elements are permitted, then, even in the integers, there are alternative factorizations such as

$$6 = 2 \cdot 3 = (-2) \cdot (-3).$$

In general, if u is a unit, meaning a number with a multiplicative inverse in O, and if p is a prime element, then up is also a prime element. Numbers such as p and up are said to be *associate*. In the integers, the primes p and $-p$ are associate, but only one of these is positive. Requiring that prime numbers be positive selects a unique element from among a set of associated prime elements. When K is not the rational numbers, however, there is no analog of positivity. For example, in the Gaussian integers $Z[i]$, the numbers $1 + 2i$ and $-2 + i$ are associate because the latter is the product of the former by i, but there is no way to single out one as being more canonical than the other. This leads to equations such as

$$5 = (1 + 2i)(1 - 2i) = (2 + i)(2 - i),$$

which prove that in $Z[i]$, it is not true that factorizations are unique up to the order of the factors. For this reason, one adopts the definition of unique factorization used in unique factorization domains (UFDs). In a UFD, the prime elements occurring in a factorization are only expected to be unique up to units and their ordering.

However, even with this weaker definition, many rings of integers in algebraic number fields do not admit unique factorization. There is an algebraic obstruction called the ideal class group. When the ideal class group is trivial, the ring is a UFD. When it is not, there is a distinction between a prime element and an irreducible element. An *irreducible element x* is an element such that if $x = yz$, then either y or z is a unit. These are the elements that cannot be factored any further. Every element in O admits a factorization into irreducible elements, but it may admit more than one. This is because, while all prime elements are irreducible, some irreducible elements may not be prime. For example, consider the ring $Z[\sqrt{-5}]$. In this ring, the numbers 3, $2 + \sqrt{-5}$ are $2 - \sqrt{-5}$ is irreducible. This means that the number 9 has two factorizations into irreducible elements,

$$9 = 3^2 = (2 + \sqrt{-5})(2 - \sqrt{-5}).$$

This equation shows that 3 divides the product $(2 + \sqrt{-5})(2 - \sqrt{-5}) = 9$. If 3 were a prime element, then it would divide $2 + \sqrt{-5}$ or $2 - \sqrt{-5}$, but it does not, because all elements divisible by 3 are of the form $3a + 3b\sqrt{-5}$. Similarly, $2 + \sqrt{-5}$ and $2 - \sqrt{-5}$ divide the product 3^2, but neither of these elements divides 3 itself, so neither of them are prime. As there is no sense in which the elements 3, $2 + \sqrt{-5}$ and $2 - \sqrt{-5}$ can be made equivalent, unique factorization fails in $Z[\sqrt{-5}]$. Unlike the situation with units, where uniqueness could be repaired by weakening the definition, overcoming this failure requires a new perspective.

Factorization into Prime Ideals

If I is an ideal in O, then there is always a factorization

$$I = \mathfrak{p}_1^{e_1} \cdots \mathfrak{p}_t^{e_t},$$

where each $\mathfrak{p}_i$ is a prime ideal, and where this expression is unique up to the order of the factors. In particular, this is true if I is the principal ideal generated by a single element. This is the strongest sense in which the ring of integers of a general number field admits unique factorization. In the language of ring theory, it says that rings of integers are Dedekind domains.

When O is a UFD, every prime ideal is generated by a prime element. Otherwise, there are prime ideals which are not generated by prime elements. In Z[√-5], for instance, the ideal $(2, 1 + \sqrt{-5})$ is a prime ideal which cannot be generated by a single element.

Historically, the idea of factoring ideals into prime ideals was preceded by Ernst Kummer's introduction of ideal numbers. These are numbers lying in an extension field E of K. This extension field is now known as the Hilbert class field. By the principal ideal theorem, every prime ideal of O generates a principal ideal of the ring of integers of E. A generator of this principal ideal is called an ideal number. Kummer used these as a substitute for the failure of unique factorization in cyclotomic fields. These eventually led Richard Dedekind to introduce a forerunner of ideals and to prove unique factorization of ideals.

An ideal which is prime in the ring of integers in one number field may fail to be prime when extended to a larger number field. Consider, for example, the prime numbers. The corresponding ideals pZ are prime ideals of the ring Z. However, when this ideal is extended to the Gaussian integers to get pZ[i], it may or may not be prime. For example, the factorization $2 = (1 + i)(1 − i)$ implies that

$$2\mathbf{Z}[i] = (1+i)\mathbf{Z}[i] \cdot (1-i)\mathbf{Z}[i] = ((1+i)\mathbf{Z}[i])^{2};$$

note that because $1 + i = (1 − i) \cdot i$, the ideals generated by $1 + i$ and $1 − i$ are the same. A complete answer to the question of which ideals remain prime in the Gaussian integers is provided by Fermat's theorem on sums of two squares. It implies that for an odd prime number p, pZ[i] is a prime ideal if $p \equiv 3 \pmod 4$ and is not a prime ideal if $p \equiv 1 \pmod 4$. This, together with the observation that the ideal $(1 + i)$Z[i] is prime, provides a complete description of the prime ideals in the Gaussian integers. Generalizing this simple result to more general rings of integers is a basic problem in algebraic number theory. Class field theory accomplishes this goal when K is an abelian extension of Q (i.e. a Galois extension with abelian Galois group).

Ideal Class Group

Unique factorization fails if and only if there are prime ideals that fail to be principal. The object which measures the failure of prime ideals to be principal is called the ideal class group. Defining the ideal class group requires enlarging the set of ideals in a ring of algebraic integers so that they admit a group structure. This is done by generalizing ideals to fractional ideals. A fractional ideal is an additive subgroup J of K which is closed under multiplication by elements of O, meaning that $xJ \subseteq J$ if $x \in O$. All ideals of O are also fractional ideals. If I and J are fractional ideals, then the set IJ of all products of an element in I and an element in J is also a fractional ideal. This operation makes the set of non-zero fractional ideals into a group. The group identity is the ideal (1), and the inverse of J is the fractional ideal J^{-1} whose elements are $\{ x^{-1} : x \in J \}$.

The principal fractional ideals, meaning the ones of the form Ox where $x \in K^{\times}$, form a subgroup of the group of all non-zero fractional ideals. The quotient of the group of non-zero fractional ideals by this subgroup is the ideal class group. Two fractional ideals I and J represent the same element of the ideal class group if and only if there exists an element $x \in K$ such that $xI = J$. Therefore the ideal class group makes two fractional ideals equivalent if one is as close to being principal as the other is. The ideal class group is generally denoted Cl K, Cl O, or Pic O (with the last notation identifying it with the Picard group in algebraic geometry).

The number of elements in the class group is called the class number of K. The class number of $Q(\sqrt{-5})$ is 2. This means that there are only two ideal classes, the class of principal fractional ideals, and the class of a non-principal fractional ideal such as $(2, 1 + \sqrt{-5})$.

The ideal class group has another description in terms of divisors. These are formal objects which represent possible factorizations of numbers. The divisor group Div K is defined to be the free abelian group generated by the prime ideals of O. There is a group homomorphism from $K^{\times}$, the non-zero elements of K up to multiplication, to Div K. Suppose that $x \in K$ satisfies

$$(x) = \mathfrak{p}_1^{e_1} \cdots \mathfrak{p}_t^{e_t}.$$

Then div x is defined to be the divisor

$$\operatorname{div} x = \sum_{i=1}^{t} e_i [\mathfrak{p}_i].$$

The kernel of div is the group of units in O, while the cokernel is the ideal class group. In the language of homological algebra, this says that there is an exact sequence of abelian groups (written multiplicatively),

$$1 \to O^{\times} \to K^{\times} \to DivK \to ClK \to 1.$$

Real aand Complex Embeddings

Some number fields, such as $Q(\sqrt{2})$, can be specified as subfields of the real numbers. Others, such as $Q(\sqrt{-1})$, cannot. Abstractly, such a specification corresponds to a field homomorphism $K \to R$ or $K \to C$. These are called real embeddings and complex embeddings, respectively.

A real quadratic field $Q(\sqrt{d})$ is so-called because it admits two real embeddings and no complex embeddings. These are the field homomorphisms which send $\sqrt{d}$ to $\sqrt{d}$ and to $\sqrt{-d}$, respectively. Dually, an imaginary quadratic field $Q(\sqrt{-d})$ admits no real embeddings and a conjugate pair of complex embeddings. One of these embeddings sends $\sqrt{-d}$ to $\sqrt{-d}$, while the other sends it to its complex conjugate.

Conventionally, the number of real embeddings of K is denoted r_1, while the number of conjugate pairs of complex embeddings is denoted r_2. The signature of K is the pair (r_1, r_2). It is a theorem that $r_1 + 2r_2 = d$, where d is the degree of K.

Considering all embeddings at once determines a function

$$M : K \to \mathbf{R}^{r_1} \oplus \mathbf{C}^{2r_2}.$$

This is called the Minkowski embedding. The subspace of the codomain fixed by complex conjugation is a real vector space of dimension d called Minkowski space. Because the Minkowski embedding is defined by field homomorphisms, multiplication of elements of K by an element $x \in K$ corresponds to multiplication by a diagonal matrix in the Minkowski embedding. The dot product on Minkowski space corresponds to the trace form $\langle x, y \rangle = \text{Tr}(xy)$.

The image of O in Minkowski space is a d-dimensional lattice. If B is a basis for this lattice, then $\det B^{\mathrm{T}}B$ is the discriminant of O. The discriminant is denoted Δ or D. The covolume of the image of O is $\sqrt{|\Delta|}$.

Places

Real and complex embeddings can be put on the same footing as prime ideals by adopting a perspective based on valuations. Consider, for example, the integers. In addition to the usual absolute value function $|\cdot| : Q \to R$, there are p-adic absolute value functions $|\cdot|_p : Q \to R$, defined for each prime number p, which measure divisibility by p. Ostrowski's theorem states that these are all possible absolute value functions on Q (up to equivalence). Therefore absolute values are a common language to describe both the real embedding of Q and the prime numbers.

A place of an algebraic number field is an equivalence class of absolute value functions on K. There are two types of places. There is a $\mathfrak{p}$-adic absolute value for each prime ideal $\mathfrak{p}$ of O, and, like the p-adic absolute values, it measures divisibility. These are called finite places. The other type of place is specified using a real or complex embedding of K and the standard absolute value function on R or C. These are infinite places. Because absolute values are unable to distinguish between a complex embedding and its conjugate, a complex embedding and its conjugate determine the same place. Therefore there are r_1 real places and r_2 complex places. Because places encompass the primes, places are sometimes referred to as primes. When this is done, finite places are called finite primes and infinite places are called infinite primes. If v is a valuation corresponding to an absolute value, then one frequently writes $v \mid \infty$ to mean that v is an infinite place and $v \nmid \infty$ to mean that it is a finite place.

Considering all the places of the field together produces the adele ring of the number field. The adele ring allows one to simultaneously track all the data available using absolute values. This produces significant advantages in situations where the behavior at one place can affect the behavior at other places, as in the Artin reciprocity law.

Units

The integers have only two units, 1 and −1. Other rings of integers may admit more units. The Gaussian integers have four units, the previous two as well as $\pm i$. The Eisenstein integers Z[exp($2\varpi i / 3$)] have six units. The integers in real quadratic number fields have infinitely many units. For example, in Z[$\sqrt{3}$], every power of $2 + \sqrt{3}$ is a unit, and all these powers are distinct.

In general, the group of units of O, denoted $O^\times$, is a finitely generated abelian group. The fundamental theorem of finitely generated abelian groups therefore implies that it is a direct sum of a torsion part and a free part. Reinterpreting this in the context of a number field, the torsion part consists of the roots of unity that lie in O. This group is cyclic. The free part is described by Dirichlet's unit theorem. This theorem says that rank of the free part is $r_1 + r_2 - 1$. Thus, for example, the only fields for which the rank of the free part is zero are Q and the imaginary quadratic fields.

A more precise statement giving the structure of $O^\times \otimes_Z Q$ as a Galois module for the Galois group of K/Q is also possible.

The free part of the unit group can be studied using the infinite places of K. Consider the function

$$L : K^\times \to \mathbf{R}^{r_1+r_2}$$

defined by

$$L(x) = (\log |x|_v)_v,$$

where v varies over the infinite places of K and $|\cdot|_v$ is the absolute value associated with v. The function L is a homomorphism from $K^\times$ to a real vector space. It can be shown that the image of $O^\times$ is a lattice that spans the hyperplane defined by $x_1 + \cdots + x_{r_1+r_2} = 0$. The covolume of this lattice is the regulator of the number field. One of the simplifications made possible by working with the adele ring is that there is a single object, the idele class group, that describes both the quotient by this lattice and the ideal class group.

Zeta Function

The Dedekind zeta function of a number field, analogous to the Riemann zeta function is an analytic object which describes the behavior of prime ideals in K. When K is an abelian extension of Q, Dedekind zeta functions are products of Dirichlet L-functions, with there being one factor for each Dirichlet character. The trivial character corresponds to the Riemann zeta function. When K is a Galois extension, the Dedekind zeta function is the Artin L-function of the regular representation of the Galois group of K, and it has a factorization in terms of irreducible Artin representations of the Galois group.

The zeta function is related to the other invariants described above by the class number formula.

Local Fields

Completing a number field K at a place w gives a complete field. If the valuation is archimedean, one gets R or C, if it is non-archimedean and lies over a prime p of the rationals, one gets a finite extension K_w / Q_p: a complete, discrete valued field with finite residue field. This process simplifies the arithmetic of the field and allows the local study of problems. For example, the Kronecker–Weber theorem can be deduced easily from the analogous local statement. The philosophy behind the study of local fields is largely motivated by geometric methods. In algebraic geometry, it is common to study varieties locally at a point by localizing to a maximal ideal. Global information can then be recovered by gluing together local data. This spirit is adopted in algebraic number theory. Given a prime in the ring of algebraic integers in a number field, it is desirable to study the field locally at that prime. Therefore, one localizes the ring of algebraic integers to that prime and then completes the fraction field much in the spirit of geometry.

Major Results

Finiteness of The Class Group

One of the classical results in algebraic number theory is that the ideal class group of an algebraic number field K is finite. The order of the class group is called the class number, and is often denoted by the letter h.

Dirichlet's Unit Theorem

Dirichlet's unit theorem provides a description of the structure of the multiplicative group of units $O^\times$ of the ring of integers O. Specifically, it states that $O^\times$ is isomorphic to $G \times Z^r$, where G is the finite cyclic group consisting of all the roots of unity in O, and $r = r_1 + r_2 - 1$ (where r_1 (respectively, r_2) denotes the number of real embeddings (respectively, pairs of conjugate non-real embeddings) of K). In other words, $O^\times$ is a finitely generated abelian group of rank $r_1 + r_2 - 1$ whose torsion consists of the roots of unity in O.

Reciprocity Laws

In terms of the Legendre symbol, the law of quadratic reciprocity for positive odd primes states

$$\left(\frac{p}{q}\right)\left(\frac{q}{p}\right) = (-1)^{\frac{p-1}{2}\frac{q-1}{2}}.$$

A reciprocity law is a generalization of the law of quadratic reciprocity.

There are several different ways to express reciprocity laws. The early reciprocity laws found in the 19th century were usually expressed in terms of a power residue symbol (p/q) generalizing the quadratic reciprocity symbol, that describes when a prime number is an nth power residue modulo another prime, and gave a relation between (p/q) and (q/p). Hilbert reformulated the reciprocity laws as saying that a product over p of Hilbert symbols $(a,b/p)$, taking values in roots of unity, is equal to 1. Artin's reformulated reciprocity law states that the Artin symbol from ideals (or ideles) to elements of a Galois group is trivial on a certain subgroup. Several more recent generalizations express reciprocity laws using cohomology of groups or representations of adelic groups or algebraic K-groups, and their relationship with the original quadratic reciprocity law can be hard to see.

Class Number Formula

The class number formula relates many important invariants of a number field to a special value of its Dedekind zeta function.

Related Areas

Algebraic number theory interacts with many other mathematical disciplines. It uses tools from homological algebra. Via the analogy of function fields vs. number fields, it relies on techniques and ideas from algebraic geometry. Moreover, the study of higher-dimensional schemes over Z instead of number rings is referred to as arithmetic geometry. Algebraic number theory is also used in the study of arithmetic hyperbolic 3-manifolds.

Dedekind Zeta Function

In mathematics, the Dedekind zeta function of an algebraic number field K, generally denoted $\zeta_K(s)$, is a generalization of the Riemann zeta function—which is obtained by specializing to the case where K is the rational numbers Q. In particular, it can be defined as a Dirichlet series, it has an Euler product expansion, it satisfies a functional equation, it has an analytic continuation to a meromorphic function on the complex plane C with only a simple pole at $s = 1$, and its values encode arithmetic data of K. The extended Riemann hypothesis states that if $\zeta_K(s) = 0$ and $0 < \mathrm{Re}(s) < 1$, then $\mathrm{Re}(s) = 1/2$.

The Dedekind zeta function is named for Richard Dedekind who introduced them in his supplement to Peter Gustav Lejeune Dirichlet's Vorlesungen über Zahlentheorie.

Definition and Basic Properties

Let K be an algebraic number field. Its Dedekind zeta function is first defined for complex numbers s with real part $\mathrm{Re}(s) > 1$ by the Dirichlet series

$$\zeta_K(s) = \sum_{I \subseteq \mathcal{O}_K} \frac{1}{(N_{K/\mathbf{Q}}(I))^s}$$

where I ranges through the non-zero ideals of the ring of integers O_K of K and $N_{K/Q}(I)$ denotes the absolute norm of I (which is equal to both the index $[O_K : I]$ of I in O_K or equivalently the cardinality of quotient ring O_K / I). This sum converges absolutely for all complex numbers s with real part $\mathrm{Re}(s) > 1$. In the case $K = Q$, this definition reduces to that of the Riemann zeta function.

Euler Product

The Dedekind zeta function of K has an Euler product which is a product over all the prime ideals P of O_K

$$\zeta_K(s) = \prod_{P \subseteq \mathcal{O}_K} \frac{1}{1 - (N_{K/\mathbf{Q}}(P))^{-s}}, \text{ for } \mathrm{Re}(s) > 1.$$

This is the expression in analytic terms of the uniqueness of prime factorization of the ideals I in O_K. The fact that, for $\mathrm{Re}(s) > 1$, $\zeta_K(s)$ is given by a product of non-zero numbers implies that it is non-zero in this region.

Analytic Continuation and Functional Equation

Erich Hecke first proved that $\zeta_K(s)$ has an analytic continuation to the complex plane as a meromorphic function, having a simple pole only at $s = 1$. The residue at that pole is given by the analytic class number formula and is made up of important arithmetic data involving invariants of the unit group and class group of K.

The Dedekind zeta function satisfies a functional equation relating its values at s and $1 - s$. Specifically, let Δ_K denote the discriminant of K, let r_1 (resp. r_2) denote the number of real places (resp. complex places) of K, and let

$$\Gamma_{\mathbf{R}}(s) = \pi^{-s/2}\Gamma(s/2)$$

and

$$\Gamma_C(s) = 2(2\pi)^{-s}\Gamma(s)$$

where $\Gamma(s)$ is the Gamma function. Then, the function

$$\Lambda_K(s) = |\Delta_K|^{s/2}\Gamma_R(s)^{r_1}\Gamma_C(s)^{r_2}\zeta_K(s) \qquad \Xi_K(s) = \tfrac{1}{2}(s^2 + \tfrac{1}{4})\Lambda_K(\tfrac{1}{2} + is)$$

satisfies the functional equation

$$\Lambda_K(s) = \Lambda_K(1-s). \qquad \Xi_K(-s) = \Xi_K(s)$$

Special Values

Analogously to the Riemann zeta function, the values of the Dedekind zeta function at integers encode (at least conjecturally) important arithmetic data of the field K. For example, the analytic class number formula relates the residue at $s = 1$ to the class number $h(K)$ of K, the regulator $R(K)$ of K, the number $w(K)$ of roots of unity in K, the absolute discriminant of K, and the number of real and complex places of K. Another example is at $s = 0$ where it has a zero whose order r is equal to the rank of the unit group of O_K and the leading term is given by

$$\lim_{s\to 0} s^{-r}\zeta_K(s) = -\frac{h(K)R(K)}{w(K)}.$$

Combining the functional equation and the fact that $\Gamma(s)$ is infinite at all integers less than or equal to zero yields that $\zeta_K(s)$ vanishes at all negative even integers. It even vanishes at all negative odd integers unless K is totally real (i.e. $r_2 = 0$; e.g. Q or a real quadratic field). In the totally real case, Carl Ludwig Siegel showed that $\zeta_K(s)$ is a non-zero rational number at negative odd integers. Stephen Lichtenbaum conjectured specific values for these rational numbers in terms of the algebraic K-theory of K.

Relations to Other L-Functions

For the case in which K is an abelian extension of Q, its Dedekind zeta function can be written as a product of Dirichlet L-functions. For example, when K is a quadratic field this shows that the ratio

$$\frac{\zeta_K(s)}{\zeta_Q(s)}$$

is the L-function $L(s, \chi)$, where χ is a Jacobi symbol used as Dirichlet character. That the zeta function of a quadratic field is a product of the Riemann zeta function and a certain Dirichlet L-function is an analytic formulation of the quadratic reciprocity law of Gauss.

In general, if K is a Galois extension of Q with Galois group G, its Dedekind zeta function is the Artin L-function of the regular representation of G and hence has a factorization in terms of Artin L-functions of irreducible Artin representations of G.

The relation with Artin L-functions shows that if L/K is a Galois extension then $\dfrac{\zeta_L(s)}{\zeta_K(s)}$ is holomorphic ($\zeta_K(s)$ "divides" $\zeta_L(s)$): for general extensions the result would follow from the Artin conjecture for L-functions.

Additionally, $\zeta_K(s)$ is the Hasse–Weil zeta function of Spec O_K and the motivic L-function of the motive coming from the cohomology of Spec K.

Arithmetically Equivalent Fields

Two fields are called arithmetically equivalent if they have the same Dedekind zeta function. Wieb Bosma and Bart de Smit (2002) used Gassmann triples to give some examples of pairs of non-isomorphic fields that are arithmetically equivalent. In particular some of these pairs have different class numbers, so the Dedekind zeta function of a number field does not determine its class number.

Fermat's Last Theorem

In number theory, Fermat's Last Theorem (sometimes called Fermat's conjecture, especially in older texts) states that no three positive integers a, b, and c satisfy the equation $a^n + b^n = c^n$ for any integer value of n greater than two. The cases $n = 1$ and $n = 2$ have been known to have infinitely many solutions since antiquity.

The 1670 edition of Diophantus' *Arithmetica* includes Fermat's commentary, particularly his "Last Theorem" (*Observatio Domini Petri de Fermat*).

This theorem was first conjectured by Pierre de Fermat in 1637 in the margin of a copy of *Arithmetica* where he claimed he had a proof that was too large to fit in the margin. The first successful proof was released in 1994 by Andrew Wiles, and formally published in 1995, after 358 years of effort by mathematicians. The unsolved problem stimulated the development of algebraic number theory in the 19th century and the proof of the modularity theorem in the 20th century. It is among the most notable theorems in the history of mathematics and prior to its proof, it was in the *Guinness Book of World Records* as the "most difficult mathematical problem", one of the reasons being that it has the largest number of unsuccessful proofs.

Overview

The Pythagorean equation, $x^2 + y^2 = z^2$, has an infinite number of positive integer solutions for x, y, and z; these solutions are known as Pythagorean triples. Around 1637, Fermat wrote in the margin of a book that the more general equation $a^n + b^n = c^n$ had no solutions in positive integers, if n is an integer greater than 2. Although he claimed to have a general proof of his conjecture, Fermat left no details of his proof, and no proof by him has ever been found. His claim was discovered some 30 years later, after his death. This claim, which came to be known as Fermat's Last Theorem, stood unsolved in mathematics for the following three and a half centuries.

The claim eventually became one of the most notable unsolved problems of mathematics. Attempts to prove it prompted substantial development in number theory, and over time Fermat's Last Theorem gained prominence as an unsolved problem in mathematics.

Subsequent Developments and Solution

With the special case $n = 4$ proved, it suffices to prove the theorem for exponents n that are prime numbers (this reduction is considered trivial to prove). Over the next two centuries (1637–1839), the conjecture was proved for only the primes 3, 5, and 7, although Sophie Germain innovated and proved an approach that was relevant to an entire class of primes. In the mid-19th century, Ernst Kummer extended this and proved the theorem for all regular primes, leaving irregular primes to be analyzed individually. Building on Kummer's work and using sophisticated computer studies, other mathematicians were able to extend the proof to cover all prime exponents up to four million, but a proof for all exponents was inaccessible (meaning that mathematicians generally considered a proof impossible, exceedingly difficult, or unachievable with current knowledge).

The proof of Fermat's Last Theorem in full, for all n, was finally accomplished 357 years later by Andrew Wiles in 1994, an achievement for which he was honoured and received numerous awards, including the 2016 Abel Prize. The solution came in a roundabout manner, from a completely different area of mathematics.

Around 1955, Japanese mathematicians Goro Shimura and Yutaka Taniyama suspected a link might exist between elliptic curves and modular forms, two completely different areas of mathematics. Known at the time as the Taniyama–Shimura-Weil conjecture, and (eventually) as the modularity theorem, it stood on its own, with no apparent connection to Fermat's Last Theorem. It was widely seen as significant and important in its own right, but was (like Fermat's theorem) widely considered completely inaccessible to proof.

In 1984, Gerhard Frey noticed an apparent link between the modularity theorem and Fermat's Last Theorem. This potential link was confirmed two years later by Ken Ribet, who gave a conditional proof of Fermat's Last Theorem that depended on the modularity theorem (see: *Ribet's Theorem* and *Frey curve*). On hearing this, English mathematician Andrew Wiles, who had a childhood fascination with Fermat's Last Theorem, decided to try to prove the modularity theorem as a way to prove Fermat's Last Theorem. In 1993, after six years working secretly on the problem, Wiles succeeded in proving enough of the modularity theorem to prove Fermat's Last Theorem for odd prime exponents. Wiles's paper was massive in size and scope. A flaw was discovered in one part of his original paper during peer review and required a further year and collaboration with a past student, Richard Taylor, to resolve. As a result, the final proof in 1995 was accompanied by a second, smaller, joint paper to that effect. Wiles's achievement was reported widely in the popular press, and was popularized in books and television programs. The remaining parts of the modularity theorem were subsequently proved by other mathematicians, building on Wiles's work, between 1996 and 2001.

Equivalent Statements of The Theorem

There are several simple alternative ways to state Fermat's Last Theorem that are equivalent to the one given above. In order to state them, let N be the set of natural numbers 1,2,3,..., let Z be the set of integers 0, ±1, ±2,..., and let Q be the set of rational numbers a/b where a and b are in Z with $b \neq 0$.

In what follows we will call a solution to $x^n + y^n = z^n$ where one or more of x, y, or z is zero a *trivial solution*. A solution where all three are non-zero will be called a *non-trivial* solution. For comparison's sake we start with the original formulation.

Original statement. With $n, x, y, z \in N$ and $n > 2$ the equation $x^n + y^n = z^n$ has no solutions.

Most popular domain treatments of the subject state it this way. In contrast, almost all math textbooks state it over Z:

Equivalent statement 1: $x^n + y^n = z^n$, where $n \geq 3$, has no non-trivial solutions $x, y, z \in Z$.

The equivalence is clear if n is even. If n is odd and all three of x, y, z are negative then we can replace x, y, z with $-x, -y, -z$ to obtain a solution in N. If two of them are negative, it must be x and z or y and z. If x, z are negative and y is positive, then we can rearrange to get $(-z)^n + y^n = (-x)^n$ resulting in a solution in N; the other case is dealt with analogously. Now if just one is negative, it must be x or y. If x is negative, and y and z are positive, then it can be rearranged to get $(-x)^n + z^n = y^n$ again resulting in a solution in N; if y is negative, the result follows symmetrically. Thus in all cases a nontrivial solution in Z results in a solution in N.

Equivalent statement 2: $x^n + y^n = z^n$, where $n \geq 3$, has no non-trivial solutions $x, y, z \in Q$.

This is because the exponent of x, y and z are equal (to n), so if there is a solution in Q then it can be multiplied through by an appropriate common denominator to get a solution in Z, and hence in N.

Equivalent statement 3: $x^n + y^n = 1$, where $n \geq 3$, has no non-trivial solutions $x, y \in Q$.

A non-trivial solution $a, b, c \in Z$ to $x^n + y^n = z^n$ yields the non-trivial solution $a/c, b/c \in Q$ for $v^n + w^n = 1$. Conversely, a solution $a/b, c/d \in Q$ to $v^n + w^n = 1$ yields the non-trivial solution ad, cb, bd for $x^n + y^n = z^n$.

This last formulation is particularly fruitful, because it reduces the problem from a problem about surfaces in three dimensions to a problem about curves in two dimensions. Furthermore, it allows working over the field Q, rather than over the ring Z; fields exhibit more structure than rings, which allows for deeper analysis of their elements.

Connection to elliptic curves: If a, b, c is a non-trivial solution to $x^p + y^p = z^p$, p odd prime, then $y^2 = x(x - a^p)(x + b^p)$ (Frey curve) is an elliptic curve.

Examining this elliptic curve with Ribet's theorem shows that it cannot have a modular form. The proof by Andrew Wiles shows that $y^2 = x(x - a^n)(x + b^n)$ always has a modular form. This implies that a non-trivial solution to $x^p + y^p = z^p$, p odd prime, would create a contradiction. This shows that no non-trivial solutions exist.

Mathematical History

Pythagoras and Diophantus

Pythagorean Triples

A Pythagorean triple – named for the ancient Greek Pythagoras – is a set of three integers (a, b, c) that satisfy a special case of Fermat's equation ($n = 2$)

$$a^2 + b^2 = c^2.$$

Examples of Pythagorean triples include (3, 4, 5) and (5, 12, 13). There are infinitely many such triples, and methods for generating such triples have been studied in many cultures, beginning with the Babylonians and later ancient Greek, Chinese, and Indian mathematicians. The traditional interest in Pythagorean triples connects with the Pythagorean theorem; in its converse form, it states that a triangle with sides of lengths a, b, and c has a right angle between the a and b legs when the numbers are a Pythagorean triple. Fermat's Last Theorem is an extension of this problem to higher powers, stating that no solution exists when the exponent 2 is replaced by any larger integer.

Diophantine Equations

Fermat's equation, $x^n + y^n = z^n$ with positive integer solutions, is an example of a Diophantine equation, named for the 3rd-century Alexandrian mathematician, Diophantus, who studied them and developed methods for the solution of some kinds of Diophantine equations. A typical Diophantine problem is to find two integers x and y such that their sum, and the sum of their squares, equal two given numbers A and B, respectively:

$$A = x + y$$

$$B = x^2 + y^2.$$

Diophantus's major work is the *Arithmetica*, of which only a portion has survived. Fermat's conjecture of his Last Theorem was inspired while reading a new edition of the *Arithmetica*, that was translated into Latin and published in 1621 by Claude Bachet.

Diophantine equations have been studied for thousands of years. For example, the solutions to the quadratic Diophantine equation $x^2 + y^2 = z^2$ are given by the Pythagorean triples, originally solved by the Babylonians (c. 1800 BC). Solutions to linear Diophantine equations, such as $26x + 65y = 13$, may be found using the Euclidean algorithm (c. 5th century BC). Many Diophantine equations have a form similar to the equation of Fermat's Last Theorem from the point of view of algebra, in that they have no *cross terms* mixing two letters, without sharing its particular properties. For example, it is known that there are infinitely many positive integers x, y, and z such that $x^n + y^n = z^m$ where n and m are relatively prime natural numbers.

Fermat's Conjecture

Problem II.8 of the *Arithmetica* asks how a given square number is split into two other squares; in other words, for a given rational number k, find rational numbers u and v such that $k^2 = u^2 + v^2$. Diophantus shows how to solve this sum-of-squares problem for $k = 4$ (the solutions being $u = 16/5$ and $v = 12/5$).

Problem II.8 in the 1621 edition of the *Arithmetica* of Diophantus. On the right is the margin that was too small to contain Fermat's alleged proof of his "last theorem".

Around 1637, Fermat wrote his Last Theorem in the margin of his copy of the *Arithmetica* next to Diophantus' sum-of-squares problem:

Cubum autem in duos cubos, aut quadratoquadratum in duos quadratoquadratos & generaliter nullam in infinitum ultra quadratum potestatem in duos eiusdem nominis fas est dividere cuius rei demonstrationem mirabilem sane detexi. Hanc marginis exiguitas non caperet.	*It is impossible to separate a cube into two cubes, or a fourth power into two fourth powers, or in general, any power higher than the second, into two like powers. I have discovered a truly marvelous proof of this, which this margin is too narrow to contain.*

After Fermat's death in 1665, his son Clément-Samuel Fermat produced a new edition of the book (1670) augmented with his father's comments. The margin note became known as *Fermat's Last Theorem*, as it was the last of Fermat's asserted theorems to remain unproved.

It is not known whether Fermat had actually found a valid proof for all exponents n, but it appears unlikely. Only one related proof by him has survived, namely for the case $n = 4$, as described in the section *Proofs for specific exponents*. While Fermat posed the cases of $n = 4$ and of $n = 3$ as challenges to his mathematical correspondents, such as Marin Mersenne, Blaise Pascal, and John Wallis, he never posed the general case. Moreover, in the last thirty years of his life, Fermat never again wrote of his "truly marvelous proof" of the general case, and never published it. Van der Poorten suggests that while the absence of a proof is insignificant, the lack of challenges means Fermat realised he did not have a proof; he quotes Weil as saying Fermat must have briefly deluded himself with an irretrievable idea.

The techniques Fermat might have used in such a "marvelous proof" are unknown.

Taylor and Wiles's proof relies on 20th-century techniques. Fermat's proof would have had to be elementary by comparison, given the mathematical knowledge of his time.

While Harvey Friedman's grand conjecture implies that any provable theorem (including Fermat's last theorem) can be proved using only 'elementary function arithmetic', such a proof need be 'elementary' only in a technical sense and could involve millions of steps, and thus be far too long to have been Fermat's proof.

Proofs for Specific Exponents

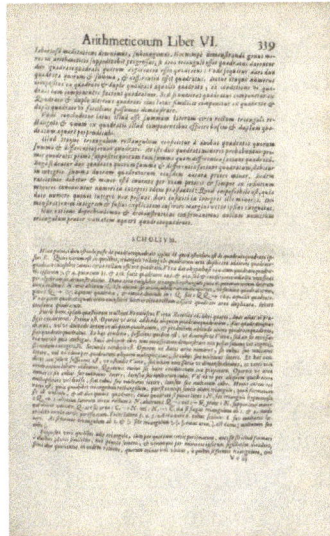

Fermat's infinite descent for Fermat's Last Theorem case n=4 in the 1670 edition of the *Arithmetica* of Diophantus (pp. 338–339).

Only one relevant proof by Fermat has survived, in which he uses the technique of infinite descent to show that the area of a right triangle with integer sides can never equal the square of an integer. His proof is equivalent to demonstrating that the equation

$$x^4 - y^4 = z^2$$

has no primitive solutions in integers (no pairwise coprime solutions). In turn, this proves Fermat's Last Theorem for the case $n = 4$, since the equation $a^4 + b^4 = c^4$ can be written as $c^4 - b^4 = (a^2)^2$.

Alternative proofs of the case $n = 4$ were developed later by Frénicle de Bessy (1676), Leonhard Euler (1738), Kausler (1802), Peter Barlow (1811), Adrien-Marie Legendre (1830), Schopis (1825), Terquem (1846), Joseph Bertrand (1851), Victor Lebesgue (1853, 1859, 1862), Theophile Pepin (1883), Tafelmacher (1893), David Hilbert (1897), Bendz (1901), Gambioli (1901), Leopold Kronecker (1901), Bang (1905), Sommer (1907), Bottari (1908), Karel Rychlík (1910), Nutzhorn (1912), Robert Carmichael (1913), Hancock (1931), and Vrănceanu (1966).

For another proof for $n=4$ by infinite descent, Infinite descent: Non-solvability of $r^2 + s^4 = t^4$. For various proofs for $n=4$ by infinite descent, see Grant and Perella (1999), Barbara (2007), and Dolan (2011).

After Fermat proved the special case $n = 4$, the general proof for all n required only that the theorem be established for all odd prime exponents. In other words, it was necessary to prove only that the equation $a^n + b^n = c^n$ has no integer solutions (a, b, c) when n is an odd prime number. This follows because a solution (a, b, c) for a given n is equivalent to a solution for all the factors of n. For illustration, let n be factored into d and e, $n = de$. The general equation

$$a^n + b^n = c^n$$

implies that (a^d, b^d, c^d) is a solution for the exponent e

$$(a^d)^e + (b^d)^e = (c^d)^e.$$

Thus, to prove that Fermat's equation has no solutions for $n > 2$, it would suffice to prove that it has no solutions for at least one prime factor of every n. Each integer $n > 2$ is divisible by 4 or an odd prime number (or both). Therefore, Fermat›s Last Theorem could be proved for all n if it could be proved for $n = 4$ and for all odd primes p.

In the two centuries following its conjecture (1637–1839), Fermat's Last Theorem was proved for three odd prime exponents $p = 3$, 5 and 7. The case $p = 3$ was first stated by Abu-Mahmud Khojandi (10th century), but his attempted proof of the theorem was incorrect. In 1770, Leonhard Euler gave a proof of $p = 3$, but his proof by infinite descent contained a major gap. However, since Euler himself had proved the lemma necessary to complete the proof in other work, he is generally credited with the first proof. Independent proofs were published by Kausler (1802), Legendre (1823, 1830), Calzolari (1855), Gabriel Lamé (1865), Peter Guthrie Tait (1872), Günther (1878), Gambioli (1901), Krey (1909), Rychlík (1910), Stockhaus (1910), Carmichael (1915), Johannes van der Corput (1915), Axel Thue (1917), and Duarte (1944). The case $p = 5$ was proved independently by Legendre and Peter Gustav Lejeune Dirichlet around 1825. Alternative proofs were developed by Carl Friedrich Gauss (1875, posthumous), Lebesgue (1843), Lamé (1847), Gambioli (1901), Werebrusow (1905), Rychlík (1910), van der Corput (1915), and Guy Terjanian (1987). The case $p = 7$ was proved by Lamé in 1839. His rather complicated proof was simplified in 1840 by Lebesgue, and still simpler proofs were published by Angelo Genocchi in 1864, 1874 and 1876. Alternative proofs were developed by Théophile Pépin (1876) and Edmond Maillet (1897).

Fermat's Last Theorem was also proved for the exponents $n = 6$, 10, and 14. Proofs for $n = 6$ were published by Kausler, Thue, Tafelmacher, Lind, Kapferer, Swift, and Breusch. Similarly, Dirichlet and Terjanian each proved the case $n = 14$, while Kapferer and Breusch each proved the case $n = 10$. Strictly speaking, these proofs are unnecessary, since these cases follow from the proofs for $n = 3$, 5, and 7, respectively. Nevertheless, the reasoning of these even-exponent proofs differs from their odd-exponent counterparts. Dirichlet›s proof for $n = 14$ was published in 1832, before Lamé›s 1839 proof for $n = 7$.

All proofs for specific exponents used Fermat's technique of infinite descent, either in its original form, or in the form of descent on elliptic curves or abelian varieties. The details and auxiliary arguments, however, were often *ad hoc* and tied to the individual exponent under consideration. Since they became ever more complicated as p increased, it seemed unlikely that the general case of Fermat's Last Theorem could be proved by building upon the proofs for individual exponents. Although some general results on Fermat's Last Theorem were published in the early 19th century by Niels Henrik Abel and Peter Barlow, the first significant work on the general theorem was done by Sophie Germain.

Sophie Germain

In the early 19th century, Sophie Germain developed several novel approaches to prove Fermat's Last Theorem for all exponents. First, she defined a set of auxiliary primes θ constructed from the prime exponent p by the equation $\theta = 2hp + 1$, where h is any integer not divisible by three. She showed that, if no integers raised to the pth power were adjacent modulo θ (the *non-consecutivity condition*), then θ must divide the product xyz. Her goal was to use mathematical induction to prove that, for any given p, infinitely many auxiliary primes θ satisfied the non-consecutivity condition and thus divided xyz; since the product xyz can have at most a finite number of prime factors, such a proof would have established Fermat's Last Theorem. Although she developed many techniques for establishing the non-consecutivity condition, she did not succeed in her strategic goal. She also worked to set lower limits on the size of solutions to Fermat's equation for a given exponent p, a modified version of which was published by Adrien-Marie Legendre. As a byproduct of this latter work, she proved Sophie Germain's theorem, which verified the first case of Fermat's Last Theorem (namely, the case in which p does not divide xyz) for every odd prime exponent less than 100. Germain tried unsuccessfully to prove the first case of Fermat's Last Theorem for all even exponents, specifically for $n = 2p$, which was proved by Guy Terjanian in 1977. In 1985, Leonard Adleman, Roger Heath-Brown and Étienne Fouvry proved that the first case of Fermat's Last Theorem holds for infinitely many odd primes p.

Ernst Kummer and The Theory of Ideals

In 1847, Gabriel Lamé outlined a proof of Fermat's Last Theorem based on factoring the equation $x^p + y^p = z^p$ in complex numbers, specifically the cyclotomic field based on the roots of the number 1. His proof failed, however, because it assumed incorrectly that such complex numbers can be factored uniquely into primes, similar to integers. This gap was pointed out immediately by Joseph Liouville, who later read a paper that demonstrated this failure of unique factorisation, written by Ernst Kummer.

Kummer set himself the task of determining whether the cyclotomic field could be generalized to include new prime numbers such that unique factorisation was restored. He succeeded in that task by developing the ideal numbers. Using the general approach outlined by Lamé, Kummer proved both cases of Fermat's Last Theorem for all regular prime numbers. However, he could not prove the theorem for the exceptional primes (irregular primes) that conjecturally occur approximately 39% of the time; the only irregular primes below 100 are 37, 59 and 67.

Mordell Conjecture

In the 1920s, Louis Mordell posed a conjecture that implied that Fermat's equation has at most a finite number of nontrivial primitive integer solutions, if the exponent n is greater than two. This conjecture was proved in 1983 by Gerd Faltings, and is now known as Faltings' theorem.

Computational Studies

In the latter half of the 20th century, computational methods were used to extend Kummer's approach to the irregular primes. In 1954, Harry Vandiver used a SWAC computer to prove Fermat's Last Theorem for all primes up to 2521. By 1978, Samuel Wagstaff had extended this to all primes less than 125,000. By 1993, Fermat's Last Theorem had been proved for all primes less than four million.

However despite these efforts and their results, no proof existed of Fermat's Last Theorem. Proofs of individual exponents by their nature could never prove the *general* case: even if all exponents were verified up to an extremely large number X, a higher exponent beyond X might still exist for which the claim was not true. (This had been the case with some other past conjectures, and it could not be ruled out in this conjecture.)

Connection With Elliptic Curves

The strategy that ultimately led to a successful proof of Fermat's Last Theorem arose from the "astounding" Taniyama–Shimura-Weil conjecture, proposed around 1955—which many mathematicians believed would be near to impossible to prove, and was linked in the 1980s by Gerhard Frey, Jean-Pierre Serre and Ken Ribet to Fermat's equation. By accomplishing a partial proof of this conjecture in 1994, Andrew Wiles ultimately succeeded in proving Fermat's Last Theorem, as well as leading the way to a full proof by others of what is now the modularity theorem.

Taniyama–Shimura–Weil Conjecture

Around 1955, Japanese mathematicians Goro Shimura and Yutaka Taniyama observed a possible link between two apparently completely distinct branches of mathematics, elliptic curves and modular forms. The resulting modularity theorem (at the time known as the Taniyama–Shimura conjecture) states that every elliptic curve is modular, meaning that it can be associated with a unique modular form.

It was initially dismissed as unlikely or highly speculative, and was taken more seriously when number theorist André Weil found evidence supporting it, but no proof; as a result the conjecture was often known as the Taniyama–Shimura-Weil conjecture. It became a part of the Langlands programme, a list of important conjectures needing proof or disproof.

Even after gaining serious attention, the conjecture was seen by contemporary mathematicians as extraordinarily difficult or perhaps inaccessible to proof. For example, Wiles's ex-supervisor John Coates states that it seemed "impossible to actually prove", and Ken Ribet considered himself "one of the vast majority of people who believed [it] was completely inaccessible", adding that "Andrew Wiles was probably one of the few people on earth who had the audacity to dream that you can actually go and prove [it]."

Ribet's Theorem for Frey Curves

In 1984, Gerhard Frey noted a link between Fermat's equation and the modularity theorem, then still a conjecture. If Fermat's equation had any solution (a, b, c) for exponent $p > 2$, then it could be shown that the elliptic curve (now known as a Frey-Hellegouarch)

$$y^2 = x\,(x - a^p)(x + b^p)$$

would have such unusual properties that it was unlikely to be modular. This would conflict with the modularity theorem, which asserted that all elliptic curves are modular. As such, Frey observed that a proof of the Taniyama–Shimura-Weil conjecture would simultaneously prove Fermat's Last Theorem and equally, a *disproof* or refutation of Fermat's Last Theorem would disprove the conjecture.

Following this strategy, a proof of Fermat's Last Theorem required two steps. First, it was necessary to prove the modularity theorem – or at least to prove it for the sub-class of cases (known as semistable elliptic curves) that included Frey's equation – and this was widely believed inaccessible to proof by contemporary mathematicians. Second, it was necessary to show that Frey's intuition was correct: that if an elliptic curve were constructed in this way, using a set of numbers that were a solution of Fermat's equation, the resulting elliptic curve could not be modular. Frey did not quite succeed in proving this rigorously; the missing piece (the so-called "epsilon conjecture", now known as Ribet's theorem) was identified by Jean-Pierre Serre and proved in 1986 by Ken Ribet.

- The modularity theorem – if proved – would mean all elliptic curves (or at least all semistable elliptic curves) are of necessity modular.

- Ribet's theorem – proved in 1986 – showed that, if a solution to Fermat's equation existed, it could be used to create a semistable elliptic curve that was not modular;

- The contradiction would imply (if the modularity theorem were correct) that *no* solutions can exist to Fermat's equation – therefore proving Fermat's Last Theorem.

Wiles's General Proof

Ribet's proof of the epsilon conjecture in 1986 accomplished the first of the two goals proposed by Frey. Upon hearing of Ribet's success, Andrew Wiles, an English mathematician with a childhood fascination with Fermat's Last Theorem, and a prior study area of elliptical equations, decided to commit himself to accomplishing the second half: proving a special case of the modularity theorem (then known as the Taniyama–Shimura conjecture) for semistable elliptic curves.

British mathematician Andrew Wiles.

Wiles worked on that task for six years in near-total secrecy, covering up his efforts by releasing prior work in small segments as separate papers and confiding only in his wife. His initial study suggested proof by induction, and he based his initial work and first significant breakthrough on Galois theory before switching to an attempt to extend horizontal Iwasawa theory for the inductive argument around 1990–91 when it seemed that there was no existing approach adequate to the problem. However, by the summer of 1991, Iwasawa theory also seemed to not be reaching the central issues in the problem. In response, he approached colleagues to seek out any hints of cutting edge research and new techniques, and discovered an Euler system recently developed by Victor Kolyvagin and Matthias Flach that seemed "tailor made" for the inductive part of his proof. Wiles studied and extended this approach, which worked. Since his work relied extensively on this approach, which was new to mathematics and to Wiles, in January 1993 he asked his Princeton colleague, Nick Katz, to check his reasoning for subtle errors. Their conclusion at the time was that the techniques Wiles used seemed to work correctly.

By mid-May 1993, Wiles felt able to tell his wife he thought he had solved the proof of Fermat's Last Theorem, and by June he felt sufficiently confident to present his results in three lectures delivered on 21–23 June 1993 at the Isaac Newton Institute for Mathematical Sciences. Specifically, Wiles presented his proof of the Taniyama–Shimura conjecture for semistable elliptic curves; together with Ribet's proof of the epsilon conjecture, this implied Fermat's Last Theorem. However, it became apparent during peer review that a critical point in the proof was incorrect. It contained an error in a bound on the order of a particular group. The error was caught by several mathematicians refereeing Wiles's manuscript including Katz (in his role as reviewer), who alerted Wiles on 23 August 1993.

The error would not have rendered his work worthless – each part of Wiles's work was highly significant and innovative by itself, as were the many developments and techniques he had created in the course of his work, and only one part was affected. However without this part proved, there was no actual proof of Fermat's Last Theorem. Wiles spent almost a year trying to repair his proof, initially by himself and then in collaboration with Richard Taylor, without success.

On 19 September 1994, on the verge of giving up, Wiles had a flash of insight that the proof could be saved by returning to his original Horizontal Iwasawa theory approach, which he had abandoned in favour of the Kolyvagin–Flach approach, this time strengthening it with expertise gained in Kolyvagin–Flach's approach. On 24 October 1994, Wiles submitted two manuscripts, "Modular elliptic curves and Fermat's Last Theorem" and "Ring theoretic properties of certain Hecke algebras", the second of which was co-authored with Taylor and proved that certain conditions were met that were needed to justify the corrected step in the main paper. The two papers were vetted and published as the entirety of the May 1995 issue of the *Annals of Mathematics*. These papers established the modularity theorem for semistable elliptic curves, the last step in proving Fermat's Last Theorem, 358 years after it was conjectured.

Subsequent Developments

The full Taniyama–Shimura–Weil conjecture was finally proved by Diamond (1996), Conrad, Diamond & Taylor (1999), and Breuil et al. (2001) who, building on Wiles's work, incrementally chipped away at the remaining cases until the full result was proved. The now fully proved conjecture became known as the modularity theorem.

Several other theorems in number theory similar to Fermat's Last Theorem also follow from the same reasoning, using the modularity theorem. For example: no cube can be written as a sum of two coprime n-th powers, $n \geq 3$. (The case $n = 3$ was already known by Euler.)

Exponents other Than Positive Integers

Reciprocal Integers (Inverse Fermat Equation)

The equation $a1 >$ can be considered the "inverse" Fermat equation. All solutions of this equation were computed by Lenstra in 1992. In the case in which the mth roots are required to be real and positive, all solutions are given by

$$a = rs^m$$

$$b = rt^m$$

$$c = r(s+t)^m$$

for positive integers r, s, t with s and t coprime.

Rational Exponents

For the Diophantine equation $a^{n/m} + b^{n/m} = c^{n/m}$ with n not equal to 1, Bennett, Glass, and Székely proved in 2004 for $n > 2$, that if n and m are coprime, then there are integer solutions if and only if 6 divides m, and $a1$, $b^{1/m}$, and $c1$ are different complex 6th roots of the same real number.

Monetary Prizes

In 1816, and again in 1850, the French Academy of Sciences offered a prize for a general proof of Fermat's Last Theorem. In 1857, the Academy awarded 3000 francs and a gold medal to Kummer for his research on ideal numbers, although he had not submitted an entry for the prize. Another prize was offered in 1883 by the Academy of Brussels.

In 1908, the German industrialist and amateur mathematician Paul Wolfskehl bequeathed 100,000 gold marks—a large sum at the time—to the Göttingen Academy of Sciences to offer as a prize for a complete proof of Fermat's Last Theorem. On 27 June 1908, the Academy published nine rules for awarding the prize. Among other things, these rules required that the proof be published in a peer-reviewed journal; the prize would not be awarded until two years after the publication; and that no prize would be given after 13 September 2007, roughly a century after the competition was begun. Wiles collected the Wolfskehl prize money, then worth $50,000, on 27 June 1997. In March 2016, Wiles was awarded the Norwegian government's Abel prize worth €600,000 for "for his stunning proof of Fermat's Last Theorem by way of the modularity conjecture for semistable elliptic curves, opening a new era in number theory."

Prior to Wiles's proof, thousands of incorrect proofs were submitted to the Wolfskehl committee, amounting to roughly 10 feet (3 meters) of correspondence. In the first year alone (1907–1908), 621 attempted proofs were submitted, although by the 1970s, the rate of submission had decreased to roughly 3–4 attempted proofs per month. According to F. Schlichting, a Wolfskehl reviewer, most of the proofs were based on elementary methods taught in schools, and often submitted by "people with a technical education but a failed career". In the words of mathematical historian Howard Eves, "Fermat's Last Theorem has the peculiar distinction of being the mathematical problem for which the greatest number of incorrect proofs have been published."

Local Field

In mathematics, a local field is a special type of field that is a locally compact topological field with respect to a non-discrete topology. Given such a field, an absolute value can be defined on it. There are two basic types of local fields: those in which the absolute value is archimedean and those in which it is not. In the first case, one calls the local field an archimedean local field, in the second case, one calls it a non-archimedean local field. Local fields arise naturally in number theory as completions of global fields.

Every local field is isomorphic (as a topological field) to one of the following:

- Archimedean local fields (characteristic zero): the real numbers R, and the complex numbers C.

- Non-archimedean local fields of characteristic zero: finite extensions of the p-adic numbers Q_p (where p is any prime number).

- Non-archimedean local fields of characteristic p (for p any given prime number): finite extensions of the field of formal Laurent series $F_q((T))$ over a finite field F_q (where q is a power of p).

There is an equivalent definition of non-archimedean local field: it is a field that is complete with respect to a discrete valuation and whose residue field is finite. However, some authors consider a more general notion, requiring only that the residue field be perfect, not necessarily finite. This article uses the former definition.

Induced Absolute Value

Given a locally compact topological field K, an absolute value can be defined as follows. First, consider the additive group of the field. As a locally compact topological group, it has a unique (up to positive scalar multiple) Haar measure μ. The absolute value is defined so as to measure the change in size of a set after multiplying it by an element of K. Specifically, define $|\cdot| : K \to$ R by

$$|a| := \frac{\mu(aX)}{\mu(X)}$$

for any measurable subset X of K (with $0 < \mu(X) < \infty$). This absolute value does not depend on X nor on the choice of Haar measure (since the same scalar multiple ambiguity will occur in both the numerator and the denominator). This definition is very similar to that of the modular function.

Given such an absolute value on K, a new induced topology can be defined on K. This topology is the same as the original topology. Explicitly, for a positive real number m, define the subset B_m of K by

$$B_m := \{a \in K : |a| \le m\}.$$

Then, the B_m make up a neighbourhood basis of 0 in K.

Non-archimedean Local Field Theory

For a non-archimedean local field F (with absolute value denoted by $|\cdot|$), the following objects are important:

- its ring of integers $\mathcal{O} = \{a \in F : |a| \leq 1\}$ which is a discrete valuation ring, is the closed unit ball of F, and is compact;

- the units in its ring of integers $\mathcal{O}^{\times} = \{a \in F : |a| = 1\}$ which forms a group and is the unit sphere of F;

- the unique non-zero prime ideal $\mathfrak{m}$ in its ring of integers which is its open unit ball $\{a \in F : |a| < 1\}$;

- a generator ϖ of $\mathfrak{m}$ called a uniformizer of F;

- its residue field $k = \mathcal{O} / \mathfrak{m}$ which is finite (since it is compact and discrete).

Every non-zero element a of F can be written as $a = \varpi^n u$ with u a unit, and n a unique integer. The normalized valuation of F is the surjective function $v : F \to \mathbb{Z} \cup \{\infty\}$ defined by sending a non-zero a to the unique integer n such that $a = \varpi^n u$ with u a unit, and by sending 0 to ∞. If q is the cardinality of the residue field, the absolute value on F induced by its structure as a local field is given by

$$|a| = q^{-v(a)}.$$

An equivalent definition of a non-archimedean local field is that it is a field that is complete with respect to a discrete valuation and whose residue field is finite.

Higher-dimensional Local Fields

It is natural to introduce non-archimedean local fields in a uniform geometric way as the field of fractions of the completion of the local ring of a one-dimensional arithmetic scheme of rank 1 at its non-singular point. For generalizations, a local field is sometimes called a *one-dimensional local field*.

For a non-negative integer n, an n-dimensional local field is a complete discrete valuation field whose residue field is an $(n - 1)$-dimensional local field. Depending on the definition of local field, a *zero-dimensional local field* is then either a finite field (with the definition used in this article), or a quasi-finite field, or a perfect field.

From the geometric point of view, n-dimensional local fields with last finite residue field are naturally associated to a complete flag of subschemes of an n-dimensional arithmetic scheme.

References

- Apostol, Tom M. (1976), Introduction to analytic number theory, Undergraduate Texts in Mathematics, New York-Heidelberg: Springer-Verlag, ISBN 978-0-387-90163-3.

- Davenport, Harold (2000), Multiplicative number theory, Graduate Texts in Mathematics, 74 (3rd revised ed.), New York: Springer-Verlag, ISBN 978-0-387-95097-6, MR 1790423

- Tenenbaum, Gérald (1995), Introduction to Analytic and Probabilistic Number Theory, Cambridge studies in advanced mathematics, 46, Cambridge University Press, ISBN 0-521-41261-7

- Nathanson, Melvyn B. (1996). Additive Number Theory: The Classical Bases. Graduate Texts in Mathematics. 164. Springer-Verlag. ISBN 0-387-94656-X. Zbl 0859.11002.

- Nathanson, Melvyn B. (1996). Additive Number Theory: Inverse Problems and the Geometry of Sumsets. Graduate Texts in Mathematics. 165. Springer-Verlag. ISBN 0-387-94655-1. Zbl 0859.11003.

- Martinet, J. (1977), "Character theory and Artin L-functions", in Fröhlich, A., Algebraic Number Fields, Proc. Symp. London Math. Soc., Univ. Durham 1975, Academic Press, pp. 1–87, ISBN 0-12-268960-7, Zbl 0359.12015

- Serre, Jean-Pierre (1995), Local Fields, Graduate texts in mathematics, 67, Berlin, Heidelberg: Springer-Verlag, ISBN 0-387-90424-7

- Weil, André (1995), Basic number theory, Classics in Mathematics, Berlin, Heidelberg: Springer-Verlag, ISBN 3-540-58655-5

- Neukirch, Jürgen (1999). Algebraic Number Theory. Grundlehren der mathematischen Wissenschaften. 322. Berlin: Springer-Verlag. ISBN 978-3-540-65399-8. Zbl 0956.11021. MR 1697859.

- "The Abel Prize citation 2016" (PDF). The Abel Prize. The Abel Prize Committee. March 2016. Retrieved 16 March 2016.

Numbers: An Overview

A number is used to count and measure objects. Apart from counting and measuring, numbers are also used for codes and mathematical abstraction. This chapter also explains theories such as natural numbers, rational numbers, integers, prime numbers, real numbers and complex numbers. This section is an overview of the subject matter incorporating all the major aspects of numbers.

Number

A number is a mathematical object used to count, measure, and label. The original examples are the natural numbers 1, 2, 3, and so forth. A notational symbol that represents a number is called a numeral. In addition to their use in counting and measuring, numerals are often used for labels (as with telephone numbers), for ordering (as with serial numbers), and for codes (as with ISBNs). In common usage, *number* may refer to a symbol, a word, or a mathematical abstraction.

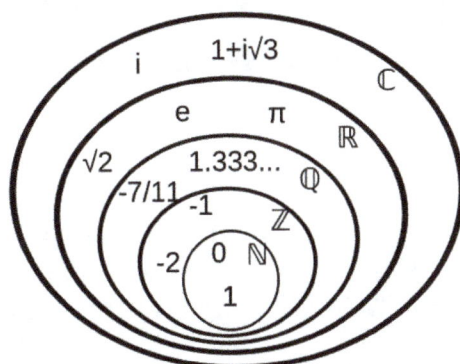

Subsets of the complex numbers.

Besides their practical uses, numbers have cultural significance throughout the world. For example, in Western society the number 13 is regarded as unlucky, and "a million" may signify "a lot." Though it is now regarded as pseudoscience, numerology, the belief in a mystical significance of numbers permeated ancient and medieval thought. Numerology heavily influenced the development of Greek mathematics, stimulating the investigation of many problems in number theory which are still of interest today.

During the 19th century, mathematicians began to develop many different abstractions which share certain properties of numbers and may be seen as extending the concept. Among the first were the hypercomplex numbers, which consist of various extensions or modifications of the complex number system. Today, number systems are considered important special examples of much more general categories such as rings and fields, and the application of the term "number" is a matter of convention, without fundamental significance.

Numerals

Numbers should be distinguished from numerals, the symbols used to represent numbers. Boyer showed that Egyptians created the first ciphered numeral system. Greeks followed by mapping their counting numbers onto Ionian and Doric alphabets. The number five can be represented by digit "5" or by the Roman numeral "V". Notations used to represent numbers are discussed in the article numeral systems. An important development in the history of numerals was the development of a positional system, like modern decimals, which have many advantages, such as representing large numbers with only a few symbols. The Roman numerals require extra symbols for larger numbers.

Main Classification

Different types of numbers have many different uses. Numbers can be classified into sets, called number systems, such as the natural numbers and the real numbers. The same number can be written in many different ways. For different methods of expressing numbers with symbols, such as the Roman numerals.

Each of these number systems may be considered as a proper subset of the next one. This means that each one is canonically isomorphic to a proper subset of the next one, and that there is generally no problem with the abuse of notation consisting of identifying each number system with a subset of the next one. This is expressed, symbolically, by writing

$$\mathbb{N} \subset \mathbb{Z} \subset \mathbb{Q} \subset \mathbb{R} \subset \mathbb{C}.$$

Natural Numbers

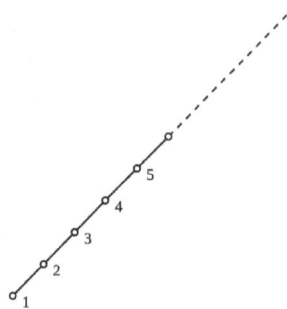

The natural numbers, starting with 1

The most familiar numbers are the natural numbers (sometimes called whole numbers or counting numbers): 1, 2, 3, and so on. Traditionally, the sequence of natural numbers started with 1 (0 was not even considered a number for the Ancient Greeks.) However, in the 19th century, set theorists and other mathematicians started including 0 (cardinality of the empty set, i.e. 0 elements, where 0 is thus the smallest cardinal number) in the set of natural numbers. Today, different mathematicians use the term to describe both sets, including 0 or not. The mathematical symbol for the set of all natural numbers is N, also written $\mathbb{N}$, and sometimes $\mathbb{N}_0$ or $\mathbb{N}_1$ when it is necessary to indicate whether the set should start with 0 or 1, respectively.

In the base 10 numeral system, in almost universal use today for mathematical operations, the symbols for natural numbers are written using ten digits: 0, 1, 2, 3, 4, 5, 6, 7, 8, and 9. The radix or base is the number of unique numerical digits, including zero, that a numeral system uses to represent numbers (for the decimal system, the radix is 10). In this base 10 system, the rightmost digit of a natural number has a place value of 1, and every other digit has a place value ten times that of the place value of the digit to its right.

In set theory, which is capable of acting as an axiomatic foundation for modern mathematics, natural numbers can be represented by classes of equivalent sets. For instance, the number 3 can be represented as the class of all sets that have exactly three elements. Alternatively, in Peano Arithmetic, the number 3 is represented as sss0, where s is the "successor" function (i.e., 3 is the third successor of 0). Many different representations are possible; all that is needed to formally represent 3 is to inscribe a certain symbol or pattern of symbols three times.

Integers

The negative of a positive integer is defined as a number that produces 0 when it is added to the corresponding positive integer. Negative numbers are usually written with a negative sign (a minus sign). As an example, the negative of 7 is written −7, and 7 + (−7) = 0. When the set of negative numbers is combined with the set of natural numbers (including 0), the result is defined as the set of integers, Z also written $\mathbb{Z}$. Here the letter Z comes from German *Zahl*, meaning "number". The set of integers forms a ring with the operations addition and multiplication.

The natural numbers form a subset of the integers. As there is no common standard for the inclusion or not of zero in the natural numbers, the natural numbers without zero are commonly referred to as positive integers, and the natural numbers with zero are referred to as non-negative integers.

Rational Numbers

A rational number is a number that can be expressed as a fraction with an integer numerator and a positive integer denominator. Negative denominators are allowed, but are commonly avoided, as every rational number is equal to a fraction with positive denominator. Fractions are written as two integers, the numerator and the denominator, with a dividing bar between them. The fraction m/n represents m parts of a whole divided into n equal parts. Two different fractions may correspond to the same rational number; for example 1/2 and 2/4 are equal, that is:

$$\frac{1}{2} = \frac{2}{4}.$$

If the absolute value of m is greater than n (supposed to be positive), then the absolute value of the fraction is greater than 1. Fractions can be greater than, less than, or equal to 1 and can also be positive, negative, or 0. The set of all rational numbers includes the integers, since every integer can be written as a fraction with denominator 1. For example −7 can be written −7/1. The symbol for the rational numbers is Q (for *quotient*), also written $\mathbb{Q}$.

Real Numbers

The real numbers include all the measuring numbers. The symbol for the real numbers is R, also written as $\mathbb{R}$. Real numbers are usually represented by using decimal numerals, in which a decimal point is placed to the right of the digit with place value 1. Each digit to the right of the decimal point has a place value one-tenth of the place value of the digit to its left. For example, 123.456 represents 123456/1000, or, in words, one hundred, two tens, three ones, four tenths, five hundredths, and six thousandths. A finite decimal representation allows us to represent exactly only the integers and those rational numbers whose denominators have only prime factors which are factors of ten. Thus one half is 0.5, one fifth is 0.2, one tenth is 0.1, and one fiftieth is 0.02. To represent the rest of the real numbers requires an infinite sequence of digits after the decimal point. Since it is impossible to write infinitely many digits, real numbers are commonly represented by rounding or truncating this sequence, or by establishing a pattern, such as 0.333..., with an ellipsis to indicate that the pattern continues. Thus 123.456 is an approximation of any real number between 1234555/10000 and 1234565/10000 (rounding) or any real number between 123456/1000 and 123457/1000 (truncation). Negative real numbers are written with a preceding minus sign: -123.456.

Every rational number is also a real number. It is not the case, however, that every real number is rational. A real number, which is not rational, is called irrational. A decimal represents a rational number if and only if it has a finite number of digits or eventually repeats for ever, after any initial finite string digits. For example, 1/2 = 0.5 and 1/3 = 0.333... (forever repeating 3s, otherwise written 0.3). On the other hand, the real number π, the ratio of the circumference of any circle to its diameter, is

$$\pi = 3.14159265358979\ldots$$

Since the decimal neither ends nor eventually repeats forever it cannot be written as a fraction, and is an example of an irrational number. Other irrational numbers include

$$\sqrt{2} = 1.41421356237\ldots$$

(the square root of 2, that is, the positive number whose square is 2).

Just as the same fraction can be written in more than one way, the same decimal may have more than one representation. 1.0 and 0.999... are two different decimal numerals representing the natural number 1. There are infinitely many other ways of representing the number 1, for example 1.00, 1.000, and so on.

Every real number is either rational or irrational. Every real number corresponds to a point on the number line. The real numbers also have an important but highly technical property called the least upper bound property.

When a real number represents a measurement, there is always a margin of error. This gives rise to rounding or truncating a decimal, so that digits that suggest a greater accuracy than the measurement itself can grant are removed. The remaining digits are called significant digits. For example, measurements with a ruler can seldom be made without a margin of error of at least 0.001 meters. If the sides of a rectangle are measured as 1.23 meters and 4.56 meters, then multiplication gives an area for the rectangle between 5.614591 square meters and 5.603011 square meters. Since not even the second digit after the decimal place is preserved, the following digits are fully *insignificant*. Therefore, sensibly, the result is usually rounded to 5.61.

In abstract algebra, it can be shown that any complete ordered field is isomorphic to the real numbers. The real numbers are not, however, an algebraically closed field, because they do not include a solution to the algebraic equation $x^2 + 1 = 0$, often addressed as the square root of minus one.

Complex Numbers

Moving to a greater level of abstraction, the real numbers can be extended to the complex numbers. This set of numbers arose historically from trying to find closed formulas for the roots of cubic and quartic polynomials. This led to expressions involving the square roots of negative numbers, and eventually to the definition of a new number: a square root of –1, denoted by i, a symbol assigned by Leonhard Euler, and called the imaginary unit. The complex numbers consist of all numbers of the form

$$a + bi$$

where a and b are real numbers. Because of this, complex numbers correspond to points on the complex plane, a vector space of two real dimensions. In the expression $a + bi$, the real number a is called the real part and b is called the imaginary part. If the real part of a complex number is 0, then the number is called an imaginary number or is referred to as *purely imaginary*; if the imaginary part is 0, then the number is a real number. Thus the real numbers are a subset of the complex numbers. If the real and imaginary parts of a complex number are both integers, then the number is called a Gaussian integer. The symbol for the complex numbers is C or $\mathbb{C}$.

In abstract algebra, the complex numbers are an example of an algebraically closed field, meaning that every polynomial with complex coefficients can be factored into linear factors. Like the real number system, the complex number system is a field and is complete, but unlike the real numbers, it is not ordered. That is, there is no consistent meaning assignable to saying that i is greater than 1, nor is there any meaning in saying that i is less than 1. In technical terms, the complex numbers lack of a total order that is compatible with field operations.

Subclasses of The Integers

Even and Odd Numbers

An even number is an integer that is "evenly divisible" by two, that is divisible by two without remainder; an odd number is an integer that is not even. (The old-fashioned term "evenly divisible" is now almost always shortened to "divisible".) Equivalently, another way of defining an odd number is that it is an integer of the form $n = 2k + 1$, where k is an integer, and an even number has the form $n = 2k$ where k is an integer.

Prime Numbers

A prime number is an integer greater than 1 that is not the product of two smaller positive integers. The first few prime numbers are 2, 3, 5, 7, and 11. The prime numbers have been widely studied for more than 2000 years and have led to many questions, only some of which have been answered. The study of these questions is called number theory. An example of a question that is still unanswered is whether every even number is the sum of two primes. This is called Goldbach's conjecture.

A question that has been answered is whether every integer greater than one is a product of primes in only one way, except for a rearrangement of the primes. This is called fundamental theorem of arithmetic. A proof appears in Euclid's Elements.

Other Classes of Integers

Many subsets of the natural numbers have been the subject of specific studies and have been named, often after the first mathematician that has studied them. Example of such sets of integers are Fibonacci numbers and perfect numbers. For more examples, see Integer sequence.

Subclasses of The Complex Numbers

Algebraic, Irrational and Transcendental Numbers

Algebraic numbers are those that are a solution to a polynomial equation with integer coefficients. Real numbers that are not rational numbers are called irrational numbers. Complex numbers which are not algebraic are called transcendental numbers. The algebraic numbers that are solutions of a monic polynomial equation with integer coefficients are called algebraic integers.

Computable Numbers

A computable number, also known as *recursive number*, is a real number such that there exists an algorithm which, given a positive number n as input, produces the first n digits of the computable number's decimal representation. Equivalent definitions can be given using μ-recursive functions, Turing machines or λ-calculus. The computable numbers are stable for all usual arithmetic operations, including the computation of the roots of a polynomial, and thus form a real closed field that contains the real algebraic numbers.

The computable numbers may be viewed as the real numbers that may be exactly represented in a computer: a computable number is exactly represented by its first digits and a program for computing further digits. However, the computable numbers are rarely used in practice. One reason is that there is no algorithm for testing the equality of two computable numbers. More precisely, there cannot exist any algorithm which takes any computable number as an input, and decides in every case if this number is equal to zero or not.

The set of computable numbers has the same cardinality as the natural numbers. Therefore, almost all real numbers are non-computable. However, it is very difficult to produce explicitly a real number that is not computable.

Extensions of The Concept

P-Adic Numbers

The *p*-adic numbers may have infinitely long expansions to the left of the decimal point, in the same way that real numbers may have infinitely long expansions to the right. The number system that results depends on what base is used for the digits: any base is possible, but a prime number base provides the best mathematical properties. The set of the *p*-adic numbers contains the rational numbers, but is not contained in the complex numbers.

The elements of an algebraic function field over a finite field and algebraic numbers have many similar properties. Therefore, they are often regarded as numbers by number theorists. The *p*-adic numbers play an important role in this analogy.

Hypercomplex Numbers

Some number systems that are not included in the complex numbers may be constructed from the real numbers in a way that generalize the construction of the complex numbers. They are sometimes called hypercomplex numbers. They include the quaternions H, introduced by Sir William Rowan Hamilton, in which multiplication is not commutative, and the octonions, in which multiplication is not associative.

Transfinite Numbers

For dealing with infinite sets, the natural numbers have been generalized to the ordinal numbers and to the cardinal numbers. The former gives the ordering of the set, while the latter gives its size. For finite sets, both ordinal and cardinal numbers are identified with the natural numbers. In the infinite case, many ordinal numbers correspond to the same cardinal number.

Nonstandard Numbers

Hyperreal numbers are used in non-standard analysis. The hyperreals, or nonstandard reals (usually denoted as *R), denote an ordered field that is a proper extension of the ordered field of real numbers R and satisfies the transfer principle. This principle allows true first-order statements about R to be reinterpreted as true first-order statements about *R.

Superreal and surreal numbers extend the real numbers by adding infinitesimally small numbers and infinitely large numbers, but still form fields.

A relation number is defined as the class of relations consisting of all those relations that are similar to one member of the class.

History

First Use of Numbers

Bones and other artifacts have been discovered with marks cut into them that many believe are tally marks. These tally marks may have been used for counting elapsed time, such as numbers of days, lunar cycles or keeping records of quantities, such as of animals.

A tallying system has no concept of place value (as in modern decimal notation), which limits its representation of large numbers. Nonetheless tallying systems are considered the first kind of abstract numeral system.

The first known system with place value was the Mesopotamian base 60 system (ca. 3400 BC) and the earliest known base 10 system dates to 3100 BC in Egypt.

Zero

The use of 0 as a number should be distinguished from its use as a placeholder numeral in place-value systems. Many ancient texts used 0. Babylonian (Modern Iraq) and Egyptian texts used it. Egyptians used the word *nfr* to denote zero balance in double entry accounting entries. Indian texts used a Sanskrit word *Shunye* or *shunya* to refer to the concept of *void*. In mathematics texts this word often refers to the number zero.

The number 605 in Khmer numerals, from an inscription from 683 AD. An early use of zero as a decimal figure.

Records show that the Ancient Greeks seemed unsure about the status of 0 as a number: they asked themselves "how can 'nothing' be something?" leading to interesting philosophical and, by the Medieval period, religious arguments about the nature and existence of 0 and the vacuum. The paradoxes of Zeno of Elea depend in large part on the uncertain interpretation of 0. (The ancient Greeks even questioned whether 1 was a number.)

The late Olmec people of south-central Mexico began to use a true zero (a shell glyph) in the New World possibly by the 4th century BC but certainly by 40 BC, which became an integral part of Maya numerals and the Maya calendar. Mayan arithmetic used base 4 and base 5 written as base 20. Sanchez in 1961 reported a base 4, base 5 "finger" abacus.

By 130 AD, Ptolemy, influenced by Hipparchus and the Babylonians, was using a symbol for 0 (a small circle with a long overbar) within a sexagesimal numeral system otherwise using alphabetic Greek numerals. Because it was used alone, not as just a placeholder, this Hellenistic zero was the first *documented* use of a true zero in the Old World. In later Byzantine manuscripts of his *Syntaxis Mathematica* (*Almagest*), the Hellenistic zero had morphed into the Greek letter omicron (otherwise meaning 70).

Another true zero was used in tables alongside Roman numerals by 525 (first known use by Dionysius Exiguus), but as a word, *nulla* meaning *nothing*, not as a symbol. When division produced 0 as a remainder, *nihil*, also meaning *nothing*, was used. These medieval zeros were used by all future medieval computists (calculators of Easter). An isolated use of their initial, N, was used in a table of Roman numerals by Bede or a colleague about 725, a true zero symbol.

An early documented use of the zero by Brahmagupta (in the *Brāhmasphuṭasiddhānta*) dates to 628. He treated 0 as a number and discussed operations involving it, including division. By this time (the 7th century) the concept had clearly reached Cambodia as Khmer numerals, and documentation shows the idea later spreading to China and the Islamic world.

Negative Numbers

The abstract concept of negative numbers was recognized as early as 100 BC – 50 BC in China. *The Nine Chapters on the Mathematical Art* contains methods for finding the areas of figures; red rods were used to denote positive coefficients, black for negative. The first reference in a Western work was in the 3rd century AD in Greece. Diophantus referred to the equation equivalent to $4x + 20 = 0$ (the solution is negative) in *Arithmetica*, saying that the equation gave an absurd result.

During the 600s, negative numbers were in use in India to represent debts. Diophantus' previous reference was discussed more explicitly by Indian mathematician Brahmagupta, in *Brāhmasphuṭasiddhānta* 628, who used negative numbers to produce the general form quadratic formula that remains in use today. However, in the 12th century in India, Bhaskara gives negative roots for quadratic equations but says the negative value "is in this case not to be taken, for it is inadequate; people do not approve of negative roots."

European mathematicians, for the most part, resisted the concept of negative numbers until the 17th century, although Fibonacci allowed negative solutions in financial problems where they could be interpreted as debts (chapter 13 of *Liber Abaci*, 1202) and later as losses (in *Flos*). At the same time, the Chinese were indicating negative numbers by drawing a diagonal stroke through the right-most non-zero digit of the corresponding positive number's numeral. The first use of negative numbers in a European work was by Nicolas Chuquet during the 15th century. He used them as exponents, but referred to them as "absurd numbers".

As recently as the 18th century, it was common practice to ignore any negative results returned by equations on the assumption that they were meaningless, just as René Descartes did with negative solutions in a Cartesian coordinate system.

Rational Numbers

It is likely that the concept of fractional numbers dates to prehistoric times. The Ancient Egyptians used their Egyptian fraction notation for rational numbers in mathematical texts such as the Rhind Mathematical Papyrus and the Kahun Papyrus. Classical Greek and Indian mathematicians made studies of the theory of rational numbers, as part of the general study of number theory. The best known of these is Euclid's *Elements*, dating to roughly 300 BC. Of the Indian texts, the most relevant is the Sthananga Sutra, which also covers number theory as part of a general study of mathematics.

The concept of decimal fractions is closely linked with decimal place-value notation; the two seem to have developed in tandem. For example, it is common for the Jain math sutra to include calculations of decimal-fraction approximations to pi or the square root of 2. Similarly, Babylonian math texts had always used sexagesimal (base 60) fractions with great frequency.

Irrational Numbers

The earliest known use of irrational numbers was in the Indian Sulba Sutras composed between 800 and 500 BC. The first existence proofs of irrational numbers is usually attributed to Pythagoras, more specifically to the Pythagorean Hippasus of Metapontum, who produced a (most likely geometrical) proof of the irrationality of the square root of 2. The story goes that Hippasus discovered irrational numbers when trying to represent the square root of 2 as a fraction. However Pythagoras believed in the absoluteness of numbers, and could not accept the existence of irrational numbers. He could not disprove their existence through logic, but he could not accept irrational numbers, so he sentenced Hippasus to death by drowning.

The 16th century brought final European acceptance of negative integral and fractional numbers. By the 17th century, mathematicians generally used decimal fractions with modern notation. It was not, however, until the 19th century that mathematicians separated irrationals into algebraic and transcendental parts, and once more undertook scientific study of irrationals. It had remained almost dormant since Euclid. In 1872, the publication of the theories of Karl Weierstrass (by his pupil Kossak), Heine (*Crelle*, 74), Georg Cantor (Annalen, 5), and Richard Dedekind was brought about. In 1869, Méray had taken the same point of departure as Heine, but the theory is generally referred to the year 1872. Weierstrass's method was completely set forth by Salvatore Pincherle (1880), and Dedekind's has received additional prominence through the author's later work (1888) and endorsement by Paul Tannery (1894). Weierstrass, Cantor, and Heine base their theories on infinite series, while Dedekind founds his on the idea of a cut (Schnitt) in the system of real numbers, separating all rational numbers into two groups having certain characteristic properties. The subject has received later contributions at the hands of Weierstrass, Kronecker (Crelle, 101), and Méray.

The search for roots of quintic and higher degree equations was an important development, the Abel–Ruffini theorem (Ruffini 1799, Abel 1824) showed that they could not be solved by radicals (formulas involving only arithmetical operations and roots). Hence it was necessary to consider the wider set of algebraic numbers (all solutions to polynomial equations). Galois (1832) linked polynomial equations to group theory giving rise to the field of Galois theory.

Continued fractions, closely related to irrational numbers (and due to Cataldi, 1613), received attention at the hands of Euler, and at the opening of the 19th century were brought into prominence through the writings of Joseph Louis Lagrange. Other noteworthy contributions have been made by Druckenmüller (1837), Kunze (1857), Lemke (1870), and Günther (1872). Ramus (1855) first connected the subject with determinants, resulting, with the subsequent contributions of Heine, Möbius, and Günther, in the theory of Kettenbruchdeterminanten.

Transcendental Numbers and Reals

The existence of transcendental numbers was first established by Liouville (1844, 1851). Hermite proved in 1873 that e is transcendental and Lindemann proved in 1882 that π is transcendental. Finally, Cantor showed that the set of all real numbers is uncountably infinite but the set of all algebraic numbers is countably infinite, so there is an uncountably infinite number of transcendental numbers.

Infinity and Infinitesimals

The earliest known conception of mathematical infinity appears in the Yajur Veda, an ancient Indian script, which at one point states, "If you remove a part from infinity or add a part to infinity, still what remains is infinity." Infinity was a popular topic of philosophical study among the Jain mathematicians c. 400 BC. They distinguished between five types of infinity: infinite in one and two directions, infinite in area, infinite everywhere, and infinite perpetually.

Aristotle defined the traditional Western notion of mathematical infinity. He distinguished between actual infinity and potential infinity—the general consensus being that only the latter had true value. Galileo Galilei's *Two New Sciences* discussed the idea of one-to-one correspondences between infinite sets. But the next major advance in the theory was made by Georg Cantor; in 1895 he published a book about his new set theory, introducing, among other things, transfinite numbers and formulating the continuum hypothesis.

In the 1960s, Abraham Robinson showed how infinitely large and infinitesimal numbers can be rigorously defined and used to develop the field of nonstandard analysis. The system of hyper-real numbers represents a rigorous method of treating the ideas about infinite and infinitesimal numbers that had been used casually by mathematicians, scientists, and engineers ever since the invention of infinitesimal calculus by Newton and Leibniz.

A modern geometrical version of infinity is given by projective geometry, which introduces "ideal points at infinity", one for each spatial direction. Each family of parallel lines in a given direction is postulated to converge to the corresponding ideal point. This is closely related to the idea of vanishing points in perspective drawing.

Complex Numbers

The earliest fleeting reference to square roots of negative numbers occurred in the work of the mathematician and inventor Heron of Alexandria in the 1st century AD, when he considered the volume of an impossible frustum of a pyramid. They became more prominent when in the 16th century closed formulas for the roots of third and fourth degree polynomials were discovered by Italian mathematicians such as Niccolò Fontana Tartaglia and Gerolamo Cardano. It was soon realized that these formulas, even if one was only interested in real solutions, sometimes required the manipulation of square roots of negative numbers.

This was doubly unsettling since they did not even consider negative numbers to be on firm ground at the time. When René Descartes coined the term "imaginary" for these quantities in 1637, he intended it as derogatory. A further source of confusion was that the equation

$$\left(\sqrt{-1}\right)^2 = \sqrt{-1}\sqrt{-1} = -1$$

seemed capriciously inconsistent with the algebraic identity

$$\sqrt{a}\sqrt{b} = \sqrt{ab},$$

which is valid for positive real numbers a and b, and was also used in complex number calculations with one of a, b positive and the other negative. The incorrect use of this identity, and the related identity

$$\frac{1}{\sqrt{a}} = \sqrt{\frac{1}{a}}$$

in the case when both a and b are negative even bedeviled Euler. This difficulty eventually led him to the convention of using the special symbol i in place of $\sqrt{-1}$ to guard against this mistake.

The 18th century saw the work of Abraham de Moivre and Leonhard Euler. De Moivre's formula (1730) states:

$$(\cos\theta + i\sin\theta)^n = \cos n\theta + i\sin n\theta$$

and to Euler (1748) Euler's formula of complex analysis:

$$\cos\theta + i\sin\theta = e^{i\theta}.$$

The existence of complex numbers was not completely accepted until Caspar Wessel described the geometrical interpretation in 1799. Carl Friedrich Gauss rediscovered and popularized it several years later, and as a result the theory of complex numbers received a notable expansion. The idea of the graphic representation of complex numbers had appeared, however, as early as 1685, in Wallis's *De Algebra tractatus*.

Also in 1799, Gauss provided the first generally accepted proof of the fundamental theorem of algebra, showing that every polynomial over the complex numbers has a full set of solutions in that realm. The general acceptance of the theory of complex numbers is due to the labors of Augustin Louis Cauchy and Niels Henrik Abel, and especially the latter, who was the first to boldly use complex numbers with a success that is well known.

Gauss studied complex numbers of the form $a + bi$, where a and b are integral, or rational (and i is one of the two roots of $x^2 + 1 = 0$). His student, Gotthold Eisenstein, studied the type $a + b\omega$, where ω is a complex root of $x^3 - 1 = 0$. Other such classes (called cyclotomic fields) of complex numbers derive from the roots of unity $x^k - 1 = 0$ for higher values of k. This generalization is largely due to Ernst Kummer, who also invented ideal numbers, which were expressed as geometrical entities by Felix Klein in 1893.

In 1850 Victor Alexandre Puiseux took the key step of distinguishing between poles and branch points, and introduced the concept of essential singular points. This eventually led to the concept of the extended complex plane.

Prime Numbers

Prime numbers have been studied throughout recorded history. Euclid devoted one book of the *Elements* to the theory of primes; in it he proved the infinitude of the primes and the fundamental theorem of arithmetic, and presented the Euclidean algorithm for finding the greatest common divisor of two numbers.

In 240 BC, Eratosthenes used the Sieve of Eratosthenes to quickly isolate prime numbers. But most further development of the theory of primes in Europe dates to the Renaissance and later eras.

In 1796, Adrien-Marie Legendre conjectured the prime number theorem, describing the asymptotic distribution of primes. Other results concerning the distribution of the primes include Euler's proof that the sum of the reciprocals of the primes diverges, and the Goldbach conjecture, which claims that any sufficiently large even number is the sum of two primes. Yet another conjecture related to the distribution of prime numbers is the Riemann hypothesis, formulated by Bernhard Riemann in 1859. The prime number theorem was finally proved by Jacques Hadamard and Charles de la Vallée-Poussin in 1896. Goldbach and Riemann's conjectures remain unproven and unrefuted.

Natural Number

In mathematics, the natural numbers are those used for counting (as in "there are *six* coins on the table") and ordering (as in "this is the *third* largest city in the country"). In common language, words used for counting are "cardinal numbers" and words used for ordering are "ordinal numbers".

Natural numbers can be used for counting (one apple, two apples, three apples, ...)

Some authors and ISO 31-11 begin the natural numbers with 0, corresponding to the non-negative integers 0, 1, 2, 3, ..., whereas others start with 1, corresponding to the positive integers 1, 2, 3, ... Texts that exclude zero from the natural numbers sometimes refer to the natural numbers together with zero as the whole numbers, but in other writings, that term is used instead for the integers (including negative integers).

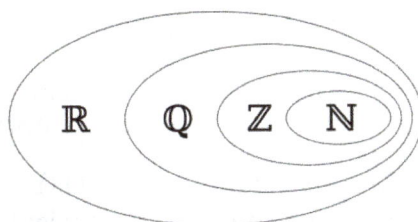

Real numbers (R) include the rational (Q), which include the integers (Z), which include the natural numbers (N)

The natural numbers are the basis from which many other number sets may be built by extension: the integers, by including an additive inverse ($-n$) for each natural number n (and zero, if it is not there already, as its own additive inverse); the rational numbers, by including a multiplicative inverse ($1/n$) for each nonzero integer n; the real numbers by including with the rationals the (converging) Cauchy sequences of rationals; the complex numbers, by including with the real numbers the unresolved square root of minus one; and so on. These chains of extensions make the natural numbers canonically embedded (identified) in the other number systems.

Properties of the natural numbers, such as divisibility and the distribution of prime numbers, are studied in number theory. Problems concerning counting and ordering, such as partitioning and enumerations, are studied in combinatorics.

In common language, for example in primary school, natural numbers may be called counting numbers to contrast the discreteness of counting to the continuity of measurement, established by the real numbers.

The natural numbers can, at times, appear as a convenient set of names (labels), that is, as what linguists call nominal numbers, foregoing many or all of the properties of being a number in a mathematical sense.

History

Ancient Roots

The most primitive method of representing a natural number is to put down a mark for each object. Later, a set of objects could be tested for equality, excess or shortage, by striking out a mark and removing an object from the set.

The Ishango bone (on exhibition at the Royal Belgian Institute of Natural Sciences) is believed to have been used 20,000 years ago for natural number arithmetic.

The first major advance in abstraction was the use of numerals to represent numbers. This allowed systems to be developed for recording large numbers. The ancient Egyptians developed a powerful system of numerals with distinct hieroglyphs for 1, 10, and all the powers of 10 up to over 1 million. A stone carving from Karnak, dating from around 1500 BC and now at the Louvre in Paris, depicts 276 as 2 hundreds, 7 tens, and 6 ones; and similarly for the number 4,622. The Babylonians had a place-value system based essentially on the numerals for 1 and 10, using base sixty, so that the symbol for sixty was the same as the symbol for one, its value being determined from context.

A much later advance was the development of the idea that 0 can be considered as a number, with its own numeral. The use of a 0 digit in place-value notation (within other numbers) dates back as early as 700 BC by the Babylonians, but they omitted such a digit when it would have been the last symbol in the number. The Olmec and Maya civilizations used 0 as a separate number as early as the 1st century BC, but this usage did not spread beyond Mesoamerica. The use of a numeral 0 in modern times originated with the Indian mathematician Brahmagupta in 628. However, 0 had been used as a number in the medieval computus (the calculation of the date of Easter), beginning with Dionysius Exiguus in 525, without being denoted by a numeral (standard Roman numerals do not have a symbol for 0); instead *nulla* (or the genitive form *nullae*) from *nullus*, the Latin word for "none", was employed to denote a 0 value.

The first systematic study of numbers as abstractions is usually credited to the Greek philosophers Pythagoras and Archimedes. Some Greek mathematicians treated the number 1 differently than larger numbers, sometimes even not as a number at all.

Independent studies also occurred at around the same time in India, China, and Mesoamerica.

Modern Definitions

In 19th century Europe, there was mathematical and philosophical discussion about the exact nature of the natural numbers. A school of Naturalism stated that the natural numbers were a direct consequence of the human psyche. Henri Poincaré was one of its advocates, as was Leopold Kronecker who summarized "God made the integers, all else is the work of man".

In opposition to the Naturalists, the constructivists saw a need to improve the logical rigor in the foundations of mathematics. In the 1860s, Hermann Grassmann suggested a recursive definition for natural numbers thus stating they were not really natural but a consequence of definitions. Later, two classes of such formal definitions were constructed; later, they were shown to be equivalent in most practical applications.

Set-theoretical definitions of natural numbers were initiated by Frege and he initially defined a natural number as the class of all sets that are in one-to-one correspondence with a particular set, but this definition turned out to lead to paradoxes including Russell's paradox. Therefore, this formalism was modified so that a natural number is defined as a particular set, and any set that can be put into one-to-one correspondence with that set is said to have that number of elements.

The second class of definitions was introduced by Giuseppe Peano and is now called Peano arithmetic. It is based on an axiomatization of the properties of ordinal numbers: each natural number has a successor and every non-zero natural number has a unique predecessor. Peano arithmetic is equiconsistent with several weak systems of set theory. One such system is ZFC with the axiom of infinity replaced by its negation. Theorems that can be proved in ZFC but cannot be proved using the Peano Axioms include Goodstein's theorem.

With all these definitions it is convenient to include 0 (corresponding to the empty set) as a natural number. Including 0 is now the common convention among set theorists and logicians. Other mathematicians also include 0 although many have kept the older tradition and take 1 to be the first natural number. Computer scientists often start from zero when enumerating items like loop counters and string- or array- elements.

Notation

Mathematicians use N or $\mathbb{N}$ (an N in blackboard bold) to refer to the set of all natural numbers. This set is countably infinite: it is infinite but countable by definition. This is also expressed by saying that the cardinal number of the set is aleph-naught $\aleph_0$.

To be unambiguous about whether 0 is included or not, sometimes an index (or superscript) "0" is added in the former case, and a superscript "*" or subscript "1" is added in the latter case:

N

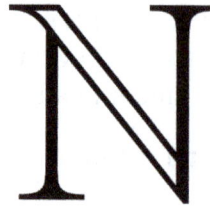

The double-struck capital N symbol, often used to denote the set of all natural numbers.

$$N^0 = N_0 = \{0, 1, 2, \ldots\}$$

$$N^* = N^+ = N_1 = N_{>0} = \{1, 2, \ldots\}.$$

Alternately natural numbers may be distinguished from positive integers with the notation, but it must be understood by context that, since both symbols are used, the natural number do contain zero.

$$N = \{0, 1, 2, \ldots\}.$$

$$Z^+ = \{1, 2, \ldots\}.$$

Properties

Addition

One can recursively define an addition on the natural numbers by setting $a + 0 = a$ and $a + S(b) = S(a + b)$ for all a, b. Here S should be read as "successor". This turns the natural numbers (N, +) into a commutative monoid with identity element 0, the so-called free object with one generator. This monoid satisfies the cancellation property and can be embedded in a group (in the mathematical sense of the word *group*). The smallest group containing the natural numbers is the integers.

If 1 is defined as $S(0)$, then $b + 1 = b + S(0) = S(b + 0) = S(b)$. That is, $b + 1$ is simply the successor of b.

Multiplication

Analogously, given that addition has been defined, a multiplication $\times$ can be defined via $a \times 0 = 0$ and $a \times S(b) = (a \times b) + a$. This turns $(N^*, \times)$ into a free commutative monoid with identity element 1; a generator set for this monoid is the set of prime numbers.

Relationship Between Addition and Multiplication

Addition and multiplication are compatible, which is expressed in the distribution law: $a \times (b + c) = (a \times b) + (a \times c)$. These properties of addition and multiplication make the natural numbers an instance of a commutative semiring. Semirings are an algebraic generalization of the natural numbers where multiplication is not necessarily commutative. The lack of additive inverses, which is equivalent to the fact that N is not closed under subtraction, means that N is *not* a ring; instead it is a semiring (also known as a *rig*).

If the natural numbers are taken as "excluding 0", and "starting at 1", the definitions of + and × are as above, except that they begin with $a + 1 = S(a)$ and $a \times 1 = a$.

Order

In this section, juxtaposed variables such as ab indicate the product $a \times b$, and the standard order of operations is assumed.

A total order on the natural numbers is defined by letting $a \leq b$ if and only if there exists another natural number c with $a + c = b$. This order is compatible with the arithmetical operations in the following sense: if a, b and c are natural numbers and $a \leq b$, then $a + c \leq b + c$ and $ac \leq bc$. An important property of the natural numbers is that they are well-ordered: every non-empty set of natural numbers has a least element. The rank among well-ordered sets is expressed by an ordinal number; for the natural numbers this is expressed as ω.

Division

In this section, juxtaposed variables such as ab indicate the product $a \times b$, and the standard order of operations is assumed.

While it is in general not possible to divide one natural number by another and get a natural number as result, the procedure of *division with remainder* is available as a substitute: for any two natural numbers a and b with $b \neq 0$ there are natural numbers q and r such that

$$a = bq + r \text{ and } r < b.$$

The number q is called the *quotient* and r is called the *remainder* of division of a by b. The numbers q and r are uniquely determined by a and b. This Euclidean division is key to several other properties (divisibility), algorithms (such as the Euclidean algorithm), and ideas in number theory.

Algebraic Properties Satisfied by The Natural Numbers

The addition (+) and multiplication (×) operations on natural numbers as defined above have several algebraic properties:

- Closure under addition and multiplication: for all natural numbers a and b, both $a + b$ and $a \times b$ are natural numbers.

- Associativity: for all natural numbers a, b, and c, $a + (b + c) = (a + b) + c$ and $a \times (b \times c) = (a \times b) \times c$.

- Commutativity: for all natural numbers a and b, $a + b = b + a$ and $a \times b = b \times a$.

- Existence of identity elements: for every natural number a, $a + 0 = a$ and $a \times 1 = a$.

- Distributivity of multiplication over addition for all natural numbers a, b, and c, $a \times (b + c) = (a \times b) + (a \times c)$.

- No nonzero zero divisors: if a and b are natural numbers such that $a \times b = 0$, then $a = 0$ or $b = 0$.

Generalizations

Two generalizations of natural numbers arise from the two uses:

- A natural number can be used to express the size of a finite set; more generally a cardinal number is a measure for the size of a set also suitable for infinite sets; this refers to a concept of "size" such that if there is a bijection between two sets they have the same size. The set of natural numbers itself and any other countably infinite set has cardinality aleph-null ($\aleph_0$).

- Linguistic ordinal numbers "first", "second", "third" can be assigned to the elements of a totally ordered finite set, and also to the elements of well-ordered countably infinite sets like the set of natural numbers itself. This can be generalized to ordinal numbers which describe the position of an element in a well-ordered set in general. An ordinal number is also used to describe the "size" of a well-ordered set, in a sense different from cardinality: if there is an order isomorphism between two well-ordered sets they have the same ordinal number. The first ordinal number that is not a natural number is expressed as ω; this is also the ordinal number of the set of natural numbers itself.

Many well-ordered sets with cardinal number $\aleph_0$ have an ordinal number greater than ω (the latter is the lowest possible). The least ordinal of cardinality $\aleph_0$ (i.e., the initial ordinal) is ω.

For finite well-ordered sets, there is one-to-one correspondence between ordinal and cardinal numbers; therefore they can both be expressed by the same natural number, the number of elements of the set. This number can also be used to describe the position of an element in a larger finite, or an infinite, sequence.

A countable non-standard model of arithmetic satisfying the Peano Arithmetic (i.e., the first-order Peano axioms) was developed by Skolem in 1933. The hypernatural numbers are an uncountable model that can be constructed from the ordinary natural numbers via the ultrapower construction.

Georges Reeb used to claim provocatively that *The naïve integers don't fill up* N. Other generalizations are discussed in the article on numbers.

Formal Definitions

Peano Axioms

Many properties of the natural numbers can be derived from the Peano axioms.

- Axiom One: 0 is a natural number.

- Axiom Two: Every natural number has a successor.

- Axiom Three: 0 is not the successor of any natural number.

- Axiom Four: If the successor of x equals the successor of y, then x equals y.

- Axiom Five (the axiom of induction): If a statement is true of 0, and if the truth of that statement for a number implies its truth for the successor of that number, then the statement is true for every natural number.

These are not the original axioms published by Peano, but are named in his honor. Some forms of the Peano axioms have 1 in place of 0. In ordinary arithmetic, the successor of x is x + 1. Replacing Axiom Five by an axiom schema one obtains a (weaker) first-order theory called *Peano Arithmetic*.

Constructions Based on Set Theory

Von Neumann Construction

In the area of mathematics called set theory, a special case of the von Neumann ordinal construction defines the natural numbers as follows:

- Set 0 = { }, the empty set,

- Define $S(a) = a \cup \{a\}$ for every set a. $S(a)$ is the successor of a, and S is called the successor function.

- By the axiom of infinity, there exists a set which contains 0 and is closed under the successor function. Such sets are said to be 'inductive'. The intersection of all such inductive sets is defined to be the set of natural numbers. It can be checked that the set of natural numbers satisfies the Peano axioms.

- It follows that each natural number is equal to the set of all natural numbers less than it:

  - 0 = { },

  - $1 = 0 \cup \{0\} = \{0\} = \{\{ \}\}$,

  - $2 = 1 \cup \{1\} = \{0, 1\} = \{\{ \}, \{\{ \}\}\}$,

  - $3 = 2 \cup \{2\} = \{0, 1, 2\} = \{\{ \}, \{\{ \}\}, \{\{ \}, \{\{ \}\}\}\}$,

  - $n = n{-}1 \cup \{n{-}1\} = \{0, 1, ..., n{-}1\} = \{\{ \}, \{\{ \}\}, ..., \{\{ \}, \{\{ \}\}, ...\}\}$, etc.

With this definition, a natural number n is a particular set with n elements, and $n \le m$ if and only if n is a subset of m.

Also, with this definition, different possible interpretations of notations like R^n (n-tuples versus mappings of n into R) coincide.

Even if one does not accept the axiom of infinity and therefore cannot accept that the set of all natural numbers exists, it is still possible to define any one of these sets.

Other Constructions

Although the standard construction is useful, it is not the only possible construction. Zermelo's construction goes as follows:

- Set 0 = { }

- Define $S(a) = \{a\}$,

- It then follows that

- $0 = \{\,\}$,
- $1 = \{0\} = \{\{\,\}\}$,
- $2 = \{1\} = \{\{\{\,\}\}\}$,
- $n = \{n-1\} = \{\{\{...\}\}\}$, etc.

Each natural number is then equal to the set of the natural number preceding it.

Rational Number

In mathematics, a rational number is any number that can be expressed as the quotient or fraction p/q of two integers, a numerator p and a non-zero denominator q. Since q may be equal to 1, every integer is a rational number. The set of all rational numbers, often referred to as "the rationals", is usually denoted by a boldface Q (or blackboard bold , Unicode Q); it was thus $\mathbb{Q}$ denoted in 1895 by Giuseppe Peano after *quoziente*, Italian for "quotient".

The decimal expansion of a rational number always either terminates after a finite number of digits or begins to repeat the same finite sequence of digits over and over. Moreover, any repeating or terminating decimal represents a rational number. These statements hold true not just for base 10, but also for any other integer base (e.g. binary, hexadecimal).

A real number that is not rational is called irrational. Irrational numbers include $\sqrt{2}$, π, e, and φ. The decimal expansion of an irrational number continues without repeating. Since the set of rational numbers is countable, and the set of real numbers is uncountable, almost all real numbers are irrational.

The rational numbers can be formally defined as the equivalence classes of the quotient set $(Z \times (Z \setminus \{0\})) / \sim$, where the cartesian product $Z \times (Z \setminus \{0\})$ is the set of all ordered pairs (m,n) where m and n are integers, n is not 0 ($n \neq 0$), and "$\sim$" is the equivalence relation defined by $(m_1,n_1) \sim (m_2,n_2)$ if, and only if, $m_1 n_2 - m_2 n_1 = 0$.

In abstract algebra, the rational numbers together with certain operations of addition and multiplication form the archetypical field of characteristic zero. As such, it is characterized as having no proper subfield or, alternatively, being the field of fractions for the ring of integers. Finite extensions of Q are called algebraic number fields, and the algebraic closure of Q is the field of algebraic numbers.

In mathematical analysis, the rational numbers form a dense subset of the real numbers. The real numbers can be constructed from the rational numbers by completion, using Cauchy sequences, Dedekind cuts, or infinite decimals.

Zero divided by any other integer equals zero; therefore, zero is a rational number (but division by zero is undefined).

Terminology

The term *rational* in reference to the set Q refers to the fact that a rational number represents a *ratio* of two integers. In mathematics, "rational" is often used as a noun abbreviating "rational number". The adjective *rational* sometimes means that the coefficients are rational numbers. For example, a rational point is a point with rational coordinates (that is a point whose coordinates are rational numbers; a *rational matrix* is a matrix of rational numbers; a *rational polynomial* may be a polynomial with rational coefficients, although the term "polynomial over the rationals" is generally preferred, for avoiding confusion with "rational expression" and "rational function" (a polynomial is a rational expression and defines a rational function, even if its coefficients are not rational numbers). However, a rational curve *is not* a curve defined over the rationals, but a curve which can be parameterized by rational functions.

Arithmetic

Embedding of Integers

Any integer n can be expressed as the rational number $n/1$.

Equality

$$\frac{a}{b} = \frac{c}{d} \text{ if and only if } ad = bc.$$

Ordering

Where both denominators are positive:

$$\frac{a}{b} < \frac{c}{d} \text{ if and only if } ad < bc.$$

If either denominator is negative, the fractions must first be converted into equivalent forms with positive denominators, through the equations:

$$\frac{-a}{-b} = \frac{a}{b} a$$

and

$$\frac{a}{-b} = \frac{-a}{b}.$$

Addition

Two fractions are added as follows:

$$\frac{a}{b} + \frac{c}{d} = \frac{ad + bc}{bd}.$$

Subtraction

$$\frac{a}{b} - \frac{c}{d} = \frac{ad - bc}{bd}.$$

Multiplication

The rule for multiplication is:

$$\frac{a}{b} \cdot \frac{c}{d} = \frac{ac}{bd}.$$

Division

Where $c \neq 0$:

$$\frac{a}{b} \div \frac{c}{d} = \frac{ad}{bc}.$$

Note that division is equivalent to multiplying by the reciprocal of the divisor fraction:

$$\frac{ad}{bc} = \frac{a}{b} \times \frac{d}{c}.$$

Inverse

Additive and multiplicative inverses exist in the rational numbers:

$$-\left(\frac{a}{b}\right) = \frac{-a}{b} = \frac{a}{-b} \quad \text{and} \quad \left(\frac{a}{b}\right)^{-1} = \frac{b}{a} \text{ if } a \neq 0.$$

Exponentiation to Integer Power

If n is a non-negative integer, then

$$\left(\frac{a}{b}\right) \quad \frac{a}{b}$$

and (if $a \neq 0$):

$$\left(\frac{a}{b}\right)^{-n} = \frac{b^n}{a^n}.$$

Continued Fraction Representation

A finite continued fraction is an expression such as

$$a_0 + \cfrac{1}{a_1 + \cfrac{1}{a_2 + \cfrac{1}{\ddots + \cfrac{1}{a_n}}}},$$

where a_n are integers. Every rational number a/b has two closely related expressions as a finite continued fraction, whose coefficients a_n can be determined by applying the Euclidean algorithm to (a,b).

Formal Construction

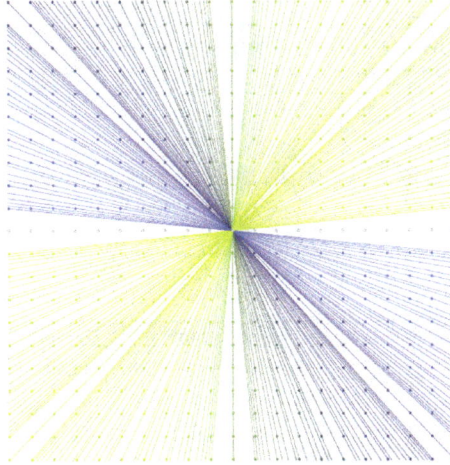

A diagram showing a representation of the equivalent classes of pairs of integers

Mathematically we may construct the rational numbers as equivalence classes of ordered pairs of integers (m,n), with $n \neq 0$. This space of equivalence classes is the quotient space $(\mathbb{Z} \times (\mathbb{Z} \setminus \{0\})) / \sim$, where $(m_1,n_1) \sim (m_2,n_2)$ if, and only if, $m_1 n_2 - m_2 n_1 = 0$. We can define addition and multiplication of these pairs with the following rules:

$$\left(m_1,n_1\right) + \left(m_2,n_2\right) \equiv \left(m_1 n_2 + n_1 m_2, n_1 n_2\right)$$

$$\left(m_1,n_1\right) \times \left(m_2,n_2\right) \equiv \left(m_1 m_2, n_1 n_2\right)$$

and, if $m_2 \neq 0$, division by

$$\frac{\left(m_1,n_1\right)}{\left(m_2,n_2\right)} \equiv \left(m_1 n_2, n_1 m_2\right).$$

The equivalence relation $(m_1,n_1) \sim (m_2,n_2)$ if, and only if, $m_1 n_2 - m_2 n_1 = 0$ is a congruence relation, i.e. it is compatible with the addition and multiplication defined above, and we may define $\mathbb{Q}$ to be the quotient set $(\mathbb{Z} \times (\mathbb{Z} \setminus \{0\})) / \sim$, i.e. we identify two pairs (m_1,n_1) and (m_2,n_2) if they are equivalent in the above sense. We denote by $[(m_1,n_1)]$ the equivalence class containing (m_1,n_1). If $(m_1,n_1) \sim (m_2,n_2)$ then, by definition, (m_1,n_1) belongs to $[(m_2,n_2)]$ and (m_2,n_2) belongs to $[(m_1,n_1)]$; in this case we can write $[(m_1,n_1)] = [(m_2,n_2)]$. Given any equivalence class $[(m,n)]$ there are a countably infinite number of representation, since

$$\cdots = [(-2m,-2n)] = [(-m,-n)] = [(m,n)] = [(2m,2n)] = \cdots.$$

The canonical choice for $[(m,n)]$ is chosen so that n is positive and $\gcd(m,n) = 1$, i.e. m and n share no common factors, i.e. m and n are coprime. For example, we would write $[(1,2)]$ instead of $[(2,4)]$

or $[(-12,-24)]$, even though $[(1,2)] = [(2,4)] = [(-12,-24)]$.

We can also define a total order on Q. Let $\wedge$ be the *and*-symbol and $\vee$ be the *or*-symbol. We say that $[(m_1,n_1)] \le [(m_2,n_2)]$ if:

$$(n_1 n_2 > 0 \wedge m_1 n_2 \le n_1 m_2) \vee (n_1 n_2 < 0 \wedge m_1 n_2 \ge n_1 m_2).$$

The integers may be considered to be rational numbers by the embedding that maps m to $[(m,1)]$.

Properties

A diagram illustrating the countability of the rationals

The set Q, together with the addition and multiplication operations shown above, forms a field, the field of fractions of the integers Z.

The rationals are the smallest field with characteristic zero: every other field of characteristic zero contains a copy of Q. The rational numbers are therefore the prime field for characteristic zero.

The algebraic closure of Q, i.e. the field of roots of rational polynomials, is the algebraic numbers.

The set of all rational numbers is countable. Since the set of all real numbers is uncountable, we say that almost all real numbers are irrational, in the sense of Lebesgue measure, i.e. the set of rational numbers is a null set.

The rationals are a densely ordered set: between any two rationals, there sits another one, and, therefore, infinitely many other ones. For example, for any two fractions such that

$$\frac{a}{b} < \frac{c}{d}$$

(where b, d are positive), we have

$$\frac{a}{b} < \frac{ad+bc}{2bd} < \frac{c}{d}.$$

Any totally ordered set which is countable, dense (in the above sense), and has no least or greatest element is order isomorphic to the rational numbers.

Real Numbers and Topological Properties

The rationals are a dense subset of the real numbers: every real number has rational numbers arbitrarily close to it. A related property is that rational numbers are the only numbers with finite expansions as regular continued fractions.

By virtue of their order, the rationals carry an order topology. The rational numbers, as a subspace of the real numbers, also carry a subspace topology. The rational numbers form a metric space by using the absolute difference metric $d(x,y) = |x - y|$, and this yields a third topology on Q. All three topologies coincide and turn the rationals into a topological field. The rational numbers are an important example of a space which is not locally compact. The rationals are characterized topologically as the unique countable metrizable space without isolated points. The space is also totally disconnected. The rational numbers do not form a complete metric space; the real numbers are the completion of Q under the metric $d(x,y) = |x - y|$, above.

P-Adic Numbers

In addition to the absolute value metric mentioned above, there are other metrics which turn Q into a topological field:

Let p be a prime number and for any non-zero integer a, let $|a|_p = p^{-n}$, where p^n is the highest power of p dividing a.

In addition set $|0|_p = 0$. For any rational number a/b, we set $|a/b|_p = |a|_p / |b|_p$.

Then $d_p(x,y) = |x - y|_p$ defines a metric on Q.

The metric space (Q,d_p) is not complete, and its completion is the p-adic number field Q_p. Ostrowski's theorem states that any non-trivial absolute value on the rational numbers Q is equivalent to either the usual real absolute value or a p-adic absolute value.

Integer

An integer is a number that can be written without a fractional component. For example, 21, 4, 0, and −2048 are integers, while 9.75, 5 $\frac{1}{2}$, and $\sqrt{2}$ are not.

The set of integers consists of zero (0), the natural numbers (1, 2, 3, ...), also called *whole numbers* or *counting numbers*, and their additive inverses (the negative integers, i.e. −1, −2, −3, ...). This is often denoted by a boldface Z ("Z") or blackboard bold $\mathbb{Z}$ (Unicode U+2124 Z) standing for the German word *Zahlen* (['tsacaiblən], "numbers"). Z is a subset of the sets of rational and real numbers and, like the natural numbers, is countably infinite.

The integers form the smallest group and the smallest ring containing the natural numbers. In algebraic number theory, the integers are sometimes called rational integers to distinguish them from the more general algebraic integers. In fact, the (rational) integers are the algebraic integers that are also rational numbers.

Algebraic Properties

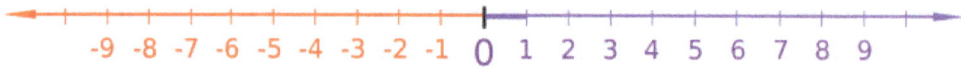

$$-9 \;\; -8 \;\; -7 \;\; -6 \;\; -5 \;\; -4 \;\; -3 \;\; -2 \;\; -1 \;\; 0 \;\; 1 \;\; 2 \;\; 3 \;\; 4 \;\; 5 \;\; 6 \;\; 7 \;\; 8 \;\; 9$$

Integers can be thought of as discrete, equally spaced points on an infinitely long number line. In the above, non-negative integers are shown in purple and negative integers in red.

Like the natural numbers, Z is closed under the operations of addition and multiplication, that is, the sum and product of any two integers is an integer. However, with the inclusion of the negative natural numbers, and, importantly, 0, Z (unlike the natural numbers) is also closed under subtraction. The integers form a unital ring which is the most basic one, in the following sense: for any unital ring, there is a unique ring homomorphism from the integers into this ring. This universal property, namely to be an initial object in the category of rings, characterizes the ring Z.

Z is not closed under division, since the quotient of two integers (e.g. 1 divided by 2), need not be an integer. Although the natural numbers are closed under exponentiation, the integers are not (since the result can be a fraction when the exponent is negative).

The following lists some of the basic properties of addition and multiplication for any integers a, b and c.

	Properties of addition and multiplication on integers	
	Addition	**Multiplication**
Closure:	$a + b$ is an integer	$a \times b$ is an integer
Associativity:	$a + (b + c) = (a + b) + c$	$a \times (b \times c) = (a \times b) \times c$
Commutativity:	$a + b = b + a$	$a \times b = b \times a$
Existence of an identity element:	$a + 0 = a$	$a \times 1 = a$
Existence of inverse elements:	$a + (-a) = 0$	The only invertible rational integers (called units) are -1 and 1.
Distributivity:	$a \times (b + c) = (a \times b) + (a \times c)$ and $(a + b) \times c = (a \times c) + (b \times c)$	
No zero divisors: (*)		If $a \times b = 0$, then $a = 0$ or $b = 0$ (or both)

In the language of abstract algebra, the first five properties listed above for addition say that Z under addition is an abelian group. As a group under addition, Z is a cyclic group, since every non-zero integer can be written as a finite sum $1 + 1 + ... + 1$ or $(-1) + (-1) + ... + (-1)$. In fact, Z under addition is the *only* infinite cyclic group, in the sense that any infinite cyclic group is isomorphic to Z.

The first four properties listed above for multiplication say that Z under multiplication is a commutative monoid. However, not every integer has a multiplicative inverse; e.g. there is no integer x such that $2x = 1$, because the left hand side is even, while the right hand side is odd. This means that Z under multiplication is not a group.

All the rules from the above property table, except for the last, taken together say that Z together with addition and multiplication is a commutative ring with unity. It is the prototype of all objects of such algebraic structure. Only those equalities of expressions are true in Z for all values of variables, which are true in any unital commutative ring. Note that certain non-zero integers map to zero in certain rings.

At last, the property (*) says that the commutative ring Z is an integral domain. In fact, Z provides the motivation for defining such a structure.

The lack of multiplicative inverses, which is equivalent to the fact that Z is not closed under division, means that Z is *not* a field. The smallest field with the usual operations containing the integers is the field of rational numbers. The process of constructing the rationals from the integers can be mimicked to form the field of fractions of any integral domain. And back, starting from an algebraic number field (an extension of rational numbers), its ring of integers can be extracted, which includes Z as its subring.

Moreover, Z is a principal ring.

Although ordinary division is not defined on Z, the division "with remainder" is defined on them. It is called Euclidean division and possesses the following important property: that is, given two integers a and b with $b \neq 0$, there exist unique integers q and r such that $a = q \times b + r$ and $0 \leq r < |b|$, where $|b|$ denotes the absolute value of b. The integer q is called the *quotient* and r is called the *remainder* of the division of a by b. The Euclidean algorithm for computing greatest common divisors works by a sequence of Euclidean divisions.

Again, in the language of abstract algebra, the above says that Z is a Euclidean domain. This implies that Z is a principal ideal domain and any positive integer can be written as the products of primes in an essentially unique way. This is the fundamental theorem of arithmetic.

Order-heoretic Properties

Z is a totally ordered set without upper or lower bound. The ordering of Z is given by: :... $-3 < -2 < -1 < 0 < 1 < 2 < 3 < $ ... An integer is *positive* if it is greater than zero and *negative* if it is less than zero. Zero is defined as neither negative nor positive.

The ordering of integers is compatible with the algebraic operations in the following way:

1. if $a < b$ and $c < d$, then $a + c < b + d$

2. if $a < b$ and $0 < c$, then $ac < bc$.

It follows that Z together with the above ordering is an ordered ring.

The integers are the only nontrivial totally ordered abelian group whose positive elements are well-ordered. This is equivalent to the statement that any Noetherian valuation ring is either a field or a discrete valuation ring.

Construction

In elementary school teaching, integers are often intuitively defined as the (positive) natural numbers, zero, and the negations of the natural numbers. However, this style of definition leads to many different cases (each arithmetic operation needs to be defined on each combination of types of integer) and makes it tedious to prove that these operations obey the laws of arithmetic. Therefore, in modern set-theoretic mathematics a more abstract construction, which allows one to define the arithmetical operations without any case distinction, is often used instead. The integers can thus be formally constructed as the equivalence classes of ordered pairs of natural numbers (a,b).

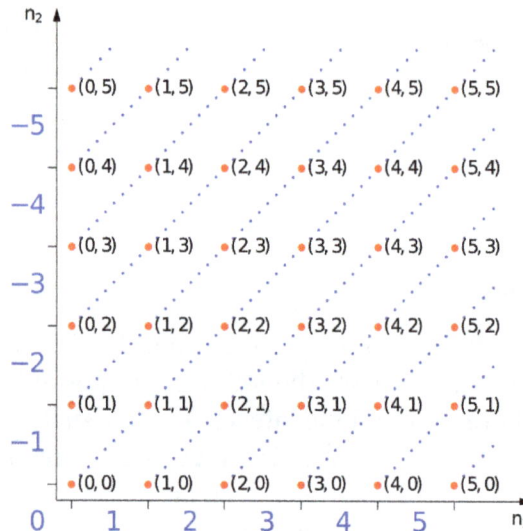

Red points represent ordered pairs of natural numbers. Linked red points are equivalence classes representing the blue integers at the end of the line.

The intuition is that (a,b) stands for the result of subtracting b from a. To confirm our expectation that $1 - 2$ and $4 - 5$ denote the same number, we define an equivalence relation $\sim$ on these pairs with the following rule:

$$(a,b) \sim (c,d)$$

precisely when

$$a + d = b + c.$$

Addition and multiplication of integers can be defined in terms of the equivalent operations on the natural numbers; denoting by $[(a,b)]$ the equivalence class having (a,b) as a member, one has:

$$[(a,b)] + [(c,d)] := [(a+c,b+d)].$$

$$[(a,b)] \cdot [(c,d)] := [(ac+bd,ad+bc)].$$

The negation (or additive inverse) of an integer is obtained by reversing the order of the pair:

$$-[(a,b)] := [(b,a)].$$

Hence subtraction can be defined as the addition of the additive inverse:

$$[(a,b)] - [(c,d)] := [(a+d,b+c)].$$

The standard ordering on the integers is given by:

$$[(a,b)] < [(c,d)] \text{ iff } a+d < b+c.$$

It is easily verified that these definitions are independent of the choice of representatives of the equivalence classes.

Every equivalence class has a unique member that is of the form $(n,0)$ or $(0,n)$ (or both at once). The natural number n is identified with the class $[(n,0)]$ (in other words the natural numbers are embedded into the integers by map sending n to $[(n,0)]$), and the class $[(0,n)]$ is denoted $-n$ (this covers all remaining classes, and gives the class $[(0,0)]$ a second time since $-0 = 0$.

Thus, $[(a,b)]$ is denoted by

$$\begin{cases} a-b, & \text{if } a \geq b \\ -(b-a), & \text{if } a < b. \end{cases}$$

If the natural numbers are identified with the corresponding integers (using the embedding mentioned above), this convention creates no ambiguity.

This notation recovers the familiar representation of the integers as $\{..., -2, -1, 0, 1, 2, ...\}$.

Some examples are:

$$[(0,0)] = [(1,1)] = \cdots = [(k,k)]$$
$$1 = [(1,0)] = [(2,1)] = \cdots = [(k+1,k)]$$
$$-1 = [(0,1)] = [(1,2)] = \cdots = [(k,k+1)]$$
$$2 = [(2,0)] = [(3,1)] = \cdots = [(k+2,k)]$$
$$-2 = [(0,2)] = [(1,3)] = \cdots = [(k,k+2)]$$

Computer Science

An integer is often a primitive data type in computer languages. However, integer data types can only represent a subset of all integers, since practical computers are of finite capacity. Also, in the common two's complement representation, the inherent definition of sign distinguishes between "negative" and "non-negative" rather than "negative, positive, and 0". (It is, however, certainly possible for a computer to determine whether an integer value is truly positive.) Fixed length integer approximation data types (or subsets) are denoted *int* or Integer in several programming languages (such as Algol68, C, Java, Delphi, etc.).

Variable-length representations of integers, such as bignums, can store any integer that fits in the computer's memory. Other integer data types are implemented with a fixed size, usually a number of bits which is a power of 2 (4, 8, 16, etc.) or a memorable number of decimal digits (e.g., 9 or 10).

Cardinality

The cardinality of the set of integers is equal to $\aleph_0$ (aleph-null). This is readily demonstrated by the construction of a bijection, that is, a function that is injective and surjective from Z to N. If N = {0, 1, 2, ...} then consider the function:

$$f(x) = \begin{cases} 2|x|, & \text{if } x \le 0 \\ 2x-1, & \text{if } x > 0. \end{cases}$$

{... (−4,8) (−3,6) (−2,4) (−1,2) (0,0) (1,1) (2,3) (3,5) ...}

If N = {1, 2, 3, ...} then consider the function:

$$g(x) = \begin{cases} 2|x|, & \text{if } x < 0 \\ 2x+1, & \text{if } x \ge 0. \end{cases}$$

{... (−4,8) (−3,6) (−2,4) (−1,2) (0,1) (1,3) (2,5) (3,7) ...}

If the domain is restricted to Z then each and every member of Z has one and only one corresponding member of N and by the definition of cardinal equality the two sets have equal cardinality.

Prime Number

A prime number (or a prime) is a natural number greater than 1 that has no positive divisors other than 1 and itself. A natural number greater than 1 that is not a prime number is called a composite number. For example, 5 is prime because 1 and 5 are its only positive integer factors, whereas 6 is composite because it has the divisors 2 and 3 in addition to 1 and 6. The fundamental theorem of arithmetic establishes the central role of primes in number theory: any integer greater than 1 can be expressed as a product of primes that is unique up to ordering. The uniqueness in this theorem requires excluding 1 as a prime because one can include arbitrarily many instances of 1 in any factorization, e.g., 3, 1 · 3, 1 · 1 · 3, etc. are all valid factorizations of 3.

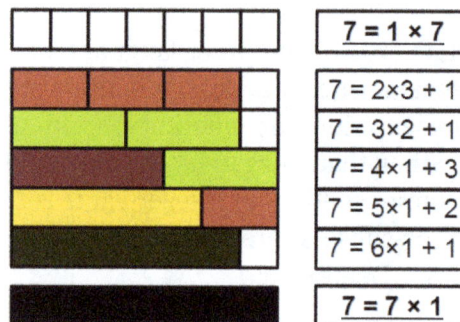

Demonstration, with Cuisenaire rods, that the number 7 is prime, being divisible only by 1 and 7

The property of being prime (or not) is called primality. A simple but slow method of verifying the primality of a given number n is known as trial division. It consists of testing whether n is a multiple of any integer between 2 and $\sqrt{n}$. Algorithms much more efficient than trial division have been devised to test the primality of large numbers. These include the Miller–Rabin primality test, which is fast but has a small probability of error, and the AKS primality test, which always produces the correct answer in polynomial time but is too slow to be practical. Particularly fast methods are available for numbers of special forms, such as Mersenne numbers. As of January 2016, the largest known prime number has 22,338,618 decimal digits.

There are infinitely many primes, as demonstrated by Euclid around 300 BC. There is no known simple formula that separates prime numbers from composite numbers. However, the distribution of primes, that is to say, the statistical behaviour of primes in the large, can be modelled. The first result in that direction is the prime number theorem, proven at the end of the 19th century, which says that the probability that a given, randomly chosen number n is prime is inversely proportional to its number of digits, or to the logarithm of n.

Many questions regarding prime numbers remain open, such as Goldbach's conjecture (that every even integer greater than 2 can be expressed as the sum of two primes), and the twin prime conjecture (that there are infinitely many pairs of primes whose difference is 2). Such questions spurred the development of various branches of number theory, focusing on analytic or algebraic aspects of numbers. Primes are used in several routines in information technology, such as public-key cryptography, which makes use of properties such as the difficulty of factoring large numbers into their prime factors. Prime numbers give rise to various generalizations in other mathematical domains, mainly algebra, such as prime elements and prime ideals.

Definition and Examples

A natural number (i.e. 1, 2, 3, 4, 5, 6, etc.) is called a prime number (or a prime) if it has exactly two positive divisors, 1 and the number itself. Natural numbers greater than 1 that are not prime are called *composite*.

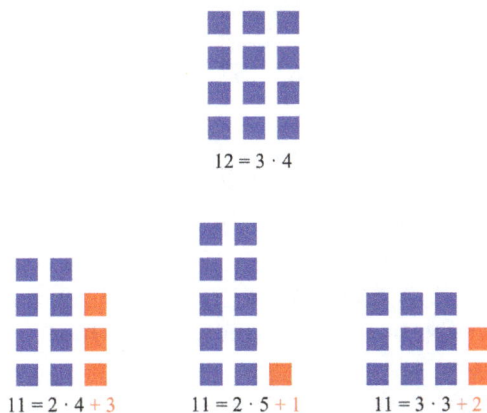

$$12 = 3 \cdot 4$$

$$11 = 2 \cdot 4 + 3 \qquad 11 = 2 \cdot 5 + 1 \qquad 11 = 3 \cdot 3 + 2$$

The number 12 is not a prime, as 12 items can be placed into 3 equal-size columns of 4 each (among other ways). 11 items cannot be all placed into several equal-size columns of more than 1 item each without some extra items leftover (a remainder). Therefore, the number 11 is a prime.

Among the numbers 1 to 6, the numbers 2, 3, and 5 are the prime numbers, while 1, 4, and 6 are not prime. 1 is excluded as a prime number, for reasons explained below. 2 is a prime number, since the only natural numbers dividing it are 1 and 2. Next, 3 is prime, too: 1 and 3 do divide 3 without remainder, but 3 divided by 2 gives remainder 1. Thus, 3 is prime. However, 4 is composite, since 2 is another number (in addition to 1 and 4) dividing 4 without remainder:

$$4 = 2 \cdot 2.$$

5 is again prime: none of the numbers 2, 3, or 4 divide 5. Next, 6 is divisible by 2 or 3, since

$$6 = 2 \cdot 3.$$

Hence, 6 is not prime. The image at the right illustrates that 12 is not prime: $12 = 3 \cdot 4$. No even number greater than 2 is prime because by definition, any such number n has at least three distinct divisors, namely 1, 2, and n. This implies that n is not prime. Accordingly, the term *odd prime* refers to any prime number greater than 2. Similarly, when written in the usual decimal system, all prime numbers larger than 5 end in 1, 3, 7, or 9, since even numbers are multiples of 2 and numbers ending in 0 or 5 are multiples of 5.

If n is a natural number, then 1 and n divide n without remainder. Therefore, the condition of being a prime can also be restated as: a number is prime if it is greater than one and if none of

$$2, 3, ..., n - 1$$

divides n (without remainder). Yet another way to say the same is: a number $n > 1$ is prime if it cannot be written as a product of two integers a and b, both of which are larger than 1:

$$n = a \cdot b.$$

In other words, n is prime if n items cannot be divided up into smaller equal-size groups of more than one item.

The set of all primes is often denoted by P.

The first 168 prime numbers (all the prime numbers less than 1000) are:

2, 3, 5, 7, 11, 13, 17, 19, 23, 29, 31, 37, 41, 43, 47, 53, 59, 61, 67, 71, 73, 79, 83, 89, 97, 101, 103, 107, 109, 113, 127, 131, 137, 139, 149, 151, 157, 163, 167, 173, 179, 181, 191, 193, 197, 199, 211, 223, 227, 229, 233, 239, 241, 251, 257, 263, 269, 271, 277, 281, 283, 293, 307, 311, 313, 317, 331, 337, 347, 349, 353, 359, 367, 373, 379, 383, 389, 397, 401, 409, 419, 421, 431, 433, 439, 443, 449, 457, 461, 463, 467, 479, 487, 491, 499, 503, 509, 521, 523, 541, 547, 557, 563, 569, 571, 577, 587, 593, 599, 601, 607, 613, 617, 619, 631, 641, 643, 647, 653, 659, 661, 673, 677, 683, 691, 701, 709, 719, 727, 733, 739, 743, 751, 757, 761, 769, 773, 787, 797, 809, 811, 821, 823, 827, 829, 839, 853, 857, 859, 863, 877, 881, 883, 887, 907, 911, 919, 929, 937, 941, 947, 953, 967, 971, 977, 983, 991, 997 (sequence A000040 in the OEIS).

Fundamental Theorem of Arithmetic

The crucial importance of prime numbers to number theory and mathematics in general stems from the *fundamental theorem of arithmetic*, which states that every integer larger than 1 can be written as a product of one or more primes in a way that is unique except for the order of the prime factors. Primes can thus be considered the "basic building blocks" of the natural numbers. For example:

23244	$= 2 \cdot 2 \cdot 3 \cdot 13 \cdot 149$
	$= 2^2 \cdot 3 \cdot 13 \cdot 149.$ (2^2 denotes the square or second power of 2.)

As in this example, the same prime factor may occur multiple times. A decomposition:

$$n = p_1 \cdot p_2 \cdot ... \cdot p_t$$

of a number n into (finitely many) prime factors p_1, p_2, ... to p_t is called *prime factorization* of n. The fundamental theorem of arithmetic can be rephrased so as to say that any factorization into primes will be identical except for the order of the factors. So, albeit there are many prime factorization algorithms to do this in practice for larger numbers, they all have to yield the same result.

If p is a prime number and p divides a product ab of integers, then p divides a or p divides b. This proposition is known as Euclid's lemma. It is used in some proofs of the uniqueness of prime factorizations.

Primality of One

Most early Greeks did not even consider 1 to be a number, so they could not consider it to be a prime. By the Middle Ages and Renaissance many mathematicians included 1 as the first prime number. In the mid-18th century Christian Goldbach listed 1 as the first prime in his famous correspondence with Leonhard Euler; however, Euler himself did not consider 1 to be a prime number. In the 19th century many mathematicians still considered the number 1 to be a prime. For example, Derrick Norman Lehmer's list of primes up to 10,006,721, reprinted as late as 1956, started with 1 as its first prime. Henri Lebesgue is said to be the last professional mathematician to call 1 prime. By the early 20th century, mathematicians began to arrive at the consensus that 1 is not a prime number, but rather forms its own special category as a "unit".

A large body of mathematical work would still be valid when calling 1 a prime, but Euclid's fundamental theorem of arithmetic (mentioned above) would not hold as stated. For example, the number 15 can be factored as $3 \cdot 5$ and $1 \cdot 3 \cdot 5$; if 1 were admitted as a prime, these two presentations would be considered different factorizations of 15 into prime numbers, so the statement of that theorem would have to be modified. Similarly, the sieve of Eratosthenes would not work correctly if 1 were considered a prime: a modified version of the sieve that considers 1 as prime would eliminate all multiples of 1 (that is, all other numbers) and produce as output only the single number 1. Furthermore, the prime numbers have several properties that the number 1 lacks, such as the relationship of the number to its corresponding value of Euler's totient function or the sum of divisors function.

History

The Sieve of Eratosthenes is a simple algorithm for finding all prime numbers up to a specified integer. It was created in the 3rd century BC by Eratosthenes, an ancient Greek mathematician.

There are hints in the surviving records of the ancient Egyptians that they had some knowledge of prime numbers: the Egyptian fraction expansions in the Rhind papyrus, for instance, have quite different forms for primes and for composites. However, the earliest surviving records of the explicit study of prime numbers come from the Ancient Greeks. Euclid's Elements (circa 300 BC) contain important theorems about primes, including the infinitude of primes and the fundamental theorem of arithmetic. Euclid also showed how to construct a perfect number from a Mersenne prime. The Sieve of Eratosthenes, attributed to Eratosthenes, is a simple method to compute primes, although the large primes found today with computers are not generated this way.

After the Greeks, little happened with the study of prime numbers until the 17th century. In 1640 Pierre de Fermat stated (without proof) Fermat's little theorem (later proved by Leibniz and Euler). Fermat also conjectured that all numbers of the form $2^{2n} + 1$ are prime (they are called Fermat numbers) and he verified this up to $n = 4$ (or $2^{16} + 1$). However, the very next Fermat number $2^{32} + 1$ is composite (one of its prime factors is 641), as Euler discovered later, and in fact no further Fermat numbers are known to be prime. The French monk Marin Mersenne looked at primes of the form $2^p - 1$, with p a prime. They are called Mersenne primes in his honor.

Euler's work in number theory included many results about primes. He showed the infinite series $1/2 + 1/3 + 1/5 + 1/7 + 1/11 + \ldots$ is divergent. In 1747 he showed that the even perfect numbers are precisely the integers of the form $2^{p-1}(2^p - 1)$, where the second factor is a Mersenne prime.

At the start of the 19th century, Legendre and Gauss independently conjectured that as x tends to infinity, the number of primes up to x is asymptotic to $x/\ln(x)$, where $\ln(x)$ is the natural logarithm of x. Ideas of Riemann in his 1859 paper on the zeta-function sketched a program that would lead to a proof of the prime number theorem. This outline was completed by Hadamard and de la Vallée Poussin, who independently proved the prime number theorem in 1896.

Proving a number is prime is not done (for large numbers) by trial division. Many mathematicians have worked on primality tests for large numbers, often restricted to specific number forms. This includes Pépin's test for Fermat numbers (1877), Proth's theorem (around 1878), the Lucas–Lehmer primality test (originated 1856), and the generalized Lucas primality test. More recent algorithms like APRT-CL, ECPP, and AKS work on arbitrary numbers but remain much slower.

For a long time, prime numbers were thought to have extremely limited application outside of pure mathematics. This changed in the 1970s when the concepts of public-key cryptography were invented, in which prime numbers formed the basis of the first algorithms such as the RSA cryptosystem algorithm.

Since 1951 all the largest known primes have been found by computers. The search for ever larger primes has generated interest outside mathematical circles. The Great Internet Mersenne Prime Search and other distributed computing projects to find large primes have become popular, while mathematicians continue to struggle with the theory of primes.

Number of Prime Numbers

There are infinitely many prime numbers. Another way of saying this is that the sequence

$$2, 3, 5, 7, 11, 13, \ldots$$

of prime numbers never ends. This statement is referred to as *Euclid's theorem* in honor of the ancient Greek mathematician Euclid, since the first known proof for this statement is attributed to him. Many more proofs of the infinitude of primes are known, including an analytical proof by Euler, Goldbach's proof based on Fermat numbers, Furstenberg's proof using general topology, and Kummer's elegant proof.

Euclid's Proof

Euclid's proof (Book IX, Proposition 20) considers any finite set S of primes. The key idea is to consider the product of all these numbers plus one:

$$N = 1 + \prod_{p \in S} p.$$

Like any other natural number, N is divisible by at least one prime number (it is possible that N itself is prime).

None of the primes by which N is divisible can be members of the finite set S of primes with which we started, because dividing N by any one of these leaves a remainder of 1. Therefore, the primes by which N is divisible are additional primes beyond the ones we started with. Thus any finite set of primes can be extended to a larger finite set of primes.

It is often erroneously reported that Euclid begins with the assumption that the set initially considered contains all prime numbers, leading to a contradiction, or that it contains precisely the n smallest primes rather than any arbitrary finite set of primes. Today, the product of the smallest n primes plus 1 is conventionally called the nth Euclid number.

Euler's Analytical Proof

Euler's proof uses the partial sums of the reciprocals of primes,

$$S(p) = \frac{1}{2} + \frac{1}{3} + \frac{1}{5} + \frac{1}{7} + \cdots + \frac{1}{p}.$$

For any arbitrary real number x, there exists a prime p for which this partial sum is bigger than x. This shows that there are infinitely many primes, because if there were finitely many primes the sum would reach its maximum value at the biggest prime rather than being unbounded. More precisely, the growth rate of $S(p)$ is doubly logarithmic, as quantified by Mertens' second theorem. For comparison, the sum

$$\frac{1}{1^2} + \frac{1}{2^2} + \frac{1}{3^2} + \cdots + \frac{1}{n^2} = \sum_{i=1}^{n} \frac{1}{i^2}$$

does not grow to infinity as n goes to infinity. In this sense, prime numbers occur more often than squares of natural numbers. Brun's theorem states that the sum of the reciprocals of twin primes,

$$\left(\frac{1}{3}+\frac{1}{5}\right)+\left(\frac{1}{5}+\frac{1}{7}\right)+\left(\frac{1}{11}+\frac{1}{13}\right)+\cdots=\sum_{\substack{p\ \text{prime,}\\ p+2\ \text{prime}}}\left(\frac{1}{p}+\frac{1}{p+2}\right),$$

is finite. Because of Brun's theorem, it is not possible to use Euler's method to solve the twin prime conjecture, that there exist infinitely many twin primes.

Testing Primality and Integer Factorization

There are various methods to determine whether a given number n is prime. The most basic routine, trial division, is of little practical use because of its slowness. One group of modern primality tests is applicable to arbitrary numbers, while more efficient tests are available for particular numbers. Most such methods only tell whether n is prime or not. Routines also yielding one (or all) prime factors of n are called factorization algorithms.

Trial Division

The most basic method of checking the primality of a given integer n is called *trial division*. This routine consists of dividing n by each integer m that is greater than 1 and less than or equal to the square root of n. If the result of any of these divisions is an integer, then n is not a prime, otherwise it is a prime. Indeed, if $\sqrt{n}$ is composite (with a and $b \neq 1$) then one of the factors a or b is necessarily at most $n = 37$. For example, for $\sqrt{n}$, the trial divisions are by $m = 2, 3, 4, 5,$ and 6. None of these numbers divides 37, so 37 is prime. This routine can be implemented more efficiently if a complete list of primes up to $\sqrt{n}$ is known—then trial divisions need to be checked only for those m that are prime. For example, to check the primality of 37, only three divisions are necessary ($m = 2, 3,$ and 5), given that 4 and 6 are composite.

While a simple method, trial division quickly becomes impractical for testing large integers because the number of possible factors grows too rapidly as n increases. According to the prime number theorem explained below, the number of prime numbers less than $\sqrt{n}$ is approximately given by $\sqrt{n} / \ln(\sqrt{n})$,, so the algorithm may need up to this number of trial divisions to check the primality of n. For $n = 10^{20}$, this number is 450 million—too large for many practical applications.

Sieves

An algorithm yielding all primes up to a given limit, such as required in the primes-only trial division method, is called a prime number sieve. The oldest example, the sieve of Eratosthenes is still the most commonly used. The sieve of Atkin is another option. Before the advent of computers, lists of primes up to bounds like 10^7 were also used.

Primality Testing Versus Primality Proving

Modern primality tests for general numbers n can be divided into two main classes, probabilistic (or "Monte Carlo") and deterministic algorithms. Deterministic algorithms provide a way to tell for sure whether a given number is prime or not. For example, trial division is a deterministic algorithm because, if performed correctly, it will always identify a prime number as prime and a composite number as composite. Probabilistic algorithms are normally faster, but do not completely prove that a number is prime. These tests rely on testing a given number in a partly random way. For example, a given test might pass all the time if applied to a prime number, but pass only with probability p if applied to a composite number. If we repeat the test n times and pass every time, then the probability that our number is composite is $1/(1-p)^n$, which decreases exponentially with the number of tests, so we can be as sure as we like (though never perfectly sure) that the number is prime. On the other hand, if the test ever fails, then we know that the number is composite.

A particularly simple example of a probabilistic test is the Fermat primality test, which relies on the fact (Fermat's little theorem) that $n^p \equiv n \pmod{p}$ for any n if p is a prime number. If we have a number b that we want to test for primality, then we work out $n^b \pmod{b}$ for a random value of n as our test. A flaw with this test is that there are some composite numbers (the Carmichael numbers) that satisfy the Fermat identity even though they are not prime, so the test has no way of distinguishing between prime numbers and Carmichael numbers. Carmichael numbers are substantially rarer than prime numbers, though, so this test can be useful for practical purposes. More powerful extensions of the Fermat primality test, such as the Baillie-PSW, Miller-Rabin, and Solovay-Strassen tests, are guaranteed to fail at least some of the time when applied to a composite number.

Deterministic algorithms do not erroneously report composite numbers as prime. In practice, the fastest such method is known as elliptic curve primality proving. Analyzing its run time is based on heuristic arguments, as opposed to the rigorously proven complexity of the more recent AKS primality test. Deterministic methods are typically slower than probabilistic ones, so the latter ones are typically applied first before a more time-consuming deterministic routine is employed.

The following table lists a number of prime tests. The running time is given in terms of n, the number to be tested and, for probabilistic algorithms, the number k of tests performed. Moreover, ε is an arbitrarily small positive number, and log is the logarithm to an unspecified base. The big O notation means that, for example, elliptic curve primality proving requires a time that is bounded by a factor (not depending on n, but on ε) times $\log^{5+\varepsilon}(n)$.

Test	Developed in	Type	Running time	Notes
AKS primality test	2002	deterministic	$O(\log^{6+\varepsilon}(n))$	
Elliptic curve primality proving	1977	deterministic	$O(\log^{5+\varepsilon}(n))$ *heuristically*	
Baillie-PSW primality test	1980	probabilistic	$O(\log^3 n)$	no known counterexamples
Miller–Rabin primality test	1980	probabilistic	$O(k \cdot \log^{2+\varepsilon}(n))$	error probability 4^{-k}
Solovay–Strassen primality test	1977	probabilistic	$O(k \cdot \log^3 n)$	error probability 2^{-k}
Fermat primality test		probabilistic	$O(k \cdot \log^{2+\varepsilon}(n))$	fails for Carmichael numbers

Special-purpose Algorithms and The Largest Known Prime

In addition to the aforementioned tests applying to any natural number n, a number of much more efficient primality tests is available for special numbers. For example, to run Lucas' primality test requires the knowledge of the prime factors of $n - 1$, while the Lucas–Lehmer primality test needs the prime factors of $n + 1$ as input. For example, these tests can be applied to check whether

$$n! \pm 1 = 1 \cdot 2 \cdot 3 \cdot \ldots \cdot n \pm 1$$

are prime. Prime numbers of this form are known as factorial primes. Other primes where either $p + 1$ or $p - 1$ is of a particular shape include the Sophie Germain primes (primes of the form $2p + 1$ with p prime), primorial primes, Fermat primes and Mersenne primes, that is, prime numbers that are of the form $2^p - 1$, where p is an arbitrary prime. The Lucas–Lehmer test is particularly fast for numbers of this form. This is why the largest *known* prime has almost always been a Mersenne prime since the dawn of electronic computers.

The following table gives the largest known primes of the mentioned types. Some of these primes have been found using distributed computing. In 2009, the Great Internet Mersenne Prime Search project was awarded a US$100,000 prize for first discovering a prime with at least 10 million digits. The Electronic Frontier Foundation also offers $150,000 and $250,000 for primes with at least 100 million digits and 1 billion digits, respectively. Some of the largest primes not known to have any particular form (that is, no simple formula such as that of Mersenne primes) have been found by taking a piece of semi-random binary data, converting it to a number n, multiplying it by 256^k for some positive integer k, and searching for possible primes within the interval $[256^k n + 1, 256^k(n + 1) - 1]$.

Type	Prime	Number of decimal digits	Date	Found by
Mersenne prime	$2^{74,207,281} - 1$	22,338,618	January 7, 2016	Great Internet Mersenne Prime Search
not a Mersenne prime (Proth number)	$19,249 \times 2^{13,018,586} + 1$	3,918,990	March 26, 2007	Seventeen or Bust
factorial prime	$150209! + 1$	712,355	October 2011	PrimeGrid
primorial prime	$1098133\# - 1$	476,311	March 2012	PrimeGrid
twin primes	$3756801695685 \times 2^{666669} \pm 1$	200,700	December 2011	PrimeGrid

Integer Factorization

Given a composite integer n, the task of providing one (or all) prime factors is referred to as *factorization* of n. Elliptic curve factorization is an algorithm relying on arithmetic on an elliptic curve.

Distribution

In 1975, number theorist Don Zagier commented that primes both

grow like weeds among the natural numbers, seeming to obey no other law than that of chance [but also] exhibit stunning regularity [and] that there are laws governing their behavior, and that they obey these laws with almost military precision.

The distribution of primes in the large, such as the question how many primes are smaller than a given, large threshold, is described by the prime number theorem, but no efficient formula for the n-th prime is known.

There are arbitrarily long sequences of consecutive non-primes, as for every positive integer n the

$(n+1)!+2$ consecutive integers from $(n+1)!+n+1$ to $(n+1)!+n+1$ (inclusive) are all composite (as $(n+1)!+k$ is divisible by k for k between 2 and $n+1$).

Dirichlet's theorem on arithmetic progressions, in its basic form, asserts that linear polynomials

$$p(n) = a + bn$$

with coprime integers a and b take infinitely many prime values. Stronger forms of the theorem state that the sum of the reciprocals of these prime values diverges, and that different such polynomials with the same b have approximately the same proportions of primes.

The corresponding question for quadratic polynomials is less well-understood.

Formulas for Primes

There is no known efficient formula for primes. For example, Mills' theorem and a theorem of Wright assert that there are real constants $A > 1$ and μ such that

$$\left\lfloor A^{3^n} \right\rfloor \text{ and } \left\lfloor 2^{\cdots^{2^{2^{2^\mu}}}} \right\rfloor$$

are prime for any natural number n. Here $\lfloor - \rfloor$ represents the floor function, i.e., largest integer not greater than the number in question. The latter formula can be shown using Bertrand's postulate (proven first by Chebyshev), which states that there always exists at least one prime number p with $n < p < 2n - 2$, for any natural number $n > 3$. However, computing A or μ requires the knowledge of infinitely many primes to begin with. Another formula is based on Wilson's theorem and generates the number 2 many times and all other primes exactly once.

There is no non-constant polynomial, even in several variables, that takes *only* prime values. However, there is a set of Diophantine equations in 9 variables and one parameter with the following property: the parameter is prime if and only if the resulting system of equations has a solution over the natural numbers. This can be used to obtain a single formula with the property that all its *positive* values are prime.

Number of Prime Numbers Below A Given Number

The prime counting function $\pi(n)$ is defined as the number of primes not greater than n. For example, $\pi(11) = 5$, since there are five primes less than or equal to 11. There are known algorithms to compute exact values of $\pi(n)$ faster than it would be possible to compute each prime up to n. The *prime number theorem* states that $\pi(n)$ is approximately given by

$$\pi(n) \approx \frac{n}{\ln n},$$

in the sense that the ratio of $\pi(n)$ and the right hand fraction approaches 1 when n grows to infinity. This implies that the likelihood that a number less than n is prime is (approximately) inversely proportional to the number of digits in n. A more accurate estimate for $\pi(n)$ is given by the offset logarithmic integral

$$\mathrm{Li}(n) = \int_2^n \frac{dt}{\ln t}.$$

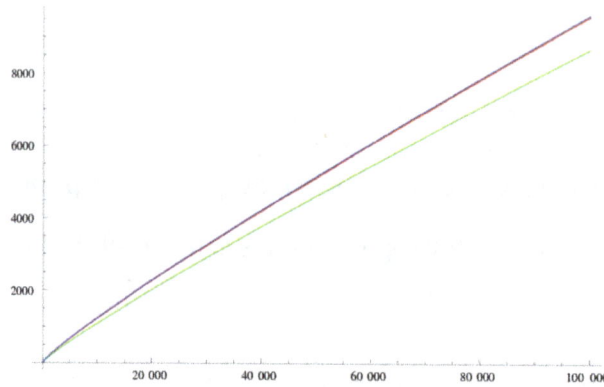

A chart depicting $\pi(n)$ (blue), $n / \ln (n)$ (green) and $\mathrm{Li}(n)$ (red)

The prime number theorem also implies estimates for the size of the n-th prime number p_n (i.e., $p_1 = 2$, $p_2 = 3$, etc.): up to a bounded factor, p_n grows like $n \log(n)$. In particular, the prime gaps, i.e. the differences $p_n - p_{n-1}$ of two consecutive primes, become arbitrarily large. This latter statement can also be seen in a more elementary way by noting that the sequence $n! + 2$, $n! + 3$, ..., $n! + n$ (for the notation $n!$ read factorial) consists of $n - 1$ composite numbers, for any natural number n.

Arithmetic Progressions

An arithmetic progression is the set of natural numbers that give the same remainder when divided by some fixed number q called modulus. For example,

$$3, 12, 21, 30, 39, \ldots,$$

is an arithmetic progression modulo $q = 9$. Except for 3, none of these numbers is prime, since $3 + 9n = 3(1 + 3n)$ so that the remaining numbers in this progression are all composite. (In general terms, all prime numbers above q are of the form $q\#\cdot n + m$, where $0 < m < q\#$, and m has no prime factor $\leq q$.) Thus, the progression

$$a, a + q, a + 2q, a + 3q, \ldots$$

can have infinitely many primes only when a and q are coprime, i.e., their greatest common divisor is one. If this necessary condition is satisfied, *Dirichlet's theorem on arithmetic progressions* asserts that the progression contains infinitely many primes. The picture below illustrates this with $q = 9$: the numbers are "wrapped around" as soon as a multiple of 9 is passed. Primes are highlighted in red. The rows (=progressions) starting with $a = 3$, 6, or 9 contain at most one prime number. In all other rows ($a = 1, 2, 4, 5, 7$, and 8) there are infinitely many prime numbers. What is more, the primes are distributed equally among those rows in the long run—the density of all primes congruent a modulo 9 is 1/6.

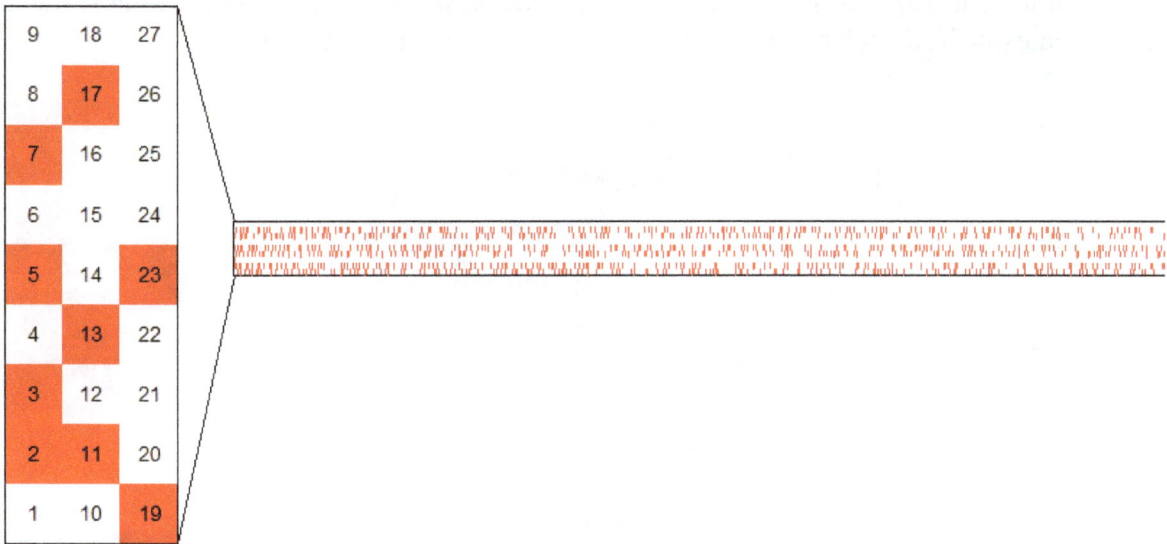

The Green–Tao theorem shows that there are arbitrarily long arithmetic progressions consisting of primes. An odd prime p is expressible as the sum of two squares, $p = x^2 + y^2$, exactly if p is congruent 1 modulo 4 (Fermat's theorem on sums of two squares).

Prime Values of Quadratic Polynomials

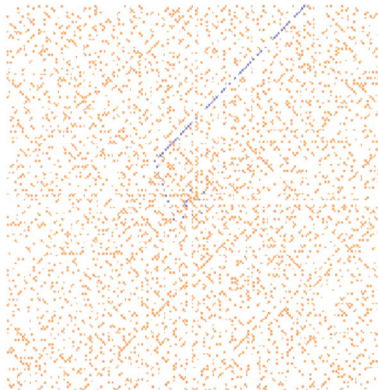

The Ulam spiral. Red pixels show prime numbers. Primes of the form $4n^2 - 2n + 41$ are highlighted in blue.

Euler noted that the function

$$n^2 + n + 41$$

yields prime numbers for $0 \leq n < 40$, a fact leading into deep algebraic number theory: more specifically, Heegner numbers. For greater n, the expression also produces composite values. The Hardy-Littlewood conjecture F makes an asymptotic prediction about the density of primes among the values of quadratic polynomials (with integer coefficients a, b, and c),

$$f(n) = ax^2 + bx + c,$$

in terms of Li(n) and the coefficients a, b, and c. However, progress has been difficult. No quadratic polynomial (with $a \neq 0$) is known to take infinitely many prime values. The Ulam spiral depicts all natural numbers in a spiral-like way. Primes cluster on certain diagonals and not others, suggesting that some quadratic polynomials take prime values more often than others.

Open Questions

Zeta Function and The Riemann Hypothesis

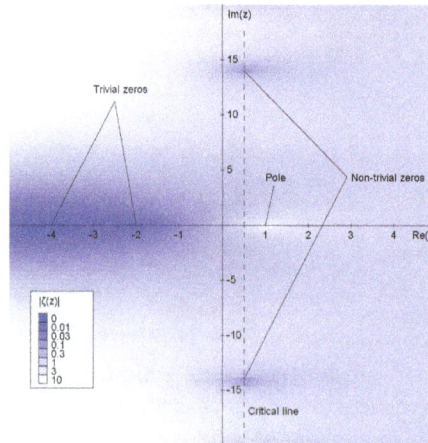

Plot of the zeta function ζ(s). At s=1, the function has a pole, that is to say, it tends to infinity.

The Riemann zeta function $\zeta(s)$ is defined as an infinite sum

$$\zeta(s) = \sum_{n=1}^{\infty} \frac{1}{n^s},$$

where s is a complex number with real part bigger than 1. It is a consequence of the fundamental theorem of arithmetic that this sum agrees with the infinite product

$$\prod_{p \text{ prime}} \frac{1}{1-p^{-s}}.$$

The zeta function is closely related to prime numbers. For example, the aforementioned fact that there are infinitely many primes can also be seen using the zeta function: if there were only finitely many primes then $\zeta(1)$ would have a finite value. However, the harmonic series $1 + 1/2 + 1/3 + 1/4 + \ldots$ diverges (i.e., exceeds any given number), so there must be infinitely many primes. Another example of the richness of the zeta function and a glimpse of modern algebraic number theory is the following identity (Basel problem), due to Euler,

$$\zeta(2) = \prod_{p} \frac{1}{1-p^{-2}} = \frac{\pi^2}{6}.$$

The reciprocal of $\zeta(2)$, $6/\pi^2$, is the probability that two numbers selected at random are relatively prime.

The unproven *Riemann hypothesis*, dating from 1859, states that except for $s = -2, -4, \ldots$, all zeroes of the ζ-function have real part equal to 1/2. The connection to prime numbers is that it essentially says that the primes are as regularly distributed as possible. From a physical viewpoint, it roughly states that the irregularity in the distribution of primes only comes from random noise. From a mathematical viewpoint, it roughly states that the asymptotic distribution of primes (about $x/\log x$ of numbers less than x are primes, the prime number theorem) also holds for much shorter intervals of length about the square root of x (for intervals near x). This hypothesis is generally believed to be correct. In particular, the simplest assumption is that primes should have no significant irregularities without good reason.

Other Conjectures

In addition to the Riemann hypothesis, many more conjectures revolving about primes have been posed. Often having an elementary formulation, many of these conjectures have withstood a proof for decades: all four of Landau's problems from 1912 are still unsolved. One of them is Goldbach's conjecture, which asserts that every even integer n greater than 2 can be written as a sum of two primes. As of February 2011, this conjecture has been verified for all numbers up to $n = 2 \cdot 10^{17}$. Weaker statements than this have been proven, for example Vinogradov's theorem says that every sufficiently large odd integer can be written as a sum of three primes. Chen's theorem says that every sufficiently large even number can be expressed as the sum of a prime and a semiprime, the product of two primes. Also, any even integer can be written as the sum of six primes. The branch of number theory studying such questions is called additive number theory.

Other conjectures deal with the question whether an infinity of prime numbers subject to certain constraints exists. It is conjectured that there are infinitely many Fibonacci primes and infinitely many Mersenne primes, but not Fermat primes. It is not known whether or not there are an infinite number of Wieferich primes and of prime Euclid numbers.

A third type of conjectures concerns aspects of the distribution of primes. It is conjectured that there are infinitely many twin primes, pairs of primes with difference 2 (twin prime conjecture). Polignac's conjecture is a strengthening of that conjecture, it states that for every positive integer n, there are infinitely many pairs of consecutive primes that differ by $2n$. It is conjectured there are infinitely many primes of the form $n^2 + 1$. These conjectures are special cases of the broad Schinzel's hypothesis H. Brocard's conjecture says that there are always at least four primes between the squares of consecutive primes greater than 2. Legendre's conjecture states that there is a prime number between n^2 and $(n + 1)^2$ for every positive integer n. It is implied by the stronger Cramér's conjecture.

Applications

For a long time, number theory in general, and the study of prime numbers in particular, was seen as the canonical example of pure mathematics, with no applications outside of the self-interest of studying the topic with the exception of use of prime numbered gear teeth to distribute wear evenly. In particular, number theorists such as British mathematician G. H. Hardy prided themselves on doing work that had absolutely no military significance. However, this vision was shattered in the 1970s, when it was publicly announced that prime numbers could be used as the basis for the creation of public key cryptography algorithms. Prime numbers are also used for hash tables and pseudorandom number generators.

Some rotor machines were designed with a different number of pins on each rotor, with the number of pins on any one rotor either prime, or coprime to the number of pins on any other rotor. This helped generate the full cycle of possible rotor positions before repeating any position.

The International Standard Book Numbers work with a check digit, which exploits the fact that 11 is a prime.

Arithmetic Modulo A Prime and Finite Fields

Modular arithmetic modifies usual arithmetic by only using the numbers

$$\{0, 1, 2, \ldots, n-1\},$$

where n is a fixed natural number called modulus. Calculating sums, differences and products is done as usual, but whenever a negative number or a number greater than $n - 1$ occurs, it gets replaced by the remainder after division by n. For instance, for $n = 7$, the sum $3 + 5$ is 1 instead of 8, since 8 divided by 7 has remainder 1. This is referred to by saying "$3 + 5$ is congruent to 1 modulo 7" and is denoted

Similarly, $6 + 1 \equiv 0 \pmod 7$, $2 - 5 \equiv 4 \pmod 7$, since $-3 + 7 = 4$, and $3 \cdot 4 \equiv 5 \pmod 7$ as 12 has remainder 5. Standard properties of addition and multiplication familiar from the integers remain valid in modular arithmetic. In the parlance of abstract algebra, the above set of integers, which is also denoted Z/nZ, is therefore a commutative ring for any n. Division, however, is not in general possible in this setting. For example, for $n = 6$, the equation

$$3 + 5 \equiv 1 \pmod 7.$$

a solution x of which would be an analogue of 2/3, cannot be solved, as one can see by calculating 3 · 0, ..., 3 · 5 modulo 6. The distinctive feature of prime numbers is the following: division *is* possible in modular arithmetic if and only if n is a prime. Equivalently, n is prime if and only if all integers m satisfying $2 \le m \le n - 1$ are *coprime* to n, i.e. their only common divisor is one. Indeed, for $n = 7$, the equation

$$3 \cdot x \equiv 2 \pmod 6,$$

has a unique solution, $x = 3$. Because of this, for any prime p, Z/pZ (also denoted F_p) is called a field or, more specifically, a finite field since it contains finitely many, namely p, elements.

A number of theorems can be derived from inspecting F_p in this abstract way. For example, Fermat's little theorem, stating

$$3 \cdot x \equiv 2 \pmod 7,$$

for any integer a not divisible by p, may be proved using these notions. This implies

$$a^{p-1} \equiv 1 \pmod p$$

Giuga's conjecture says that this equation is also a sufficient condition for p to be prime. Another consequence of Fermat's little theorem is the following: if p is a prime number other than 2 and 5, $1/_p$ is always a recurring decimal, whose period is $p-1$ or a divisor of $p-1$. The fraction $1/_p$ expressed likewise in base q (rather than base 10) has similar effect, provided that p is not a prime factor of q. Wilson's theorem says that an integer $p > 1$ is prime if and only if the factorial $(p-1)! + 1$ is divisible by p. Moreover, an integer $n > 4$ is composite if and only if $(n-1)!$ is divisible by n.

Other Mathematical Occurrences of Primes

Many mathematical domains make great use of prime numbers. An example from the theory of finite groups are the Sylow theorems: if G is a finite group and p^n is the highest power of the prime p that divides the order of G, then G has a subgroup of order p^n. Also, any group of prime order is cyclic (Lagrange's theorem).

Public-key Cryptography

Several public-key cryptography algorithms, such as RSA and the Diffie–Hellman key exchange, are based on large prime numbers (2048-bit primes are common). RSA relies on the assumption that it is much easier (i.e., more efficient) to perform the multiplication of two (large) numbers x and y than to calculate x and y (assumed coprime) if only the product xy is known. The Diffie–Hellman key exchange relies on the fact that there are efficient algorithms for modular exponentiation, while the reverse operation the discrete logarithm is thought to be a hard problem.

Prime Numbers in Nature

The evolutionary strategy used by cicadas of the genus *Magicicada* make use of prime numbers. These insects spend most of their lives as grubs underground. They only pupate and then emerge from their burrows after 7, 13 or 17 years, at which point they fly about, breed, and then die after a few weeks at most. The logic for this is believed to be that the prime number intervals between emergences make it very difficult for predators to evolve that could specialize as predators on *Magicicadas*. If *Magicicadas* appeared at a non-prime number intervals, say every 12 years, then predators appearing every 2, 3, 4, 6, or 12 years would be sure to meet them. Over a 200-year period, average predator populations during hypothetical outbreaks of 14- and 15-year cicadas would be up to 2% higher than during outbreaks of 13- and 17-year cicadas. Though small, this advantage appears to have been enough to drive natural selection in favour of a prime-numbered life-cycle for these insects.

There is speculation that the zeros of the zeta function are connected to the energy levels of complex quantum systems.

Generalizations

The concept of prime number is so important that it has been generalized in different ways in various branches of mathematics. Generally, "prime" indicates minimality or indecomposability, in an appropriate sense. For example, the prime field is the smallest subfield of a field F containing both 0 and 1. It is either Q or the finite field with p elements, whence the name. Often a second, additional meaning is intended by using the word prime, namely that any object can be, essentially uniquely, decomposed into its prime components. For example, in knot theory, a prime knot is a knot that is indecomposable in the sense that it cannot be written as the knot sum of two nontrivial knots. Any knot can be uniquely expressed as a connected sum of prime knots. Prime models and prime 3-manifolds are other examples of this type.

Prime Elements in Rings

Prime numbers give rise to two more general concepts that apply to elements of any commutative ring R, an algebraic structure where addition, subtraction and multiplication are defined: *prime elements* and *irreducible elements*. An element p of R is called prime element if it is neither zero nor a unit (i.e., does not have a multiplicative inverse) and satisfies the following requirement: given x and y in R such that p divides the product xy, then p divides x or y. An element is irreducible if it is not a unit and cannot be written as a product of two ring elements that are not units. In the ring Z of integers, the set of prime elements equals the set of irreducible elements, which is

$$\{\ldots, -11, -7, -5, -3, -2, 2, 3, 5, 7, 11, \ldots\}.$$

In any ring R, any prime element is irreducible. The converse does not hold in general, but does hold for unique factorization domains.

The fundamental theorem of arithmetic continues to hold in unique factorization domains. An example of such a domain is the Gaussian integers Z[i], that is, the set of complex numbers of the form $a + bi$ where i denotes the imaginary unit and a and b are arbitrary integers. Its prime elements are known as Gaussian primes. Not every prime (in Z) is a Gaussian prime: in the bigger ring Z[i], 2 factors into the product of the two Gaussian primes $(1 + i)$ and $(1 - i)$. Rational primes (i.e. prime elements in Z) of the form $4k + 3$ are Gaussian primes, whereas rational primes of the form $4k + 1$ are not.

Prime Ideals

In ring theory, the notion of number is generally replaced with that of ideal. *Prime ideals*, which generalize prime elements in the sense that the principal ideal generated by a prime element is a prime ideal, are an important tool and object of study in commutative algebra, algebraic number theory and algebraic geometry. The prime ideals of the ring of integers are the ideals (0), (2), (3), (5), (7), (11), ... The fundamental theorem of arithmetic generalizes to the Lasker–Noether theorem, which expresses every ideal in a Noetherian commutative ring as an intersection of primary ideals, which are the appropriate generalizations of prime powers.

Prime ideals are the points of algebro-geometric objects, via the notion of the spectrum of a ring. Arithmetic geometry also benefits from this notion, and many concepts exist in both geometry and number theory. For example, factorization or ramification of prime ideals when lifted to an extension field, a basic problem of algebraic number theory, bears some resemblance with ramification in geometry. Such ramification questions occur even in number-theoretic questions solely concerned with integers. For example, prime ideals in the ring of integers of quadratic number fields can be used in proving quadratic reciprocity, a statement that concerns the solvability of quadratic equations

$$x^2 \equiv p \pmod{q},$$

where x is an integer and p and q are (usual) prime numbers. Early attempts to prove Fermat's Last Theorem climaxed when Kummer introduced regular primes, primes satisfying a certain requirement concerning the failure of unique factorization in the ring consisting of expressions

$$a_0 + a_1 \zeta + \cdots + a_{p-1} \zeta^{p-1},$$

where $a_0, \ldots, a_{p-1}$ are integers and ζ is a complex number such that $\zeta^p = 1$.

Valuations

Valuation theory studies certain functions from a field K to the real numbers R called valuations. Every such valuation yields a topology on K, and two valuations are called equivalent if they yield the same topology. A *prime of K* (sometimes called a *place of K*) is an equivalence class of valuations. For example, the p-adic valuation of a rational number q is defined to be the integer $v_p(q)$, such that

$$q = p^{v_p(q)} \frac{r}{s},$$

where both r and s are not divisible by p. For example, $v_3(18/7) = 2$. The p-adic norm is defined as

$$|q|_p := p^{-v_p(q)}.$$

In particular, this norm gets smaller when a number is multiplied by p, in sharp contrast to the usual absolute value (also referred to as the infinite prime). While completing Q (roughly, filling the gaps) with respect to the absolute value yields the field of real numbers, completing with respect to the p-adic norm $|-|_p$ yields the field of p-adic numbers. These are essentially all possible ways to complete Q, by Ostrowski's theorem. Certain arithmetic questions related to Q or more general global fields may be transferred back and forth to the completed (or local) fields. This local-global principle again underlines the importance of primes to number theory.

In The Arts and Literature

Prime numbers have influenced many artists and writers. The French composer Olivier Messiaen used prime numbers to create ametrical music through "natural phenomena". In works such as *La Nativité du Seigneur* (1935) and *Quatre études de rythme* (1949–50), he simultaneously employs motifs with lengths given by different prime numbers to create unpredictable rhythms: the primes 41, 43, 47 and 53 appear in the third étude, "Neumes rythmiques". According to Messiaen this way of composing was "inspired by the movements of nature, movements of free and unequal durations".

In his science fiction novel *Contact*, NASA scientist Carl Sagan suggested that prime numbers could be used as a means of communicating with aliens, an idea that he had first developed informally with American astronomer Frank Drake in 1975. In the novel *The Curious Incident of the Dog in the Night-Time* by Mark Haddon, the narrator arranges the sections of the story by consecutive prime numbers.

Many films, such as *Cube, Sneakers, The Mirror Has Two Faces* and *A Beautiful Mind* reflect a popular fascination with the mysteries of prime numbers and cryptography. Prime numbers are used as a metaphor for loneliness and isolation in the Paolo Giordano novel *The Solitude of Prime Numbers*, in which they are portrayed as "outsiders" among integers.

Real Number

In mathematics, a real number is a value that represents a quantity along a line. The adjective *real* in this context was introduced in the 17th century by Descartes, who distinguished between real and imaginary roots of polynomials.

A symbol of the set of **real numbers** (R)

The real numbers include all the rational numbers, such as the integer −5 and the fraction 4/3, and all the irrational numbers, such as √2 (1.41421356..., the square root of 2, an irrational algebraic number). Included within the irrationals are the transcendental numbers, such as π (3.14159265...). Real numbers can be thought of as points on an infinitely long line called the number line or real line, where the points corresponding to integers are equally spaced. Any real number can be determined by a possibly infinite decimal representation, such as that of 8.632, where each consecutive digit is measured in units one tenth the size of the previous one. The real line can be thought of as a part of the complex plane, and complex numbers include real numbers.

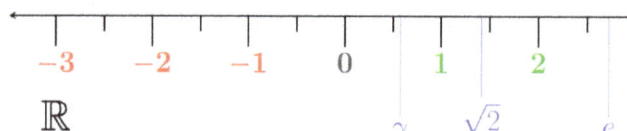

Real numbers can be thought of as points on an infinitely long number line

These descriptions of the real numbers are not sufficiently rigorous by the modern standards of pure mathematics. The discovery of a suitably rigorous definition of the real numbers – indeed, the realization that a better definition was needed – was one of the most important developments of 19th century mathematics. The current standard axiomatic definition is that real numbers form the unique complete totally ordered field (R ; + ; · ; <), up to an isomorphism, whereas popular constructive definitions of real numbers include declaring them as equivalence classes of Cauchy sequences of rational numbers, Dedekind cuts, or infinite decimal representations, together with precise interpretations for the arithmetic operations and the order relation. All these definitions satisfy the axiomatic definition and are thus equivalent.

The reals are uncountable; that is: while both the set of all natural numbers and the set of all real numbers are infinite sets, there can be no one-to-one function from the real numbers to the natural numbers: the cardinality of the set of all real numbers (denoted $\mathfrak{c}$ and called cardinality of the continuum) is strictly greater than the cardinality of the set of all natural numbers (denoted $\aleph_0$ 'aleph-naught'). The statement that there is no subset of the reals with cardinality strictly greater than $\aleph_0$ and strictly smaller than c is known as the continuum hypothesis (CH). It is known to be neither provable nor refutable using the axioms of Zermelo–Fraenkel set theory and the axiom of choice (ZFC), the standard foundation of modern mathematics, in the sense that some models of ZFC satisfy CH, while others violate it.

History

Simple fractions were used by the Egyptians around 1000 BC; the Vedic "Sulba Sutras" ("The rules of chords") in, c. 600 BC, include what may be the first "use" of irrational numbers. The concept of irrationality was implicitly accepted by early Indian mathematicians since Manava (c. 750–690 BC), who were aware that the square roots of certain numbers such as 2 and 61 could not be exactly determined. Around 500 BC, the Greek mathematicians led by Pythagoras realized the need for irrational numbers, in particular the irrationality of the square root of 2.

The Middle Ages brought the acceptance of zero, negative, integral, and fractional numbers, first by Indian and Chinese mathematicians, and then by Arabic mathematicians, who were also the first to treat irrational numbers as algebraic objects, which was made possible by the development of algebra. Arabic mathematicians merged the concepts of "number" and "magnitude" into a more general idea of real numbers. The Egyptian mathematician Abū Kāmil Shujā ibn Aslam (c. 850–930) was the first to accept irrational numbers as solutions to quadratic equations or as coefficients in an equation, often in the form of square roots, cube roots and fourth roots.

In the 16th century, Simon Stevin created the basis for modern decimal notation, and insisted that there is no difference between rational and irrational numbers in this regard.

In the 17th century, Descartes introduced the term "real" to describe roots of a polynomial, distinguishing them from "imaginary" ones.

In the 18th and 19th centuries, there was much work on irrational and transcendental numbers. Johann Heinrich Lambert (1761) gave the first flawed proof that π cannot be rational; Adrien-Marie Legendre (1794) completed the proof, and showed that π is not the square root of a rational number. Paolo Ruffini (1799) and Niels Henrik Abel (1842) both constructed proofs of the Abel–Ruffini theorem: that the general quintic or higher equations cannot be solved by a general formula involving only arithmetical operations and roots.

Évariste Galois (1832) developed techniques for determining whether a given equation could be solved by radicals, which gave rise to the field of Galois theory. Joseph Liouville (1840) showed that neither e nor e^2 can be a root of an integer quadratic equation, and then established the existence of transcendental numbers; Georg Cantor (1873) extended and greatly simplified this proof. Charles Hermite (1873) first proved that e is transcendental, and Ferdinand von Lindemann (1882), showed that π is transcendental. Lindemann's proof was much simplified by Weierstrass (1885), still further by David Hilbert (1893), and has finally been made elementary by Adolf Hurwitz and Paul Gordan.

The development of calculus in the 18th century used the entire set of real numbers without having defined them cleanly. The first rigorous definition was given by Georg Cantor in 1871. In 1874, he showed that the set of all real numbers is uncountably infinite but the set of all algebraic numbers is countably infinite. Contrary to widely held beliefs, his first method was not his famous diagonal argument, which he published in 1891.

Definition

The real number system $(\mathbb{R}; +; \cdot; <)$ can be defined axiomatically up to an isomorphism, which is described hereafter. There are also many ways to construct "the" real number system, for example, starting from natural numbers, then defining rational numbers algebraically, and finally defining real numbers as equivalence classes of their Cauchy sequences or as Dedekind cuts, which are certain subsets of rational numbers. Another possibility is to start from some rigorous axiomatization of Euclidean geometry (Hilbert, Tarski, etc.) and then define the real number system geometrically. From the structuralist point of view all these constructions are on equal footing.

Axiomatic Approach

Let R denote the set of all real numbers. Then:

- The set R is a field, meaning that addition and multiplication are defined and have the usual properties.

- The field R is ordered, meaning that there is a total order $\geq$ such that, for all real numbers x, y and z:

    o   if $x \geq y$ then $x + z \geq y + z$;

    o   if $x \geq 0$ and $y \geq 0$ then $xy \geq 0$.

- The order is Dedekind-complete; that is: every non-empty subset S of R with an upper bound in R has a least upper bound (also called supremum) in R.

The last property is what differentiates the reals from the rationals. For example, the set of rationals with square less than 2 has a rational upper bound (e.g., 1.5) but no rational least upper bound, because the square root of 2 is not rational.

The real numbers are uniquely specified by the above properties. More precisely, given any two Dedekind-complete ordered fields R_1 and R_2, there exists a unique field isomorphism from R_1 to R_2, allowing us to think of them as essentially the same mathematical object.

For another axiomatization of R, see Tarski's axiomatization of the reals.

Construction from The Rational Numbers

The real numbers can be constructed as a completion of the rational numbers in such a way that a sequence defined by a decimal or binary expansion like (3; 3.1; 3.14; 3.141; 3.1415; ...) converges to a unique real number, in this case π. For details and other constructions of real numbers, see construction of the real numbers.

Properties

Basic Properties

A real number may be either rational or irrational; either algebraic or transcendental; and either positive, negative, or zero. Real numbers are used to measure continuous quantities. They may be expressed by decimal representations that have an infinite sequence of digits to the right of the decimal point; these are often represented in the same form as 324.823122147... The ellipsis (three dots) indicates that there would still be more digits to come.

More formally, real numbers have the two basic properties of being an ordered field, and having the least upper bound property. The first says that real numbers comprise a field, with addition and multiplication as well as division by non-zero numbers, which can be totally ordered on a number line in a way compatible with addition and multiplication. The second says that, if a non-empty set of real numbers has an upper bound, then it has a real least upper bound. The second condition distinguishes the real numbers from the rational numbers: for example, the set of rational numbers whose square is less than 2 is a set with an upper bound (e.g. 1.5) but no (rational) least upper bound: hence the rational numbers do not satisfy the least upper bound property.

Completeness

A main reason for using real numbers is that the reals contain all limits. More precisely, every sequence of real numbers having the property that consecutive terms of the sequence become arbitrarily close to each other necessarily has the property that after some term in the sequence the remaining terms are arbitrarily close to some specific real number. In mathematical terminology, this means that the reals are complete (in the sense of metric spaces or uniform spaces, which is a different sense than the Dedekind completeness of the order in the previous section). This is formally defined in the following way:

A sequence (x_n) of real numbers is called a *Cauchy sequence* if for any $\varepsilon > 0$ there exists an integer N (possibly depending on ε) such that the distance $|x_n - x_m|$ is less than ε for all n and m that are both greater than N. In other words, a sequence is a Cauchy sequence if its elements x_n eventually come and remain arbitrarily close to each other.

A sequence (x_n) *converges to the limit* x if for any $\varepsilon > 0$ there exists an integer N (possibly depending on ε) such that the distance $|x_n - x|$ is less than ε provided that n is greater than N. In other words, a sequence has limit x if its elements eventually come and remain arbitrarily close to x.

Notice that every convergent sequence is a Cauchy sequence. The converse is also true:

> Every Cauchy sequence of real numbers is convergent to a real number.

That is: the reals are complete.

Note that the rationals are not complete. For example, the sequence (1; 1.4; 1.41; 1.414; 1.4142; 1.41421...), where each term adds a digit of the decimal expansion of the positive square root of 2, is Cauchy but it does not converge to a rational number. (In the real numbers, in contrast, it converges to the positive square root of 2.)

The existence of limits of Cauchy sequences is what makes calculus work and is of great practical use. The standard numerical test to determine if a sequence has a limit is to test if it is a Cauchy sequence, as the limit is typically not known in advance.

For example, the standard series of the exponential function

$$e^x = \sum_{n=0}^{\infty} \frac{x^n}{n!}$$

converges to a real number because for every x the sums

$$\sum_{n=N}^{M} \frac{x^n}{n!}$$

can be made arbitrarily small by choosing N sufficiently large. This proves that the sequence is Cauchy, so we know that the sequence converges even if the limit is not known in advance.

"The Complete Ordered Field"

The real numbers are often described as "the complete ordered field", a phrase that can be interpreted in several ways.

First, an order can be lattice-complete. It is easy to see that no ordered field can be lattice-complete, because it can have no largest element (given any element z, $z + 1$ is larger), so this is not the sense that is meant.

Additionally, an order can be Dedekind-complete, as defined in the section Axioms. The uniqueness result at the end of that section justifies using the word "the" in the phrase "complete ordered field" when this is the sense of "complete" that is meant. This sense of completeness is most closely related to the construction of the reals from Dedekind cuts, since that construction starts from an ordered field (the rationals) and then forms the Dedekind-completion of it in a standard way.

These two notions of completeness ignore the field structure. However, an ordered group (in this case, the additive group of the field) defines a uniform structure, and uniform structures have a notion of completeness (topology); the description in the previous section Completeness is a special case. (We refer to the notion of completeness in uniform spaces rather than the related and better known notion for metric spaces, since the definition of metric space relies on already having a characterization of the real numbers.) It is not true that R is the *only* uniformly complete ordered field, but it is the only uniformly complete *Archimedean field*, and indeed one often hears the phrase "complete Archimedean field" instead of "complete ordered field". Every uniformly complete Archimedean field must also be Dedekind-complete (and vice versa, of course), justifying using "the" in the phrase "the complete Archimedean field". This sense of completeness is most closely related to the construction of the reals from Cauchy sequences, since it starts with an Archimedean field (the rationals) and forms the uniform completion of it in a standard way.

But the original use of the phrase "complete Archimedean field" was by David Hilbert, who meant still something else by it. He meant that the real numbers form the *largest* Archimedean field in the sense that every other Archimedean field is a subfield of R. Thus R is "complete" in the sense that nothing further can be added to it without making it no longer an Archimedean field. This sense of completeness is most closely related to the construction of the reals from surreal numbers, since that construction starts with a proper class that contains every ordered field (the surreals) and then selects from it the largest Archimedean subfield.

Advanced Properties

The reals are uncountable; that is: there are strictly more real numbers than natural numbers, even though both sets are infinite. In fact, the cardinality of the reals equals that of the set of subsets (i.e. the power set) of the natural numbers, and Cantor's diagonal argument states that the latter set's cardinality is strictly greater than the cardinality of N. Since the set of algebraic numbers is countable, almost all real numbers are transcendental. The non-existence of a subset of the reals with cardinality strictly between that of the integers and the reals is known as the continuum hypothesis. The continuum hypothesis can neither be proved nor be disproved; it is independent from the axioms of set theory.

As a topological space, the real numbers are separable. This is because the set of rationals, which is countable, is dense in the real numbers. The irrational numbers are also dense in the real numbers, however they are uncountable and have the same cardinality as the reals.

The real numbers form a metric space: the distance between x and y is defined as the absolute value $|x - y|$. By virtue of being a totally ordered set, they also carry an order topology; the topology arising from the metric and the one arising from the order are identical, but yield different presentations for the topology – in the order topology as ordered intervals, in the metric topology as epsilon-balls. The Dedekind cuts construction uses the order topology presentation, while the Cauchy sequences construction uses the metric topology presentation. The reals are a contractible (hence connected and simply connected), separable and complete metric space of Hausdorff dimension 1. The real numbers are locally compact but not compact. There are various properties that uniquely specify them; for instance, all unbounded, connected, and separable order topologies are necessarily homeomorphic to the reals.

Every nonnegative real number has a square root in R, although no negative number does. This shows that the order on R is determined by its algebraic structure. Also, every polynomial of odd degree admits at least one real root: these two properties make R the premier example of a real closed field. Proving this is the first half of one proof of the fundamental theorem of algebra.

The reals carry a canonical measure, the Lebesgue measure, which is the Haar measure on their structure as a topological group normalized such that the unit interval [0;1] has measure 1. There exist sets of real numbers that are not Lebesgue measurable, e.g. Vitali sets.

The supremum axiom of the reals refers to subsets of the reals and is therefore a second-order logical statement. It is not possible to characterize the reals with first-order logic alone: the Löwenheim–Skolem theorem implies that there exists a countable dense subset of the real numbers satisfying exactly the same sentences in first-order logic as the real numbers themselves. The set of hyperreal numbers satisfies the same first order sentences as R. Ordered fields that satisfy the same first-order sentences as R are called nonstandard models of R. This is what makes nonstandard analysis work; by proving a first-order statement in some nonstandard model (which may be easier than proving it in R), we know that the same statement must also be true of R.

The field R of real numbers is an extension field of the field Q of rational numbers, and R can therefore be seen as a vector space over Q. Zermelo–Fraenkel set theory with the axiom of choice guarantees the existence of a basis of this vector space: there exists a set B of real numbers such that every real number can be written uniquely as a finite linear combination of elements of this set, using rational coefficients only, and such that no element of B is a rational linear combination of the others. However, this existence theorem is purely theoretical, as such a base has never been explicitly described.

The well-ordering theorem implies that the real numbers can be well-ordered if the axiom of choice is assumed: there exists a total order on R with the property that every non-empty subset of R has a least element in this ordering. (The standard ordering $\leq$ of the real numbers is not a well-ordering since e.g. an open interval does not contain a least element in this ordering.) Again, the existence of such a well-ordering is purely theoretical, as it has not been explicitly described. If V=L is assumed in addition to the axioms of ZF, a well ordering of the real numbers can be shown to be explicitly definable by a formula.

Applications and Connections to Other Areas

Real Numbers and Logic

The real numbers are most often formalized using the Zermelo–Fraenkel axiomatization of set theory, but some mathematicians study the real numbers with other logical foundations of mathematics. In particular, the real numbers are also studied in reverse mathematics and in constructive mathematics.

The hyperreal numbers as developed by Edwin Hewitt, Abraham Robinson and others extend the set of the real numbers by introducing infinitesimal and infinite numbers, allowing for building infinitesimal calculus in a way closer to the original intuitions of Leibniz, Euler, Cauchy and others.

Edward Nelson's internal set theory enriches the Zermelo–Fraenkel set theory syntactically by introducing a unary predicate "standard". In this approach, infinitesimals are (non-"standard") elements of the set of the real numbers (rather than being elements of an extension thereof, as in Robinson's theory).

The continuum hypothesis posits that the cardinality of the set of the real numbers is $\aleph_1$; i.e. the smallest infinite cardinal number after $\aleph_0$, the cardinality of the integers. Paul Cohen proved in 1963 that it is an axiom independent of the other axioms of set theory; that is: one may choose either the continuum hypothesis or its negation as an axiom of set theory, without contradiction.

In Physics

In the physical sciences, most physical constants such as the universal gravitational constant, and physical variables, such as position, mass, speed, and electric charge, are modeled using real numbers. In fact, the fundamental physical theories such as classical mechanics, electromagnetism, quantum mechanics, general relativity and the standard model are described using mathematical structures, typically smooth manifolds or Hilbert spaces, that are based on the real numbers, although actual measurements of physical quantities are of finite accuracy and precision.

In some recent developments of theoretical physics stemming from the holographic principle, the Universe is seen fundamentally as an information store, essentially zeroes and ones, organized in much less geometrical fashion and manifesting itself as space-time and particle fields only on a more superficial level. This approach removes the real number system from its foundational role in physics and even prohibits the existence of infinite precision real numbers in the physical universe by considerations based on the Bekenstein bound.

In Computation

With some exceptions, most calculators do not operate on real numbers. Instead, they work with finite-precision approximations called floating-point numbers. In fact, most scientific computation uses floating-point arithmetic. Real numbers satisfy the usual rules of arithmetic, but floating-point numbers do not.

Computers cannot directly store arbitrary real numbers with infinitely many digits.

The precision is limited by the number of bits allocated to store a number, whether as floating-point numbers or arbitrary precision numbers. However, computer algebra systems can operate on irrational quantities exactly by manipulating formulas for them (such as $\sqrt{2}$, $\arcsin\left(\dfrac{2}{23}\right)$, or $\int_0^1 x^x \, dx$) rather than their rational or decimal approximation; however, it is not in general possible to determine whether two such expressions are equal (the constant problem).

A real number is called *computable* if there exists an algorithm that yields its digits. Because there are only countably many algorithms, but an uncountable number of reals, almost all real numbers fail to be computable. Moreover, the equality of two computable numbers is an undecidable problem. Some constructivists accept the existence of only those reals that are computable. The set of definable numbers is broader, but still only countable.

"Reals" in Set Theory

In set theory, specifically descriptive set theory, the Baire space is used as a surrogate for the real numbers since the latter have some topological properties (connectedness) that are a technical inconvenience. Elements of Baire space are referred to as "reals".

Vocabulary and Notation

Mathematicians use the symbol R, or, alternatively, R, the letter "R" in blackboard bold (encoded in Unicode as U+211D R DOUBLE-STRUCK CAPITAL R (HTML ℝ)), to represent the set of all real numbers. As this set is naturally endowed with the structure of a field, the expression *field of real numbers* is frequently used when its algebraic properties are under consideration.

The sets of positive real numbers and negative real numbers are often noted R^+ and R^-, respectively; R_+ and R_- are also used. The non-negative real numbers can be noted $R_{\geq 0}$ but one often sees this set noted $R^+ \cup \{0\}$. In French mathematics, the *positive real numbers* and *negative real numbers* commonly include zero, and these sets are noted respectively R_+ and R_-. In this understanding, the respective sets without zero are called strictly positive real numbers and strictly negative real numbers, and are noted R_+^* and R_-^*.

The notation R^n refers to the cartesian product of n copies of R, which is an n-dimensional vector space over the field of the real numbers; this vector space may be identified to the n-dimensional space of Euclidean geometry as soon as a coordinate system has been chosen in the latter. For example, a value from R^3 consists of three real numbers and specifies the coordinates of a point in 3dimensional space.

In mathematics, *real* is used as an adjective, meaning that the underlying field is the field of the real numbers (or *the real field*). For example, *real matrix*, *real polynomial* and *real Lie algebra*. The word is also used as a noun, meaning a real number (as in "the set of all reals").

Generalizations and Extensions

The real numbers can be generalized and extended in several different directions:

- The complex numbers contain solutions to all polynomial equations and hence are an algebraically closed field unlike the real numbers. However, the complex numbers are not an ordered field.

- The affinely extended real number system adds two elements $+\infty$ and $-\infty$. It is a compact space. It is no longer a field, not even an additive group, but it still has a total order; moreover, it is a complete lattice.

- The real projective line adds only one value ∞. It is also a compact space. Again, it is no longer a field, not even an additive group. However, it allows division of a non-zero element by zero. It has cyclic order described by a separation relation.

- The long real line pastes together $\aleph_1^* + \aleph_1$ copies of the real line plus a single point (here $\aleph_1^*$ denotes the reversed ordering of $\aleph_1$) to create an ordered set that is "locally" identical to the real numbers, but somehow longer; for instance, there is an order-preserving embedding of $\aleph_1$ in the long real line but not in the real numbers. The long real line is the largest ordered set that is complete and locally Archimedean. As with the previous two examples, this set is no longer a field or additive group.

- Ordered fields extending the reals are the hyperreal numbers and the surreal numbers; both of them contain infinitesimal and infinitely large numbers and are therefore non-Archimedean ordered fields.

- Self-adjoint operators on a Hilbert space (for example, self-adjoint square complex matrices) generalize the reals in many respects: they can be ordered (though not totally ordered), they are complete, all their eigenvalues are real and they form a real associative algebra. Positive-definite operators correspond to the positive reals and normal operators correspond to the complex numbers.

Complex Number

A complex number is a number that can be expressed in the form $a + bi$, where a and b are real numbers and i is the imaginary unit, satisfying the equation $i^2 = -1$. In this expression, a is the *real part* and b is the *imaginary part* of the complex number.

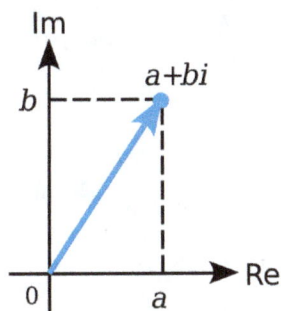

A complex number can be visually represented as a pair of numbers (a, b) forming a vector on a diagram called an Argand diagram, representing the complex plane. "Re" is the real axis, "Im" is the imaginary axis, and i is the imaginary unit which satisfies $i^2 = -1$.

Complex numbers extend the concept of the one-dimensional number line to the two-dimensional complex plane by using the horizontal axis for the real part and the vertical axis for the imaginary part. The complex number $a + bi$ can be identified with the point (a, b) in the complex plane. A complex number whose real part is zero is said to be purely imaginary, whereas a complex number whose imaginary part is zero is a real number. In this way, the complex numbers are a field extension of the ordinary real numbers, in order to solve problems that cannot be solved with real numbers alone.

As well as their use within mathematics, complex numbers have practical applications in many fields, including physics, chemistry, biology, economics, electrical engineering, and statistics. The Italian mathematician Gerolamo Cardano is the first known to have introduced complex numbers. He called them "fictitious" during his attempts to find solutions to cubic equations in the 16th century.

Overview

Complex numbers allow solutions to certain equations that have no solutions in real numbers. For example, the equation

$$(x+1)^2 = -9$$

has no real solution, since the square of a real number cannot be negative. Complex numbers provide a solution to this problem. The idea is to extend the real numbers with the imaginary unit i where $i^2 = -1$, so that solutions to equations like the preceding one can be found. In this case the solutions are $-1 + 3i$ and $-1 - 3i$, as can be verified using the fact that $i^2 = -1$:

$$((-1+3i)+1)^2 = (3i)^2 = (3^2)(i^2) = 9(-1) = -9,$$

$$((-1-3i)+1)^2 = (-3i)^2 = (-3)^2(i^2) = 9(-1) = -9.$$

According to the fundamental theorem of algebra, all polynomial equations with real or complex coefficients in a single variable have a solution in complex numbers.

Definition

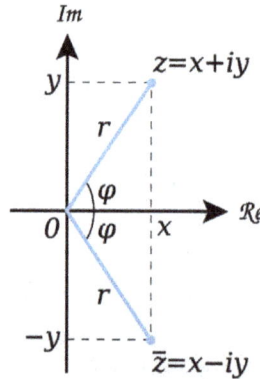

An illustration of the complex plane. The real part of a complex number
$z = x + iy$ is x, and its imaginary part is y.

A complex number is a number of the form $a + bi$, where a and b are real numbers and i is the *imaginary unit*, satisfying $i^2 = -1$. For example, $-3.5 + 2i$ is a complex number.

The real number a is called the *real part* of the complex number $a + bi$; the real number b is called the *imaginary part* of $a + bi$. By this convention the imaginary part does not include the imaginary unit: hence b, not bi, is the imaginary part. The real part of a complex number z is denoted by $\operatorname{Re}(z)$ or $R(z)$; the imaginary part of a complex number z is denoted by $\operatorname{Im}(z)$ or $I(z)$. For example,

$$\operatorname{Re}(-3.5 + 2i) = -3.5 \operatorname{Im}(-3.5 + 2i) = 2$$

Hence, in terms of its real and imaginary parts, a complex number z is equal to . This expression $\operatorname{Re}(z) + \operatorname{Im}(z) \cdot i$ is sometimes known as the Cartesian form of z.

A real number a can be regarded as a complex number $a + 0i$ whose imaginary part is 0. A purely imaginary number bi is a complex number $0 + bi$ whose real part is zero. It is common to write a for $a + 0i$ and bi for $0 + bi$. Moreover, when the imaginary part is negative, it is common to write $a - bi$ with $b > 0$ instead of $a + (-b)i$, for example $3 - 4i$ instead of $3 + (-4)i$.

The set of all complex numbers is denoted by C, **C** or $\mathbb{C}$.

Notation

Some authors write $a + ib$ instead of $a + bi$, particularly when b is a radical. In some disciplines, in particular electromagnetism and electrical engineering, j is used instead of i, since i is frequently used for electric current. In these cases complex numbers are written as $a + bj$ or $a + jb$.

Complex Plane

A complex number can be viewed as a point or position vector in a two-dimensional Cartesian coordinate system called the complex plane or Argand diagram, named after Jean-Robert Argand. The numbers are conventionally plotted using the real part as the horizontal component, and imaginary part as vertical (see Figure). These two values used to identify a given complex number are therefore called its *Cartesian*, *rectangular*, or *algebraic form*.

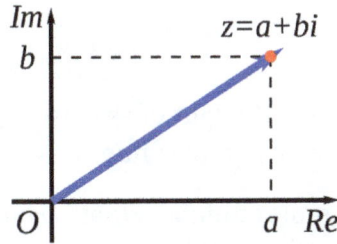

Figure: A complex number plotted as a point (red) and position vector (blue) on an Argand diagram; $a+bi$ is the *rectangular* expression of the point.

A position vector may also be defined in terms of its magnitude and direction relative to the origin. These are emphasized in a complex number's *polar form*. Using the polar form of the complex number in calculations may lead to a more intuitive interpretation of mathematical results. Notably, the operations of addition and multiplication take on a very natural geometric character when complex numbers are viewed as position vectors: addition corresponds to vector addition while multiplication corresponds to multiplying their magnitudes and adding their arguments (i.e. the angles they make with the x axis). Viewed in this way the multiplication of a complex number by i corresponds to rotating the position vector counterclockwise by a quarter turn (90°) about the origin: $(a+bi)i = ai+bi^2 = -b+ai$.

History in Brief

The solution in radicals (without trigonometric functions) of a general cubic equation contains the square roots of negative numbers when all three roots are real numbers, a situation that cannot be rectified by factoring aided by the rational root test if the cubic is irreducible (the so-called casus irreducibilis). This conundrum led Italian mathematician Gerolamo Cardano to conceive of complex numbers in around 1545, though his understanding was rudimentary.

Work on the problem of general polynomials ultimately led to the fundamental theorem of algebra, which shows that with complex numbers, a solution exists to every polynomial equation of degree one or higher. Complex numbers thus form an algebraically closed field, where any polynomial equation has a root.

Many mathematicians contributed to the full development of complex numbers. The rules for addition, subtraction, multiplication, and division of complex numbers were developed by the Italian mathematician Rafael Bombelli. A more abstract formalism for the complex numbers was further developed by the Irish mathematician William Rowan Hamilton, who extended this abstraction to the theory of quaternions.

Relations

Equality

Two complex numbers are equal if and only if both their real and imaginary parts are equal. In symbols:

$$z_1 = z_2 \leftrightarrow (\text{Re}(z_1) = \text{Re}(z_2) \wedge \text{Im}(z_1) = \text{Im}(z_2)).$$

Ordering

Because complex numbers are naturally thought of as existing on a two-dimensional plane, there is no natural linear ordering on the set of complex numbers.

There is no linear ordering on the complex numbers that is compatible with addition and multiplication. Formally, we say that the complex numbers cannot have the structure of an ordered field. This is because any square in an ordered field is at least 0, but $i^2 = -1$.

Elementary Operations

Conjugate

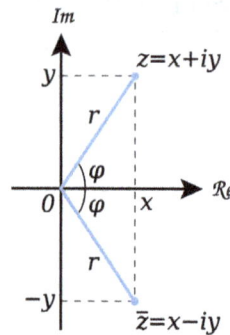

Geometric representation of z and its conjugate $\overline{z}$ in the complex plane

The *complex conjugate* of the complex number $z = x + yi$ is defined to be $x - yi$. It is denoted by either $\overline{z}$ or z^*.

Formally, for any complex number z:

$$\overline{z} = \text{Re}(z) - \text{Im}(z) \cdot i.$$

Geometrically, $\overline{z}$ is the "reflection" of z about the real axis. Conjugating twice gives the original complex number: $\overline{\overline{z}} = z..$

The real and imaginary parts of a complex number z can be extracted using the conjugate:

$$\text{Re}(z) = \tfrac{1}{2}(z + \overline{z}),$$

$$\text{Im}(z) = \tfrac{1}{2i}(z - \overline{z}).$$

Moreover, a complex number is real if and only if it equals its conjugate.

Conjugation distributes over the standard arithmetic operations:

$$\overline{z + w} = \overline{z} + \overline{w},$$

$$\overline{z - w} = \overline{z} - \overline{w},$$

$$\overline{zw} = \overline{z}\,\overline{w},$$

$$\overline{(z\,/\,w)} = \overline{z}\,/\,\overline{w}.$$

Addition and Subtraction

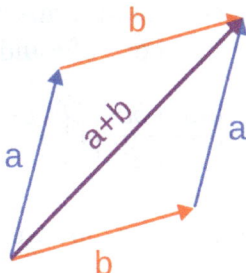

Addition of two complex numbers can be done geometrically by constructing a parallelogram.

Complex numbers are added by separately adding the real and imaginary parts of the summands. That is to say:

$$(a+bi)+(c+di)=(a+c)+(b+d)i.$$

Similarly, subtraction is defined by

$$(a+bi)-(c+di)=(a-c)+(b-d)i.$$

Using the visualization of complex numbers in the complex plane, the addition has the following geometric interpretation: the sum of two complex numbers A and B, interpreted as points of the complex plane, is the point X obtained by building a parallelogram, three of whose vertices are O, A and B. Equivalently, X is the point such that the triangles with vertices O, A, B, and X, B, A, are congruent.

Multiplication and Division

The multiplication of two complex numbers is defined by the following formula:

$$(a+bi)(c+di)=(ac-bd)+(bc+ad)i.$$

In particular, the square of the imaginary unit is −1:

$$i^2 = i \times i = -1.$$

The preceding definition of multiplication of general complex numbers follows naturally from this fundamental property of the imaginary unit. Indeed, if i is treated as a number so that di means d times i, the above multiplication rule is identical to the usual rule for multiplying two sums of two terms.

$$(a+bi)(c+di)=ac+bci+adi+bidi \qquad \text{(distributive property)}$$

$= ac + bidi + bci + adi$ (commutative property of addition—the order of the summands can be changed)

$= ac + bdi^2 + (bc + ad)i$ (commutative and distributive properties)

$= (ac - bd) + (bc + ad)i$ (fundamental property of the imaginary unit).

The division of two complex numbers is defined in terms of complex multiplication, which is described above, and real division. When at least one of c and d is non-zero, we have

$$\frac{a + bi}{c + di} = \left(\frac{ac + bd}{c^2 + d^2}\right) + \left(\frac{bc - ad}{c^2 + d^2}\right)i.$$

Division can be defined in this way because of the following observation:

$$\frac{a + bi}{c + di} = \frac{(a + bi) \cdot (c - di)}{(c + di) \cdot (c - di)} = \left(\frac{ac + bd}{c^2 + d^2}\right) + \left(\frac{bc - ad}{c^2 + d^2}\right)i.$$

As shown earlier, $c - di$ is the complex conjugate of the denominator $c + di$. At least one of the real part c and the imaginary part d of the denominator must be nonzero for division to be defined. This is called "rationalization" of the denominator (although the denominator in the final expression might be an irrational real number).

Reciprocal

The reciprocal of a nonzero complex number $z = x + yi$ is given by

$$\frac{1}{z} = \frac{\bar{z}}{z\bar{z}} = \frac{\bar{z}}{x^2 + y^2} = \frac{x}{x^2 + y^2} - \frac{y}{x^2 + y^2}i.$$

This formula can be used to compute the multiplicative inverse of a complex number if it is given in rectangular coordinates. Inversive geometry, a branch of geometry studying reflections more general than ones about a line, can also be expressed in terms of complex numbers. In the network analysis of electrical circuits, the complex conjugate is used in finding the equivalent impedance when the maximum power transfer theorem is used.

Square Root

The square roots of $a + bi$ (with $b \neq 0$) are $\pm(\gamma + \delta i)$, where

$$\gamma = \sqrt{\frac{a + \sqrt{a^2 + b^2}}{2}}$$

and

$$\delta = \text{sgn}(b)\sqrt{\dfrac{-a+\sqrt{a^2+b^2}}{2}},$$

where sgn is the signum function. This can be seen by squaring $\pm(\gamma + \delta i)$ to obtain $a + bi$. Here $\sqrt{a^2+b^2}$ is called the modulus of $a + bi$, and the square root sign indicates the square root with non-negative real part, called the principal square root; also $\sqrt{a^2+b^2} = \sqrt{z\bar{z}}$, , where $z = a + bi$.

Polar Form

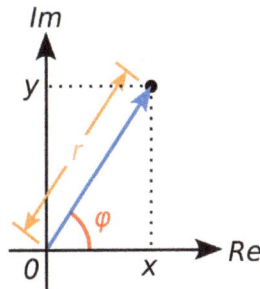

Figure 2: The argument φ and modulus r locate a point on an Argand diagram; $r(\cos\varphi + i\sin\varphi)$ or $re^{i\varphi}$ are *polar* expressions of the point.

Absolute Value and Argument

An alternative way of defining a point P in the complex plane, other than using the x- and y-coordinates, is to use the distance of the point from O, the point whose coordinates are $(0,0)$ (the origin), together with the angle subtended between the positive real axis and the line segment OP in a counterclockwise direction. This idea leads to the polar form of complex numbers.

The *absolute value* (or *modulus* or *magnitude*) of a complex number $z = x + yi$ is

$$r = |z| = \sqrt{x^2 + y^2}.$$

If z is a real number (i.e., $y = 0$), then $r = |x|$. In general, by Pythagoras' theorem, r is the distance of the point P representing the complex number z to the origin. The square of the absolute value is

$$|z|^2 = z\bar{z} = x^2 + y^2.$$

where $\bar{z}$ is the complex conjugate of .

The *argument* of z (in many applications referred to as the "phase") is the angle of the radius OP with the positive real axis, and is written as $\arg(z)$. As with the modulus, the argument can be found from the rectangular form $x + yi$:

$$\varphi = \arg(z) = \begin{cases} \arctan(\dfrac{y}{x}) & \text{if } x > 0 \\[2mm] \arctan(\dfrac{y}{x}) + \pi & \text{if } x < 0 \text{ and } y \geq 0 \\[2mm] \arctan(\dfrac{y}{x}) - \pi & \text{if } x < 0 \text{ and } y < 0 \\[2mm] \dfrac{\pi}{2} & \text{if } x = 0 \text{ and } y > 0 \\[2mm] -\dfrac{\pi}{2} & \text{if } x = 0 \text{ and } y < 0 \\[2mm] \text{indeterminate} & \text{if } x = 0 \text{ and } y = 0. \end{cases}$$

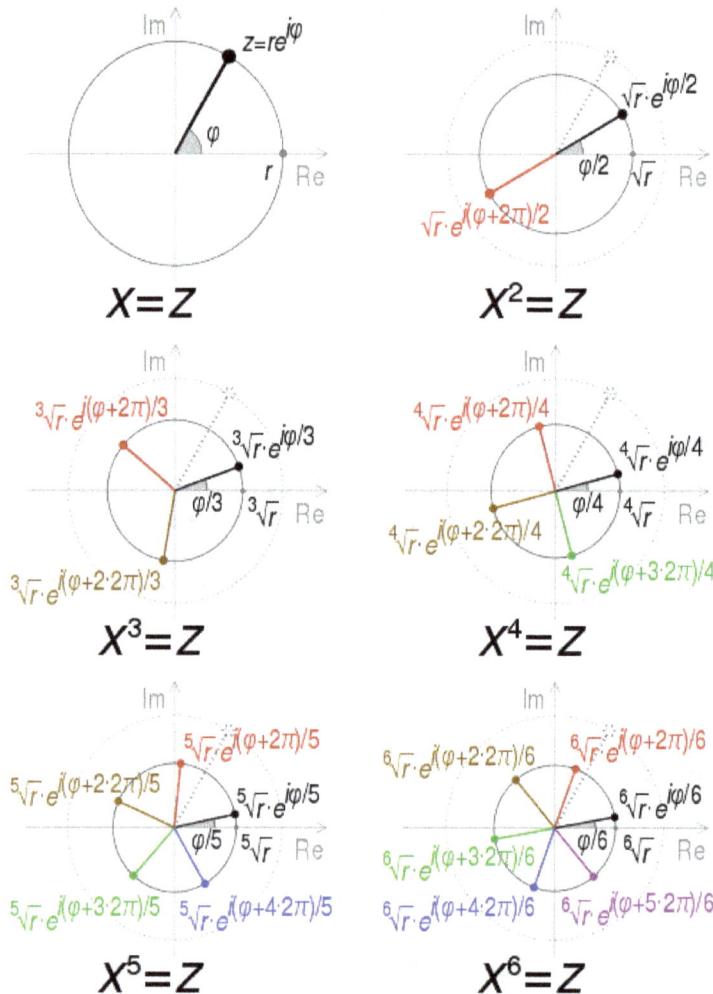

$X = Z$

$X^2 = Z$

$X^3 = Z$

$X^4 = Z$

$X^5 = Z$

$X^6 = Z$

Visualisation of the square to sixth roots of a complex number z, in polar form $re^{i\varphi}$ where $\varphi = \arg z$ and $r = |z|$ — if z is real, $\varphi = 0$ or π. Principal roots are in black.

Normally, as given above, the principal value in the interval $(-\pi, \pi]$ is chosen. Values in the range $[0, 2\pi)$ are obtained by adding 2π if the value is negative. The value of φ is expressed in radians in this article. It can increase by any integer multiple of 2π and still give the same angle. Hence, the arg function is sometimes considered as multivalued. The polar angle for the complex number o is indeterminate, but arbitrary choice of the angle o is common.

The value of φ equals the result of atan2: $\varphi = \text{atan2}(\text{imaginary}, \text{real})$.

Together, r and φ give another way of representing complex numbers, the *polar form*, as the combination of modulus and argument fully specify the position of a point on the plane. Recovering the original rectangular co-ordinates from the polar form is done by the formula called *trigonometric form*

$$z = r(\cos\varphi + i\sin\varphi).$$

Using Euler's formula this can be written as

$$z = re^{i\varphi}.$$

Using the cis function, this is sometimes abbreviated to

$$z = r\,\text{cis}\,\varphi.$$

In angle notation, often used in electronics to represent a phasor with amplitude r and phase φ, it is written as

$$z = r\angle\varphi.$$

Multiplication and Division in Polar Form

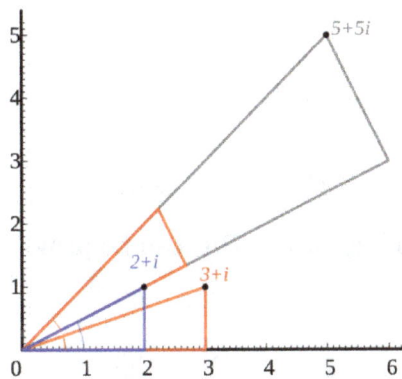

Multiplication of $2 + i$ (blue triangle) and $3 + i$ (red triangle). The red triangle is rotated to match the vertex of the blue one and stretched by $\sqrt{5}$, the length of the hypotenuse of the blue triangle.

Formulas for multiplication, division and exponentiation are simpler in polar form than the corresponding formulas in Cartesian coordinates. Given two complex numbers $z_1 = r_1(\cos\varphi_1 + i\sin\varphi_1)$ and $z_2 = r_2(\cos\varphi_2 + i\sin\varphi_2)$, because of the well-known trigonometric identities

$$\cos(a)\cos(b) - \sin(a)\sin(b) = \cos(a+b)$$

$$\cos(a)\sin(b) + \sin(a)\cos(b) = \sin(a+b)$$

we may derive

$$z_1 z_2 = r_1 r_2 (\cos(\varphi_1 + \varphi_2) + i \sin(\varphi_1 + \varphi_2)).$$

In other words, the absolute values are multiplied and the arguments are added to yield the polar form of the product. For example, multiplying by i corresponds to a quarter-turn counter-clockwise, which gives back $i^2 = -1$. The picture at the right illustrates the multiplication of

$$(2+i)(3+i) = 5+5i.$$

Since the real and imaginary part of $5 + 5i$ are equal, the argument of that number is 45 degrees, or $\pi/4$ (in radian). On the other hand, it is also the sum of the angles at the origin of the red and blue triangles are arctan(1/3) and arctan(1/2), respectively. Thus, the formula

$$\frac{\pi}{4} = \arctan\frac{1}{2} + \arctan\frac{1}{3}$$

holds. As the arctan function can be approximated highly efficiently, formulas like this—known as Machin-like formulas—are used for high-precision approximations of π.

Similarly, division is given by

$$\frac{z_1}{z_2} = \frac{r_1}{r_2}\left(\cos(\varphi_1 - \varphi_2) + i \sin(\varphi_1 - \varphi_2)\right).$$

Exponentiation

Euler's Formula

Euler's formula states that, for any real number x,

$$e^{ix} = \cos x + i \sin x,,$$

where e is the base of the natural logarithm. This can be proved through induction by observing that

$$i^0 = 1, \quad i^1 = i, \quad i^2 = -1, \quad i^3 = -i, i^4 = 1, \quad i^5 = i, \quad i^6 = -1, \quad i^7 = -i$$

and so on, and by considering the Taylor series expansions of e^{ix}, $\cos x$ and $\sin x$:

$$e^{ix} = 1 + ix + \frac{(ix)^2}{2!} + \frac{(ix)^3}{3!} + \frac{(ix)^4}{4!} + \frac{(ix)^5}{5!} + \frac{(ix)^6}{6!} + \frac{(ix)^7}{7!} + \frac{(ix)^8}{8!} + \cdots [8pt]$$

$$= 1 + ix - \frac{x^2}{2!} - \frac{ix^3}{3!} + \frac{x^4}{4!} + \frac{ix^5}{5!} - \frac{x^6}{6!} - \frac{ix^7}{7!} + \frac{x^8}{8!} + \cdots [8pt]$$

$$= \left(1 - \frac{x^2}{2!} + \frac{x^4}{4!} - \frac{x^6}{6!} + \frac{x^8}{8!} - \cdots\right) + i\left(x - \frac{x^3}{3!} + \frac{x^5}{5!} - \frac{x^7}{7!} + \cdots\right)[8pt] = \cos x + i \sin x$$

The rearrangement of terms is justified because each series is absolutely convergent.

Natural Logarithm

Euler's formula allows us to observe that, for any complex number

$$z = r(\cos\varphi + i\sin\varphi).$$

where r is a non-negative real number, one possible value for z's natural logarithm is

$$\ln(z) = \ln(r) + \varphi i$$

Because cos and sin are periodic functions, the natural logarithm may be considered a multi-valued function, with:

$$\ln(z) = \left\{\ln(r) + (\varphi + 2\pi k)i \mid k \in \mathbb{Z}\right\}$$

Integer and Fractional Exponents

We may use the identity

$$\ln(a^b) = b\ln(a)$$

to define complex exponentiation, which is likewise multi-valued:

$$\ln(z^n) = \ln((r(\cos\varphi + i\sin\varphi))^n)$$

$$= n\ln(r(\cos\varphi + i\sin\varphi))$$

$$= \{n(\ln(r) + (\varphi + k2\pi)i) \mid k \in \mathbb{Z}\}$$

$$= \{n\ln(r) + n\varphi i + nk2\pi i \mid k \in \mathbb{Z}\}.$$

When n is an integer, this simplifies to de Moivre's formula:

$$z^n = (r(\cos\varphi + i\sin\varphi))^n = r^n(\cos n\varphi + i\sin n\varphi).$$

The nth roots of z are given by

$$\sqrt[n]{z} = \sqrt[n]{r}\left(\cos\left(\frac{\varphi + 2k\pi}{n}\right) + i\sin\left(\frac{\varphi + 2k\pi}{n}\right)\right)$$

for any integer k satisfying $0 \le k \le n - 1$. Here $\sqrt[n]{r}$ is the usual (positive) nth root of the positive real number r. While the nth root of a positive real number r is chosen to be the *positive* real number c satisfying $c^n = r$ there is no natural way of distinguishing one particular complex nth root of a complex number. Therefore, the nth root of z is considered as a multivalued function (in z), as opposed to a usual function f, for which $f(z)$ is a uniquely defined number. Formulas such as

$$\sqrt[n]{z^n} = z$$

(which holds for positive real numbers), do in general not hold for complex numbers.

Properties

Field Structure

The set C of complex numbers is a field. Briefly, this means that the following facts hold: first, any two complex numbers can be added and multiplied to yield another complex number. Second, for any complex number z, its additive inverse $-z$ is also a complex number; and third, every nonzero complex number has a reciprocal complex number. Moreover, these operations satisfy a number of laws, for example the law of commutativity of addition and multiplication for any two complex numbers z_1 and z_2:

$$z_1 + z_2 = z_2 + z_1,$$

$$z_1 z_2 = z_2 z_1.$$

These two laws and the other requirements on a field can be proven by the formulas given above, using the fact that the real numbers themselves form a field.

Unlike the reals, C is not an ordered field, that is to say, it is not possible to define a relation $z_1 < z_2$ that is compatible with the addition and multiplication. In fact, in any ordered field, the square of any element is necessarily positive, so $i^2 = -1$ precludes the existence of an ordering on C.

When the underlying field for a mathematical topic or construct is the field of complex numbers, the topic's name is usually modified to reflect that fact. For example: complex analysis, complex matrix, complex polynomial, and complex Lie algebra.

Solutions of Polynomial Equations

Given any complex numbers (called coefficients) $a_0, ..., a_n$, the equation

$$a_n z^n + \cdots + a_1 z + a_0 = 0$$

has at least one complex solution z, provided that at least one of the higher coefficients $a_1, ..., a_n$ is nonzero. This is the statement of the *fundamental theorem of algebra*. Because of this fact, C is called an algebraically closed field. This property does not hold for the field of rational numbers Q (the polynomial $x^2 - 2$ does not have a rational root, since $\sqrt{2}$ is not a rational number) nor the real numbers R (the polynomial $x^2 + a$ does not have a real root for $a > 0$, since the square of x is positive for any real number x).

There are various proofs of this theorem, either by analytic methods such as Liouville's theorem, or topological ones such as the winding number, or a proof combining Galois theory and the fact that any real polynomial of *odd* degree has at least one real root.

Because of this fact, theorems that hold *for any algebraically closed field*, apply to C. For example, any non-empty complex square matrix has at least one (complex) eigenvalue.

Algebraic Characterization

The field C has the following three properties: first, it has characteristic 0. This means that $1 + 1 + \cdots + 1 \neq 0$ for any number of summands (all of which equal one). Second, its transcendence degree over Q, the prime field of C, is the cardinality of the continuum. Third, it is algebraically closed. It can be shown that any field having these properties is isomorphic (as a field) to C. For example, the algebraic closure of Q_p also satisfies these three properties, so these two fields are isomorphic. Also, C is isomorphic to the field of complex Puiseux series. However, specifying an isomorphism requires the axiom of choice. Another consequence of this algebraic characterization is that C contains many proper subfields that are isomorphic to C.

Characterization as A Topological Field

The preceding characterization of C describes only the algebraic aspects of C. That is to say, the properties of nearness and continuity, which matter in areas such as analysis and topology, are not dealt with. The following description of C as a topological field (that is, a field that is equipped with a topology, which allows the notion of convergence) does take into account the topological properties. C contains a subset P (namely the set of positive real numbers) of nonzero elements satisfying the following three conditions:

- P is closed under addition, multiplication and taking inverses.

- If x and y are distinct elements of P, then either $x - y$ or $y - x$ is in P.

- If S is any nonempty subset of P, then $S + P = x + P$ for some x in C.

Moreover, C has a nontrivial involutive automorphism $x \mapsto x^*$ (namely the complex conjugation), such that xx^* is in P for any nonzero x in C.

Any field F with these properties can be endowed with a topology by taking the sets $B(x,p) = \{y \mid p - (y - x)(y - x)^* \in P\}$ as a base, where x ranges over the field and p ranges over P. With this topology F is isomorphic as a *topological* field to C.

The only connected locally compact topological fields are R and C. This gives another characterization of C as a topological field, since C can be distinguished from R because the nonzero complex numbers are connected, while the nonzero real numbers are not.

Formal Construction

Formal Development

Above, complex numbers have been defined by introducing i, the imaginary unit, as a symbol.

More rigorously, the set C of complex numbers can be defined as the set R^2 of ordered pairs (a, b) of real numbers. In this notation, the above formulas for addition and multiplication read

$$(a,b)+(c,d) = (a+c, b+d)(a,b) \cdot (c,d) = (ac-bd, bc+ad)$$

It is then just a matter of notation to express (a, b) as $a + bi$.

Though this low-level construction does accurately describe the structure of the complex numbers, the following equivalent definition reveals the algebraic nature of C more immediately. This characterization relies on the notion of fields and polynomials. A field is a set endowed with addition, subtraction, multiplication and division operations that behave as is familiar from, say, rational numbers. For example, the distributive law

$$(x+y)z = xz + yz$$

must hold for any three elements x, y and z of a field. The set R of real numbers does form a field. A polynomial $p(X)$ with real coefficients is an expression of the form

$$a_n X^n + \cdots + a_1 X + a_0,,$$

where the $a_0, ..., a_n$ are real numbers. The usual addition and multiplication of polynomials endows the set $R[X]$ of all such polynomials with a ring structure. This ring is called polynomial ring.

The quotient ring $R[X]/(X^2 + 1)$ can be shown to be a field. This extension field contains two square roots of -1, namely (the cosets of) X and $-X$, respectively. (The cosets of) 1 and X form a basis of $R[X]/(X^2 + 1)$ as a real vector space, which means that each element of the extension field can be uniquely written as a linear combination in these two elements. Equivalently, elements of the extension field can be written as ordered pairs (a, b) of real numbers. Moreover, the above formulas for addition etc. correspond to the ones yielded by this abstract algebraic approach—the two definitions of the field C are said to be isomorphic (as fields). Together with the above-mentioned fact that C is algebraically closed, this also shows that C is an algebraic closure of R.

Matrix Representation of Complex Numbers

Complex numbers $a + bi$ can also be represented by 2×2 matrices that have the following form:

$$\begin{pmatrix} a & -b \\ b & a \end{pmatrix}.$$

Here the entries a and b are real numbers. The sum and product of two such matrices is again of this form, and the sum and product of complex numbers corresponds to the sum and product of such matrices. The geometric description of the multiplication of complex numbers can also be expressed in terms of rotation matrices by using this correspondence between complex numbers and such matrices. Moreover, the square of the absolute value of a complex number expressed as a matrix is equal to the determinant of that matrix:

The conjugate $\bar{z}$ corresponds to the transpose of the matrix.

$$|z|^2 = \begin{vmatrix} a & -b \\ b & a \end{vmatrix} = (a^2) - ((-b)(b)) = a^2 + b^2.$$

Though this representation of complex numbers with matrices is the most common, many other representations arise from matrices *other than* $\begin{pmatrix} 0 & -1 \\ 1 & 0 \end{pmatrix}$ that square to the negative of the identity matrix. See the article on 2×2 real matrices for other representations of complex numbers.

Complex Analysis

The study of functions of a complex variable is known as complex analysis and has enormous practical use in applied mathematics as well as in other branches of mathematics. Often, the most natural proofs for statements in real analysis or even number theory employ techniques from complex analysis (see prime number theorem for an example). Unlike real functions, which are commonly represented as two-dimensional graphs, complex functions have four-dimensional graphs and may usefully be illustrated by color-coding a three-dimensional graph to suggest four dimensions, or by animating the complex function's dynamic transformation of the complex plane.

Color wheel graph of $\sin(1/z)$. Black parts inside refer to numbers having large absolute values.

Complex Exponential and Related Functions

The notions of convergent series and continuous functions in (real) analysis have natural analogs in complex analysis. A sequence of complex numbers is said to converge if and only if its real and imaginary parts do. This is equivalent to the (ε, δ)-definition of limits, where the absolute value of real numbers is replaced by the one of complex numbers. From a more abstract point of view, C, endowed with the metric

$$d(z_1, z_2) = |z_1 - z_2|$$

is a complete metric space, which notably includes the triangle inequality

$$|z_1 + z_2| \leq |z_1| + |z_2|$$

for any two complex numbers z_1 and z_2.

Like in real analysis, this notion of convergence is used to construct a number of elementary functions: the *exponential function* exp(z), also written e^z, is defined as the infinite series

$$\exp(z) := 1 + z + \frac{z^2}{2 \cdot 1} + \frac{z^3}{3 \cdot 2 \cdot 1} + \cdots = \sum_{n=0}^{\infty} \frac{z^n}{n!}.$$

and the series defining the real trigonometric functions sine and cosine, as well as hyperbolic functions such as sinh also carry over to complex arguments without change. *Euler's identity* states:

$$\exp(i\varphi) = \cos(\varphi) + i\sin(\varphi)$$

for any real number φ, in particular

$$\exp(i\pi) = -1$$

Unlike in the situation of real numbers, there is an infinitude of complex solutions z of the equation

$$\exp(z) = w$$

for any complex number $w \neq 0$. It can be shown that any such solution z—called complex logarithm of w—satisfies

$$\log(x + iy) = \ln|w| + i\arg(w),$$

where arg is the argument defined above, and ln the (real) natural logarithm. As arg is a multivalued function, unique only up to a multiple of 2π, log is also multivalued. The principal value of log is often taken by restricting the imaginary part to the interval $(-\pi, \pi]$.

Complex exponentiation z^ω is defined as

$$z^\omega = \exp(\omega \log z).$$

Consequently, they are in general multi-valued. For $\omega = 1/n$, for some natural number n, this recovers the non-uniqueness of nth roots mentioned above.

Complex numbers, unlike real numbers, do not in general satisfy the unmodified power and logarithm identities, particularly when naïvely treated as single-valued functions; see failure of power and logarithm identities. For example, they do not satisfy

$$a^{bc} = (a^b)^c.$$

Both sides of the equation are multivalued by the definition of complex exponentiation given here, and the values on the left are a subset of those on the right.

Holomorphic Functions

A function $f: C \rightarrow C$ is called holomorphic if it satisfies the Cauchy–Riemann equations. For example, any R-linear map $C \rightarrow C$ can be written in the form

$$f(z) = az + b\overline{z}$$

with complex coefficients a and b. This map is holomorphic if and only if $b = 0$. The second summand $b\overline{z}$ is real-differentiable, but does not satisfy the Cauchy–Riemann equations.

Complex analysis shows some features not apparent in real analysis. For example, any two holomorphic functions f and g that agree on an arbitrarily small open subset of C necessarily agree everywhere. Meromorphic functions, functions that can locally be written as $f(z)/(z - z_0)^n$ with a holomorphic function f, still share some of the features of holomorphic functions. Other functions have essential singularities, such as $\sin(1/z)$ at $z = 0$.

Applications

Complex numbers have essential concrete applications in a variety of scientific and related areas such as signal processing, control theory, electromagnetism, fluid dynamics, quantum mechanics, cartography, and vibration analysis. Some applications of complex numbers are:

Control Theory

In control theory, systems are often transformed from the time domain to the frequency domain using the Laplace transform. The system's poles and zeros are then analyzed in the *complex plane*. The root locus, Nyquist plot, and Nichols plot techniques all make use of the complex plane.

In the root locus method, it is especially important whether the poles and zeros are in the left or right half planes, i.e. have real part greater than or less than zero. If a linear, time-invariant (LTI) system has poles that are

- in the right half plane, it will be unstable,

- all in the left half plane, it will be stable,

- on the imaginary axis, it will have marginal stability.

If a system has zeros in the right half plane, it is a nonminimum phase system.

Improper Integrals

In applied fields, complex numbers are often used to compute certain real-valued improper integrals, by means of complex-valued functions. Several methods exist to do this; see methods of contour integration.

Fluid Dynamics

In fluid dynamics, complex functions are used to describe potential flow in two dimensions.

Dynamic Equations

In differential equations, it is common to first find all complex roots r of the characteristic equation of a linear differential equation or equation system and then attempt to solve the system in terms of base functions of the form $f(t) = e^{rt}$. Likewise, in difference equations, the complex roots r of the characteristic equation of the difference equation system are used, to attempt to solve the system in terms of base functions of the form $f(t) = r^t$.

Electromagnetism and Electrical Engineering

In electrical engineering, the Fourier transform is used to analyze varying voltages and currents. The treatment of resistors, capacitors, and inductors can then be unified by introducing imaginary, frequency-dependent resistances for the latter two and combining all three in a single complex number called the impedance. This approach is called phasor calculus.

In electrical engineering, the imaginary unit is denoted by j, to avoid confusion with I, which is generally in use to denote electric current, or, more particularly, i, which is generally in use to denote instantaneous electric current.

Since the voltage in an AC circuit is oscillating, it can be represented as

$$V(t) = V_0 e^{j\omega t} = V_0\left(\cos \omega t + j \sin \omega t\right),$$

To obtain the measurable quantity, the real part is taken:

$$v(t) = \mathrm{Re}(V) = \mathrm{Re}\left[V_0 e^{j\omega t}\right] = V_0 \cos \omega t.$$

The complex-valued signal $V(t)$ is called the analytic representation of the real-valued, measurable signal $v(t)$.

Signal Analysis

Complex numbers are used in signal analysis and other fields for a convenient description for periodically varying signals. For given real functions representing actual physical quantities, often in terms of sines and cosines, corresponding complex functions are considered of which the real parts are the original quantities. For a sine wave of a given frequency, the absolute value $|z|$ of the corresponding z is the amplitude and the argument $\arg(z)$ is the phase.

If Fourier analysis is employed to write a given real-valued signal as a sum of periodic functions, these periodic functions are often written as complex valued functions of the form

$$x(t) = Re\{X(t)\}$$

and

$$X(t) = Ae^{i\omega t} = ae^{i\phi}e^{i\omega t} = ae^{i(\omega t + \phi)}$$

where ω represents the angular frequency and the complex number A encodes the phase and amplitude as explained above.

This use is also extended into digital signal processing and digital image processing, which utilize digital versions of Fourier analysis (and wavelet analysis) to transmit, compress, restore, and otherwise process digital audio signals, still images, and video signals.

Another example, relevant to the two side bands of amplitude modulation of AM radio, is:

$$\cos((\omega+\alpha)t) + \cos((\omega-\alpha)t) = \mathrm{Re}\left(e^{i(\omega+\alpha)t} + e^{i(\omega-\alpha)t}\right)$$

$$= \mathrm{Re}\left((e^{i\alpha t} + e^{-i\alpha t}) \cdot e^{i\omega t}\right)$$

$$= \mathrm{Re}\left(2\cos(\alpha t) \cdot e^{i\omega t}\right)$$

$$= 2\cos(\alpha t) \cdot \mathrm{Re}\left(e^{i\omega t}\right)$$

$$= 2\cos(\alpha t) \cdot \cos(\omega t).$$

Quantum Mechanics

The complex number field is intrinsic to the mathematical formulations of quantum mechanics, where complex Hilbert spaces provide the context for one such formulation that is convenient and perhaps most standard. The original foundation formulas of quantum mechanics—the Schrödinger equation and Heisenberg's matrix mechanics—make use of complex numbers.

Relativity

In special and general relativity, some formulas for the metric on spacetime become simpler if one takes the time component of the spacetime continuum to be imaginary. (This approach is no longer standard in classical relativity, but is used in an essential way in quantum field theory.) Complex numbers are essential to spinors, which are a generalization of the tensors used in relativity.

Geometry

Fractals

Certain fractals are plotted in the complex plane, e.g. the Mandelbrot set and Julia sets.

Triangles

Every triangle has a unique Steiner inellipse—an ellipse inside the triangle and tangent to the midpoints of the three sides of the triangle. The foci of a triangle's Steiner inellipse can be found as follows, according to Marden's theorem: Denote the triangle's vertices in the complex plane as $a = x_A + y_A i$, $b = x_B + y_B i$, and $c = x_C + y_C i$. Write the cubic equation $(x-a)(x-b)(x-c) = 0$, take its derivative, and equate the (quadratic) derivative to zero. Marden's Theorem says that the solutions of this equation are the complex numbers denoting the locations of the two foci of the Steiner inellipse.

Algebraic Number Theory

Construction of a regular pentagon using straightedge and compass.

As mentioned above, any nonconstant polynomial equation (in complex coefficients) has a solution in C. A fortiori, the same is true if the equation has rational coefficients. The roots of such equations are called algebraic numbers – they are a principal object of study in algebraic number theory. Compared to Q, the algebraic closure of Q, which also contains all algebraic numbers, C has the advantage of being easily understandable in geometric terms. In this way, algebraic methods can be used to study geometric questions and vice versa. With algebraic methods, more specifically applying the machinery of field theory to the number field containing roots of unity, it can be shown that it is not possible to construct a regular nonagon using only compass and straightedge – a purely geometric problem.

Another example are Gaussian integers, that is, numbers of the form $x + iy$, where x and y are integers, which can be used to classify sums of squares.

Analytic Number Theory

Analytic number theory studies numbers, often integers or rationals, by taking advantage of the fact that they can be regarded as complex numbers, in which analytic methods can be used. This is done by encoding number-theoretic information in complex-valued functions. For example, the Riemann zeta function $\zeta(s)$ is related to the distribution of prime numbers.

History

The earliest fleeting reference to square roots of negative numbers can perhaps be said to occur in the work of the Greek mathematician Hero of Alexandria in the 1st century AD, where in his *Stereometrica* he considers, apparently in error, the volume of an impossible frustum of a pyramid to arrive at the term $\sqrt{81-144} = 3i\sqrt{7}$ in his calculations, although negative quantities were not conceived of in Hellenistic mathematics and Heron merely replaced it by its positive ($\sqrt{144-81} = 3\sqrt{7}$).

The impetus to study complex numbers proper first arose in the 16th century when algebraic solutions for the roots of cubic and quartic polynomials were discovered by Italian mathematicians. It was soon realized that these formulas, even if one was only interested in real solutions, sometimes required the manipulation of square roots of negative numbers. As an example, Tartaglia's formula for a cubic equation of the form $x^3 = px+q^{[}$ gives the solution to the equation $x^3 = x$ as

$$\frac{1}{\sqrt{3}}\left((\sqrt{-1})^{1/3} + \frac{1}{(\sqrt{-1})^{1/3}} \right).$$

At first glance this looks like nonsense. However formal calculations with complex numbers show that the equation $z^3 = i$ has solutions $-i$, $\frac{\sqrt{3}}{2} + \frac{1}{2}i$ and $\frac{-\sqrt{3}}{2} + \frac{1}{2}i$. Substituting these in turn for $\sqrt{-1}^{1/3}$ in Tartaglia's cubic formula and simplifying, one gets 0, 1 and −1 as the solutions of $x^3 - x = 0$. Of course this particular equation can be solved at sight but it does illustrate that when general formulas are used to solve cubic equations with real roots then, as later mathematicians showed rigorously, the use of complex numbers is unavoidable. Rafael Bombelli was the first to explicitly address these seemingly paradoxical solutions of cubic equations and developed the rules for complex arithmetic trying to resolve these issues.

The term "imaginary" for these quantities was coined by René Descartes in 1637, although he was at pains to stress their imaginary nature

[...] sometimes only imaginary, that is one can imagine as many as I said in each equation, but sometimes there exists no quantity that matches that which we imagine.

([...] quelquefois seulement imaginaires c'est-à-dire que l'on peut toujours en imaginer autant que j'ai dit en chaque équation, mais qu'il n'y a quelquefois aucune quantité qui corresponde à celle qu'on imagine.)

A further source of confusion was that the equation $\sqrt{-1}^2 = \sqrt{-1}\sqrt{-1} = -1$ seemed to be capriciously inconsistent with the algebraic identity $\sqrt{a}\sqrt{b} = \sqrt{ab}$, which is valid for non-negative real numbers a and b, and which was also used in complex number calculations with one of a, b positive and the other negative. The incorrect use of this identity (and the related identity $\frac{1}{\sqrt{a}} = \sqrt{\frac{1}{a}}$) in the case when both a and b are negative even bedeviled Euler. This difficulty eventually led to the convention of using the special symbol i in place of $\sqrt{-1}$ to guard against this mistake. Even so, Euler considered it natural to introduce students to complex numbers much earlier than we do today. In his elementary algebra text book, Elements of Algebra, he introduces these numbers almost at once and then uses them in a natural way throughout.

In the 18th century complex numbers gained wider use, as it was noticed that formal manipulation of complex expressions could be used to simplify calculations involving trigonometric functions. For instance, in 1730 Abraham de Moivre noted that the complicated identities relating trigonometric functions of an integer multiple of an angle to powers of trigonometric functions of that angle could be simply re-expressed by the following well-known formula which bears his name, de Moivre's formula:

$$(\cos\theta + i\sin\theta)^n = \cos n\theta + i\sin n\theta.$$

In 1748 Leonhard Euler went further and obtained Euler's formula of complex analysis:

$$\cos\theta + i\sin\theta = e^{i\theta}$$

by formally manipulating complex power series and observed that this formula could be used to reduce any trigonometric identity to much simpler exponential identities.

The idea of a complex number as a point in the complex plane (above) was first described by Caspar Wessel in 1799, although it had been anticipated as early as 1685 in Wallis's *De Algebra tractatus*.

Wessel's memoir appeared in the Proceedings of the Copenhagen Academy but went largely unnoticed. In 1806 Jean-Robert Argand independently issued a pamphlet on complex numbers and provided a rigorous proof of the fundamental theorem of algebra. Gauss had earlier published an essentially topological proof of the theorem in 1797 but expressed his doubts at the time about "the true metaphysics of the square root of −1". It was not until 1831 that he overcame these doubts and published his treatise on complex numbers as points in the plane, largely establishing modern notation and terminology. In the beginning of 19th century, other mathematicians discovered independently the geometrical representation of the complex numbers: Buée, Mourey, Warren, Français and his brother, Bellavitis.

The English mathematician G. H. Hardy remarked that Gauss was the first mathematician to use complex numbers in 'a really confident and scientific way' although mathematicians such as Niels Henrik Abel and Carl Gustav Jacob Jacobi were necessarily using them routinely before Gauss published his 1831 treatise. Augustin Louis Cauchy and Bernhard Riemann together brought the fundamental ideas of complex analysis to a high state of completion, commencing around 1825 in Cauchy's case.

The common terms used in the theory are chiefly due to the founders. Argand called $\cos\phi + i\sin\phi$ the *direction factor*, and $r = \sqrt{a^2 + b^2}$ the *modulus*; Cauchy (1828) called $\cos\phi + i\sin\phi$ the *reduced form* (l'expression réduite) and apparently introduced the term *argument*; Gauss used i for $\sqrt{-1}$, introduced the term *complex number* for $a + bi$, and called $a^2 + b^2$ the *norm*. The expression *direction coefficient*, often used for $\cos\phi + i\sin\phi$, is due to Hankel (1867), and *absolute value*, for *modulus*, is due to Weierstrass.

Later classical writers on the general theory include Richard Dedekind, Otto Hölder, Felix Klein, Henri Poincaré, Hermann Schwarz, Karl Weierstrass and many others.

Generalizations and Related Notions

The process of extending the field R of reals to C is known as Cayley–Dickson construction. It can be carried further to higher dimensions, yielding the quaternions H and octonions O which (as a real vector space) are of dimension 4 and 8, respectively. In this context the complex numbers have been called the binarions.

However, just as applying the construction to reals loses the property of ordering, more properties familiar from real and complex numbers vanish with increasing dimension. The quaternions are only a skew field, i.e. for some x, y: $x \cdot y \neq y \cdot x$ for two quaternions, the multiplication of octonions fails (in addition to not being commutative) to be associative: for some x, y, z: $(x \cdot y) \cdot z \neq x \cdot (y \cdot z)$.

Reals, complex numbers, quaternions and octonions are all normed division algebras over R. However, by Hurwitz's theorem they are the only ones. The next step in the Cayley–Dickson construction, the sedenions, in fact fails to have this structure.

The Cayley–Dickson construction is closely related to the regular representation of C, thought of as an R-algebra (an R-vector space with a multiplication), with respect to the basis $(1, i)$. This means the following: the R-linear map

$$\mathbb{C} \to \mathbb{C}, z \mapsto wz$$

for some fixed complex number w can be represented by a 2×2 matrix (once a basis has been chosen). With respect to the basis $(1, i)$, this matrix is

$$\begin{pmatrix} \mathrm{Re}(w) & -\mathrm{Im}(w) \\ \mathrm{Im}(w) & \mathrm{Re}(w) \end{pmatrix}$$

i.e., the one mentioned in the section on matrix representation of complex numbers above. While this is a linear representation of C in the 2 × 2 real matrices, it is not the only one. Any matrix

$$J = \begin{pmatrix} p & q \\ r & -p \end{pmatrix}, \quad p^2 + qr + 1 = 0$$

has the property that its square is the negative of the identity matrix: $J^2 = -I$. Then

$$\{z = aI + bJ : a, b \in R\}$$

is also isomorphic to the field C, and gives an alternative complex structure on R^2. This is generalized by the notion of a linear complex structure.

Hypercomplex numbers also generalize R, C, H, and O. For example, this notion contains the split-complex numbers, which are elements of the ring $R[x]/(x^2 - 1)$ (as opposed to $R[x]/(x^2 + 1)$). In this ring, the equation $a^2 = 1$ has four solutions.

The field R is the completion of Q, the field of rational numbers, with respect to the usual absolute value metric. Other choices of metrics on Q lead to the fields Q_p of p-adic numbers (for any prime number p), which are thereby analogous to R. There are no other nontrivial ways of completing Q than R and Q_p, by Ostrowski's theorem. The algebraic closures $\overline{Q_p}$ of Q_p still carry a norm, but (unlike C) are not complete with respect to it. The completion C_p of $\overline{Q_p}$ turns out to be algebraically closed. This field is called p-adic complex numbers by analogy.

The fields R and Q_p and their finite field extensions, including C, are local fields.

References

- Gouvea, Fernando Q. The Princeton Companion to Mathematics, Chapter II.1, "The Origins of Modern Mathematics", p. 82. Princeton University Press, September 28, 2008. ISBN 978-0691118802.

- Selin, Helaine, ed. (2000). Mathematics across cultures: the history of non-Western mathematics. Kluwer Academic Publishers. p. 451. ISBN 0-7923-6481-3.

- Musser, Gary L.; Peterson, Blake E.; Burger, William F. (2013), Mathematics for Elementary Teachers: A Contemporary Approach (10th ed.), Wiley Global Education, ISBN 978-1118457443

- Thomson, Brian S.; Bruckner, Judith B.; Bruckner, Andrew M. (2008), Elementary Real Analysis (Second ed.), ClassicalRealAnalysis.com, ISBN 9781434843678

- Rosen, Kenneth (2007). Discrete Mathematics and its Applications (6th ed.). New York, NY: McGraw-Hill. pp. 105, 158–160. ISBN 978-0-07-288008-3.

- Gilbert, Jimmie; Linda, Gilbert (2005). Elements of Modern Algebra (6th ed.). Belmont, CA: Thomson Brooks/Cole. pp. 243–244. ISBN 0-534-40264-X.

- Warner, Seth (2012), Modern Algebra, Dover Books on Mathematics, Courier Corporation, Theorem 20.14, p. 185, ISBN 9780486137094.

- Mendelson, Elliott (2008), Number Systems and the Foundations of Analysis, Dover Books on Mathematics, Courier Dover Publications, p. 86, ISBN 9780486457925.

- Crandall, Richard; Pomerance, Carl (2005), Prime Numbers: A Computational Perspective (2nd ed.), Berlin, New York: Springer-Verlag, ISBN 978-0-387-25282-7

Understanding Fractions

Fraction represents a part of an entire object or a number. An example of a fraction would be 5/25 and ¾. Unit fraction, dyadic rational, repeating decimal, cyclic number and Egyptian fraction are the aspects elucidated in the following chapter.

Fraction (Mathematics)

A fraction represents a part of a whole or, more generally, any number of equal parts. When spoken in everyday English, a fraction describes how many parts of a certain size there are, for exam-

ple, one-half, eight-fifths, three-quarters. A *common*, *vulgar*, or *simple* fraction (examples: $\frac{1}{2}$ and 17/3) consists of an integer numerator displayed above a line

(or before a slash), and a non-zero integer denominator, displayed below (or after) that line. Numerators and denominators are also used in fractions that are not *common*, including compound fractions, complex fractions, and mixed numerals.

A cake with one quarter (one fourth) removed. The remaining three fourths are shown. Dotted lines indicate where the cake may be cut in order to divide it into equal parts. Each fourth of the cake is denoted by the fraction ¼.

The numerator represents a number of equal parts, and the denominator, which cannot be zero, indicates how many of those parts make up a unit or a whole. For example, in the fraction 3/4, the numerator, 3, tells us that the fraction represents 3 equal parts, and the denominator, 4, tells us that 4 parts make up a whole. The picture to the right illustrates $\frac{3}{4}$ or ¾ of a cake.

Fractional numbers can also be written without using explicit numerators or denominators, by using decimals, percent signs, or negative exponents (as in 0.01, 1%, and 10^{-2} respectively, all of which are equivalent to 1/100). An integer such as the number 7 can be thought of as having an implicit denominator of one: 7 equals 7/1.

Other uses for fractions are to represent ratios and to represent division. Thus the fraction ¾ is also used to represent the ratio 3:4 (the ratio of the part to the whole) and the division 3 ÷ 4 (three divided by four).

In mathematics the set of all numbers that can be expressed in the form a/b, where a and b are integers and b is not zero, is called the set of rational numbers and is represented by the symbol Q, which stands for quotient. The test for a number being a rational number is that it can be written in that form (i.e., as a common fraction). However, the word *fraction* is also used to describe mathematical expressions that are not rational numbers, for example algebraic fractions (quotients of algebraic expressions), and expressions that contain irrational numbers, such as √2/2.

Vocabulary

In a fraction, the number of equal parts being described is the numerator (equivalent to the dividend in division) and the number of equal parts that make up a whole is the denominator (equivalent to the divisor). Informally, they may be distinguished by placement alone but in formal contexts they are always separated by a fraction bar. The fraction bar may be horizontal (as in –1/3), oblique (as in 1/5 or –1/7), or diagonal (as in $\frac{1}{9}$). These marks are respectively known as the horizontal bar, the slash (US) or stroke (UK), the division slash, and the fraction slash. In typography, horizontal fractions are also known as "en" or "nut fractions" and diagonal fractions as "em fractions", based on the width of a line they take up.

The denominators of English fractions are generally expressed as ordinal numbers, in the plural if the numerator is not one. (For example, $\frac{2}{5}$ and $\frac{3}{5}$ are both read as a number of "fifths".) Exceptions include the denominator 2, which is always read "half" or "halves", the denominator 4, which may be alternatively expressed as "quarter"/"quarters" or as "fourth"/"fourths", and the denominator 100, which may be alternatively expressed as "hundredth"/"hundredths" or "percent". When the denominator is 1, it may be expressed in terms of "wholes" but is more commonly ignored, with the numerator read out as a whole number. (For example, 3/1 may be described as "three wholes" or as simply "three".) When the numerator is one, it may be omitted. (For example, "a tenth" or "each quarter".) A fraction may be expressed as a single composition, in which case it is hyphenated, or as a number of fractions with a numerator of one, in which case they are not. (For example, "two-fifths" is the fraction 2/5 and "two fifths" is the same fraction understood as 2 instances of $\frac{1}{5}$.) Fractions should always be hyphenated when used as adjectives.

Alternatively, a fraction may be described by reading it out as the numerator "over" the denominator, with the denominator expressed as a cardinal number. (For example, 3/1 may also be expressed as "three over one".) The term "over" is used even in the case of solidus fractions, where the numbers are placed left and right of a slash mark. (For example, ½ may be read "one-half", "one half", or "one over two".) Fractions with large denominators that are *not* powers of ten are often rendered in this fashion (e.g., 1/117 as "one over a hundred and seventeen") while those with denominators divisible by ten are typically read in the normal ordinal fashion (e.g., 6/1000000 as "six-millionths", "six millionths", or "six one-millionths").

Forms of Fractions

Simple, Common, or Vulgar Fractions

A simple fraction (also known as a common fraction or vulgar fraction) is a rational number writ-

ten as a/b or $\dfrac{a}{b}$, where a and b are both integers. As with other fractions, the denominator (b)

cannot be zero. Examples include $\dfrac{1}{2}$, $-\dfrac{8}{5}$, $\dfrac{-8}{5}$, $\dfrac{8}{-5}$, and 3/17. *Simple fractions* can be positive
or negative, proper, or improper. Compound fractions, complex fractions, mixed numerals, and decimals are not *simple fractions*, though, unless irrational, they can be evaluated to a simple fraction.

Proper and Improper Fractions

Common fractions can be classified as either proper or improper. When the numerator and the denominator are both positive, the fraction is called proper if the numerator is less than the denominator, and improper otherwise. In general, a common fraction is said to be a proper fraction if the absolute value of the fraction is strictly less than one—that is, if the fraction is greater than −1 and less than 1. It is said to be an improper fraction, or sometimes top-heavy fraction, if the absolute value of the fraction is greater than or equal to 1. Examples of proper fractions are 2/3, −3/4, and 4/9; examples of improper fractions are 9/4, −4/3, and 3/3.

Mixed Numbers

A mixed numeral (often called a *mixed number*, also called a *mixed fraction*) is the sum of a non-zero integer and a proper fraction. This sum is implied without the use of any visible operator such as "+". For example, in referring to two entire cakes and three quarters of another cake, the

whole and fractional parts of the number are written next to each other: $2+\dfrac{3}{4}=2\tfrac{3}{4}$..

When two algebraic expressions are written next to each other, the operation of multiplication is said to be "understood". In algebra, $a\tfrac{b}{c}$ for example is not a mixed number. Instead, multiplication is understood where $a\tfrac{b}{c}=a\times\tfrac{b}{c}$.

To avoid confusion, the multiplication is often explicitly expressed. So $a\dfrac{b}{c}$ may be written as

$$a\times\frac{b}{c},$$

$$a\cdot\frac{b}{c},$$

or

$$a\left(\frac{b}{c}\right).$$

An improper fraction is another way to write a whole plus a part. A mixed number can be converted to an improper fraction as follows:

1. Write the mixed number $2\frac{3}{4}$ as a sum $2+\frac{3}{4}$.

2. Convert the whole number to an improper fraction with the same denominator as the fractional part, $-$.

3. Add the fractions. The resulting sum is the improper fraction. In the example,

    $$2\frac{3}{4}=\frac{8}{4}+\frac{3}{4}=\frac{11}{4}.$$

Similarly, an improper fraction can be converted to a mixed number as follows:

1. Divide the numerator by the denominator. In the example, $\frac{11}{4}$, divide 11 by 4. $11 \div 4 = 2$ with remainder 3.

2. The quotient (without the remainder) becomes the whole number part of the mixed number. The remainder becomes the numerator of the fractional part. In the example, 2 is the whole number part and 3 is the numerator of the fractional part.

3. The new denominator is the same as the denominator of the improper fraction. In the example, they are both 4. Thus $\frac{11}{4}=2\frac{3}{4}$.

Mixed numbers can also be negative, as in $-2\frac{3}{4}$, which equals $-(2+\frac{3}{4})=-2-\frac{3}{4}$.

Ratios

A ratio is a relationship between two or more numbers that can be sometimes expressed as a fraction. Typically, a number of items are grouped and compared in a ratio, specifying numerically the relationship between each group. Ratios are expressed as "group 1 to group 2 ... to group n". For example, if a car lot had 12 vehicles, of which

*   2 are white,

*   6 are red, and

*   4 are yellow,

then the ratio of red to white to yellow cars is 6 to 2 to 4. The ratio of yellow cars to white cars is 4 to 2 and may be expressed as 4:2 or 2:1.

A ratio is often converted to a fraction when it is expressed as a ratio to the whole. In the above example, the ratio of yellow cars to all the cars on the lot is 4:12 or 1:3. We can convert these ratios to a fraction and say that 4/12 of the cars or ▢ of the cars in the lot are yellow. Therefore, if a person randomly chose one car on the lot, then there is a one in three chance or probability that it would be yellow.

Reciprocals and The "Invisible Denominator"

The reciprocal of a fraction is another fraction with the numerator and denominator exchanged.

The reciprocal of $\dfrac{3}{7}$, for instance, is $\frac{7}{3}$. The product of a fraction and its reciprocal is 1, hence the reciprocal is the multiplicative inverse of a fraction.

Any integer can be written as a fraction with the number one as denominator. For example, 17 can be written as $\dfrac{17}{1}$, where 1 is sometimes referred to as the *invisible denominator*. Therefore, every fraction or integer except for zero has a reciprocal. The reciprocal of 17 is $\dfrac{1}{17}$.

Complex Fractions

In a complex fraction, either the numerator, or the denominator, or both, is a fraction or a mixed number, corresponding to division of fractions. For example, $\dfrac{\frac{1}{2}}{\frac{1}{3}}$ and $\dfrac{12\frac{3}{4}}{26}$ are complex frac

tions. To reduce a complex fraction to a simple fraction, treat the longest fraction line as representing division. For example:

$$\frac{\frac{1}{2}}{\frac{1}{3}} = \tfrac{1}{2} \times \tfrac{3}{1} = \tfrac{3}{2} = 1\tfrac{1}{2}$$

$$\frac{12\frac{3}{4}}{26} = 12\tfrac{3}{4} \cdot \tfrac{1}{26} = \tfrac{12\cdot4+3}{4} \cdot \tfrac{1}{26} = \tfrac{51}{4} \cdot \tfrac{1}{26} = \tfrac{51}{104}$$

$$\frac{\frac{3}{2}}{5} = \tfrac{3}{2} \times \tfrac{1}{5} = \tfrac{3}{10}$$

$$\frac{8}{\frac{1}{3}} = 8 \times \tfrac{3}{1} = 24.$$

If, in a complex fraction, there is no clear way to tell which fraction lines takes precedence, then the expression is obviously improperly formed, because of ambiguity. So 5/10/20/40 is a poorly constructed mathematical expression, with multiple possible meanings, e.g.:

$$\frac{5}{10 / \frac{20}{40}} = \frac{1}{4} \quad \text{or} \quad \frac{\frac{\frac{5}{10}}{20}}{40} = 1$$

Compound Fractions

A compound fraction is a fraction of a fraction, or any number of fractions connected with the word *of*, corresponding to multiplication of fractions. To reduce a compound fraction to a simple frac

tion, just carry out the multiplication. For example, $\frac{3}{4}$ of $\frac{5}{7}$ is a compound fraction, corresponding

to $\frac{3}{4} \times \frac{5}{7} = \frac{15}{28}$. The terms compound fraction and complex fraction are closely related and sometimes one is used as a synonym for the other.

Decimal Fractions and Percentages

A decimal fraction is a fraction whose denominator is not given explicitly, but is understood to be an integer power of ten. Decimal fractions are commonly expressed using decimal notation in which the implied denominator is determined by the number of digits to the right of a decimal separator, the appearance of which (e.g., a period, a raised period (•), a comma) depends on the locale. Thus for 0.75 the numerator is 75 and the implied denominator is 10 to the second power, *viz.* 100, because there are two digits to the right of the decimal separator. In decimal numbers greater than 1 (such as 3.75), the fractional part of the number is expressed by the digits to the right of the decimal (with a value of 0.75 in this case). 3.75 can be written either as an improper fraction, 375/100, or as a mixed number, $3\frac{75}{100}$.

Decimal fractions can also be expressed using scientific notation with negative exponents, such as 6.023×10^{-7}, which represents 0.0000006023. The 10^{-7} represents a denominator of 10^7. Dividing by 10^7 moves the decimal point 7 places to the left.

Decimal fractions with infinitely many digits to the right of the decimal separator represent an infinite series. For example, ⅓ = 0.333... represents the infinite series 3/10 + 3/100 + 3/1000 + ...

Another kind of fraction is the percentage (Latin *per centum* meaning "per hundred", represented by the symbol %), in which the implied denominator is always 100. Thus, 51% means 51/100. Percentages greater than 100 or less than zero are treated in the same way, e.g. 311% equals 311/100, and −27% equals −27/100.

The related concept of *permille* or *parts per thousand* has an implied denominator of 1000, while the more general parts-per notation, as in 75 parts per million, means that the proportion is 75/1,000,000.

Whether common fractions or decimal fractions are used is often a matter of taste and context. Common fractions are used most often when the denominator is relatively small. By mental calculation, it is easier to multiply 16 by 3/16 than to do the same calculation using the fraction's decimal equivalent (0.1875). And it is more accurate to multiply 15 by 1/3, for example, than it is to multiply 15 by any decimal approximation of one third. Monetary values are commonly expressed as decimal fractions, for example $3.75. However, as noted above, in pre-decimal British currency, shillings and pence were often given the form (but not the meaning) of a fraction, as, for example 3/6 (read "three and six") meaning 3 shillings and 6 pence, and having no relationship to the fraction 3/6.

Special Cases

- A unit fraction is a vulgar fraction with a numerator of 1, e.g. $\frac{1}{7}$.. Unit fractions can also be expressed using negative exponents, as in 2^{-1}, which represents 1/2, and 2^{-2}, which represents $1/(2^2)$ or 1/4.

- An Egyptian fraction is the sum of distinct positive unit fractions, for example $\frac{1}{2}+\frac{1}{3}$. This definition derives from the fact that the ancient Egyptians expressed all fractions except $\frac{1}{2}, \frac{2}{3}$ and $\frac{3}{4}$ in this manner. Every positive rational number can be expanded as an Egyptian fraction. For example, $\frac{5}{7}$ can be written as $\frac{1}{2}+\frac{1}{6}+\frac{1}{21}$. Any positive rational number can be written as a sum of unit fractions in infinitely many ways. Two ways to write $\frac{13}{17}$ are $\frac{1}{2}+\frac{1}{4}+\frac{1}{68}$ and $\frac{1}{3}+\frac{1}{4}+\frac{1}{6}+\frac{1}{68}$..

- A dyadic fraction is a vulgar fraction in which the denominator is a power of two, e.g. $\frac{1}{8}$..

Arithmetic With Fractions

Like whole numbers, fractions obey the commutative, associative, and distributive laws, and the rule against division by zero.

Equivalent Fractions

Multiplying the numerator and denominator of a fraction by the same (non-zero) number results in a fraction that is equivalent to the original fraction. This is true because for any non-zero number n, the fraction $\frac{n}{n}=1$. Therefore, multiplying by $\frac{n}{n}$ is equivalent to multiplying by one, and any number multiplied by one has the same value as the original number. By way of an example, start with the fraction $\frac{1}{2}$. When the numerator and denominator are both multiplied by 2, the result is $\frac{2}{4}$,, which has the same value (0.5) as $\frac{1}{2}$. To picture this visually, imagine cutting a cake into four pieces; two of the pieces together ($\frac{2}{4}$) make up half the cake ($\frac{1}{2}$).

Dividing the numerator and denominator of a fraction by the same non-zero number will also yield an equivalent fraction. This is called reducing or simplifying the fraction. A simple fraction in which the numerator and denominator are coprime (that is, the only positive integer that goes into both the numerator and denominator evenly is 1) is said to be irreducible, in lowest terms, or in simplest terms. For example, $\frac{3}{9}$ is not in lowest terms because both 3 and 9 can be exactly divided by 3. In contrast, $\frac{3}{8}$ *is* in lowest terms—the only positive integer that goes into both 3 and 8 evenly is 1.

Using these rules, we can show that $\frac{5}{10} = \frac{1}{2} = \frac{10}{20} = \frac{50}{100}$.

A common fraction can be reduced to lowest terms by dividing both the numerator and denominator by their greatest common divisor. For example, as the greatest common divisor of 63 and 462 is 21, the fraction $\frac{63}{462}$ can be reduced to lowest terms by dividing the numerator and denominator by 21:

$$\frac{63}{462} = \frac{63 \div 21}{462 \div 21} = \frac{3}{22}$$

The Euclidean algorithm gives a method for finding the greatest common divisor of any two positive integers.

Comparing Fractions

Comparing fractions with the same denominator only requires comparing the numerators.

$$\frac{3}{4} > \frac{2}{4} \text{ because } 3>2.$$

If two positive fractions have the same numerator, then the fraction with the smaller denominator is the larger number. When a whole is divided into equal pieces, if fewer equal pieces are needed to make up the whole, then each piece must be larger. When two positive fractions have the same numerator, they represent the same number of parts, but in the fraction with the smaller denominator, the parts are larger.

Addition

The first rule of addition is that only like quantities can be added; for example, various quantities of quarters. Unlike quantities, such as adding thirds to quarters, must first be converted to like quantities as described below: Imagine a pocket containing two quarters, and another pocket containing three quarters; in total, there are five quarters. Since four quarters is equivalent to one (dollar), this can be represented as follows:

$$\frac{2}{4} + \frac{3}{4} = \frac{5}{4} = 1\frac{1}{4}.$$

If $\frac{1}{2}$ of a cake is to be added to $\frac{1}{4}$ of a cake, the pieces need to be converted into comparable quantities, such as cake-eighths or cake-quarters.

Adding Unlike Quantities

To add fractions containing unlike quantities (e.g. quarters and thirds), it is necessary to convert all amounts to like quantities. It is easy to work out the chosen type of fraction to convert to; simply multiply together the two denominators (bottom number) of each fraction.

For adding quarters to thirds, both types of fraction are converted to twelfths, thus:

$$\frac{1}{4}+\frac{1}{3}=\frac{1\times3}{4\times3}+\frac{1\times4}{3\times4}=\frac{3}{12}+\frac{4}{12}\frac{7}{12}$$

Consider adding the following two quantities:

$$\frac{3}{5}+\frac{2}{3}$$

Now it can be seen that:

$$\frac{3}{5}+\frac{2}{3}$$

is equivalent to:

$$\frac{9}{15}+\frac{10}{15}=\frac{19}{15}=1\frac{4}{15}$$

This method can be expressed algebraically:

$$\frac{a}{b}+\frac{c}{d}=\frac{ad+cb}{bd}$$

And for expressions consisting of the addition of three fractions:

$$\frac{a}{b}+\frac{c}{d}+\frac{e}{f}=\frac{a(df)+c(bf)+e(bd)}{bdf}$$

This method always works, but sometimes there is a smaller denominator that can be used (a least common denominator). For example, to add $\frac{3}{4}$ and $\frac{5}{12}$ the denominator 48 can be used (the product of 4 and 12), but the smaller denominator 12 may also be used, being the least common multiple of 4 and 12.

$$\frac{3}{4}+\frac{5}{12}=\frac{9}{12}+\frac{5}{12}=\frac{14}{12}=\frac{7}{6}=1\frac{1}{6}$$

Subtraction

The process for subtracting fractions is, in essence, the same as that of adding them: find a common denominator, and change each fraction to an equivalent fraction with the chosen common denominator. The resulting fraction will have that denominator, and its numerator will be the result of subtracting the numerators of the original fractions. For instance,

$$\frac{2}{3} - \frac{1}{2} = \frac{4}{6} - \frac{3}{6} = \frac{1}{6}$$

Multiplication

Multiplying A Fraction by Another Fraction

To multiply fractions, multiply the numerators and multiply the denominators. Thus:

$$\frac{2}{3} \times \frac{3}{4} = \frac{6}{12}$$

To explain the process, consider one third of one quarter. Using the example of a cake, if three small slices of equal size make up a quarter, and four quarters make up a whole, twelve of these small, equal slices make up a whole. Therefore, a third of a quarter is a twelfth. Now consider the numerators. The first fraction, two thirds, is twice as large as one third. Since one third of a quarter is one twelfth, two thirds of a quarter is two twelfth. The second fraction, three quarters, is three times as large as one quarter, so two thirds of three quarters is three times as large as two thirds of one quarter. Thus two thirds times three quarters is six twelfths.

A short cut for multiplying fractions is called "cancellation". Effectively the answer is reduced to lowest terms during multiplication. For example:

$$\frac{2}{3} \times \frac{3}{4} = \frac{\cancel{2}^{1}}{\cancel{3}^{1}} \times \frac{\cancel{3}^{1}}{\cancel{4}^{2}} = \frac{1}{1} \times \frac{1}{2} = \frac{1}{2}$$

A two is a common factor in both the numerator of the left fraction and the denominator of the right and is divided out of both. Three is a common factor of the left denominator and right numerator and is divided out of both.

Multiplying A Fraction by A Whole Number

Since a whole number can be rewritten as itself divided by 1, normal fraction multiplication rules can still apply.

$6 \times \frac{3}{4} = \frac{6}{1} \times \frac{3}{4} = \frac{18}{4}$ This method works because the fraction 6/1 means six equal parts, each one of which is a whole.

Multiplying Mixed Numbers

When multiplying mixed numbers, it is considered preferable to convert the mixed number into an improper fraction. For example:

$$3 \times 2\frac{3}{4} = 3 \times \left(\frac{8}{4} + \frac{3}{4}\right) = 3 \times \frac{11}{4} = \frac{33}{4} = 8\frac{1}{4}$$

In other words, $2\frac{3}{4}$ is the same as $\frac{8}{4} + \frac{3}{4}$, making 11 quarters in total (because 2 cakes, each split into quarters makes 8 quarters total) and 33 quarters is $8\frac{1}{4}$, since 8 cakes, each made of quarters, is 32 quarters in total.

Converting Between Decimals and Fractions

To change a common fraction to a decimal, divide the denominator into the numerator. Round the answer to the desired accuracy. For example, to change ¼ to a decimal, divide 4 into 1.00, to obtain 0.25. To change 1/4 a decimal, divide 3 into 1.0000..., and stop when the desired accuracy is obtained. Note that ¼ can be written exactly with two decimal digits, while 1/3 cannot be written exactly with any finite number of decimal digits. To change a decimal to a fraction, write in the denominator a 1 followed by as many zeroes as there are digits to the right of the decimal point, and write in the numerator all the digits in the original decimal, omitting the decimal point. Thus 12.3456 = 123456/10000.

Converting Repeating Decimals to Fractions

Decimal numbers, while arguably more useful to work with when performing calculations, sometimes lack the precision that common fractions have. Sometimes an infinite repeating decimal is required to reach the same precision. Thus, it is often useful to convert repeating decimals into fractions.

The preferred way to indicate a repeating decimal is to place a bar over the digits that repeat, for example 0.789 = 0.789789789... For repeating patterns where the repeating pattern begins immediately after the decimal point, a simple division of the pattern by the same number of nines as numbers it has will suffice. For example:

0.5 = 5/9

0.62 = 62/99

0.264 = 264/999

0.6291 = 6291/9999

In case leading zeros precede the pattern, the nines are suffixed by the same number of trailing zeros:

0.05 = 5/90

0.000392 = 392/999000

0.0012 = 12/9900

In case a non-repeating set of decimals precede the pattern (such as 0.1523987), we can write it as the sum of the non-repeating and repeating parts, respectively:

0.1523 + 0.0000987

Then, convert both parts to fractions, and add them using the methods described above:

1523/10000 + 987/9990000 = 1522464/9990000

Alternatively, algebra can be used, such as below:

1. Let x = the repeating decimal:

   x = 0.1523987

2. Multiply both sides by the power of 10 just great enough (in this case 10^4) to move the decimal point just before the repeating part of the decimal number:

   10,000x = 1,523.987

3. Multiply both sides by the power of 10 (in this case 10^3) that is the same as the number of places that repeat:

   10,000,000x = 1,523,987.987

4. Subtract the two equations from each other (if $a = b$ and $c = d$, then $a - c = b - d$):

   10,000,000x − 10,000x = 1,523,987.987 − 1,523.987

5. Continue the subtraction operation to clear the repeating decimal:

   9,990,000x = 1,523,987 − 1,523

   9,990,000x = 1,522,464

6. Divide both sides to represent x as a fraction

   x = 1522464/9990000

Fractions in Abstract Mathematics

In addition to being of great practical importance, fractions are also studied by mathematicians, who check that the rules for fractions given above are consistent and reliable. Mathematicians define a fraction as an ordered pair (a, b) of integers a and $b \neq 0$, for which the operations addition, subtraction, multiplication, and division are defined as follows:

$$(a,b)+(c,d)=(ad+bc,bd)$$

$$(a,b)-(c,d)=(ad-bc,bd)$$

$$(a,b)\cdot(c,d)=(ac,bd)$$

$$(a,b)\div(c,d)=(ad,bc) \text{ (when } c \text{ } 0)$$

In addition, an equivalence relation is specified as follows: $(a,b) \sim (c,d)$ if and only if $ad = bc$.

These definitions agree in every case with the definitions given above; only the notation is different.

More generally, a and b may be elements of any integral domain R, in which case a fraction is an element of the field of fractions of R. For example, when a and b are polynomials in one indeterminate, the field of fractions is the field of rational fractions (also known as the field of rational functions). When a and b are integers, the field of fractions is the field of rational numbers.

Algebraic Fractions

An algebraic fraction is the indicated quotient of two algebraic expressions. Two examples of algebraic fractions are $\dfrac{3x}{x^2+2x-3}$ and $\dfrac{\sqrt{x+2}}{x^2-3}$. Algebraic fractions are subject to the same laws as arithmetic fractions.

If the numerator and the denominator are polynomials, as in $\dfrac{3x}{x^2+2x-3}$, the algebraic fraction is called a rational fraction (or rational expression). An irrational fraction is one that contains the variable under a fractional exponent or root, as in $\dfrac{\sqrt{x+2}}{x^2-3}$.

The terminology used to describe algebraic fractions is similar to that used for ordinary fractions. For example, an algebraic fraction is in lowest terms if the only factors common to the numerator and the denominator are 1 and −1. An algebraic fraction whose numerator or denominator, or both, contain a fraction, such as $\dfrac{1+\frac{1}{x}}{1-\frac{1}{x}}$, is called a complex fraction.

Rational numbers are the quotient field of integers. Rational expressions are the quotient field of the polynomials (over some integral domain). Since a coefficient is a polynomial of degree zero, a radical expression such as $\sqrt{2}/2$ is a rational fraction. Another example (over the reals) is $\dfrac{\pi}{2}$, the radian measure of a right angle.

The term partial fraction is used when decomposing rational expressions into sums. The goal is to write the rational expression as the sum of other rational expressions with denominators of lesser degree. For example, the rational expression $\dfrac{2x}{x^2-1}$ can be rewritten as the sum of two fractions: $\dfrac{1}{x+1}+\dfrac{1}{x-1}$. This is useful in many areas such as integral calculus and differential equations.

Radical Expressions

A fraction may also contain radicals in the numerator and/or the denominator. If the denominator contains radicals, it can be helpful to rationalize it (compare Simplified form of a radical expression), especially if further operations, such as adding or comparing that fraction to another, are to be carried out. It is also more convenient if division is to be done manually. When the denominator is a monomial square root, it can be rationalized by multiplying both the top and the bottom of the fraction by the denominator:

$$\frac{3}{\sqrt{7}} = \frac{3}{\sqrt{7}} \cdot \frac{\sqrt{7}}{\sqrt{7}} = \frac{3\sqrt{7}}{7}$$

The process of rationalization of binomial denominators involves multiplying the top and the bottom of a fraction by the conjugate of the denominator so that the denominator becomes a rational number. For example:

$$\frac{3}{3-2\sqrt{5}} = \frac{3}{3-2\sqrt{5}} \cdot \frac{3+2\sqrt{5}}{3+2\sqrt{5}} = \frac{3(3+2\sqrt{5})}{3^2-(2\sqrt{5})^2} = \frac{3(3+2\sqrt{5})}{9-20} = -\frac{9+6\sqrt{5}}{11}$$

$$\frac{3}{3+2\sqrt{5}} = \frac{3}{3+2\sqrt{5}} \cdot \frac{3-2\sqrt{5}}{3-2\sqrt{5}} = \frac{3(3-2\sqrt{5})}{3^2-(2\sqrt{5})^2} = \frac{3(3-2\sqrt{5})}{9-20} = -\frac{9-6\sqrt{5}}{11}$$

Even if this process results in the numerator being irrational, like in the examples above, the process may still facilitate subsequent manipulations by reducing the number of irrationals one has to work with in the denominator.

Typographical Variations

In computer displays and typography, simple fractions are sometimes printed as a single character, e.g. ½ (one half). See the article on Number Forms for information on doing this in Unicode.

Scientific publishing distinguishes four ways to set fractions, together with guidelines on use:

- special fractions: fractions that are presented as a single character with a slanted bar, with roughly the same height and width as other characters in the text. Generally used for simple fractions, such as: ½, ▯, ▯, ¼, and ¾. Since the numerals are smaller, legibility can be an issue, especially for small-sized fonts. These are not used in modern mathematical notation, but in other contexts.

- case fractions: similar to special fractions, these are rendered as a single typographical character, but with a horizontal bar, thus making them *upright*. An example would be $\frac{1}{2}$, but rendered with the same height as other characters. Some sources include all rendering of fractions as *case fractions* if they take only one typographical space, regardless of the direction of the bar.

- shilling or solidus fractions: 1/2, so called because this notation was used for pre-decimal British currency (£sd), as in 2/6 for a half crown, meaning two shillings and six pence. While the notation "two shillings and six pence" did not represent a fraction, the forward slash is now used in fractions, especially for fractions inline with prose (rather than displayed), to avoid uneven lines. It is also used for fractions within fractions (complex fractions) or within exponents to increase legibility. Fractions written this way, also known as *piece fractions*, are written all on one typographical line, but take 3 or more typographical spaces.

- built-up fractions: $\frac{1}{2}$. This notation uses two or more lines of ordinary text, and results in a variation in spacing between lines when included within other text. While large and legible, these can be disruptive, particularly for simple fractions or within complex fractions.

History

The earliest fractions were reciprocals of integers: ancient symbols representing one part of two, one part of three, one part of four, and so on. The Egyptians used Egyptian fractions c. 1000 BC. About 4000 years ago, Egyptians divided with fractions using slightly different methods. They used least common multiples with unit fractions. Their methods gave the same answer as modern methods. The Egyptians also had a different notation for dyadic fractions in the Akhmim Wooden Tablet and several Rhind Mathematical Papyrus problems.

The Greeks used unit fractions and (later) continued fractions. Followers of the Greek philosopher Pythagoras (c. 530 BC) discovered that the square root of two cannot be expressed as a fraction of integers. (This is commonly though probably erroneously ascribed to Hippasus of Metapontum, who is said to have been executed for revealing this fact.) In 150 BC Jain mathematicians in India wrote the "Sthananga Sutra", which contains work on the theory of numbers, arithmetical operations, and operations with fractions.

The modern expression of fractions seems to have originated in India in the work of Aryabhatta (c. AD 500), Brahmagupta (c. 628), and Bhaskara (c. 1150). Their works form fractions by placing the numerators (Sanskrit: *amsa*) over the denominators (*cheda*), but without a bar between them. In Sanskrit literature, fractions were always expressed as an addition to or subtraction from an integer. The integer was written on one line and the fraction in its two parts on the next line. If the fraction was marked by a small circle ⟨.⟩ or cross ⟨+⟩, it is subtracted from the integer; if no such sign appears, it is understood to be added. For example, Bhaskara I writes

६	१	२
१	१	१.
४	५	१

which is the equivalent of

6	1	2
1	1	−1
4	5	9

and would be written in modern notation as 61/4, 11/5, and 2−1/9 (i.e., 18/9).

The horizontal fraction bar is first attested in the work of Al-Hassār (fl. 1200), a Muslim mathematician from Fez, Morocco, who specialized in Islamic inheritance jurisprudence. In his discussion he writes, "... for example, if you are told to write three-fifths and a third of a fifth, write thus, $\frac{3}{5} \frac{1}{3}$ " The same fractional notation—with the fraction given before the integer—appears soon after in the work of Leonardo Fibonacci in the 13th century.

In discussing the origins of decimal fractions, Dirk Jan Struik states:

"The introduction of decimal fractions as a common computational practice can be dated back to the Flemish pamphlet *De Thiende*, published at Leyden in 1585, together with a French translation, *La Disme*, by the Flemish mathematician Simon Stevin (1548–1620), then settled in the Northern Netherlands. It is true that decimal fractions were used by the Chinese many centuries before Stevin and that the Persian astronomer Al-Kāshī used both decimal and sexagesimal fractions with great ease in his *Key to arithmetic* (Samarkand, early fifteenth century)."

While the Persian mathematician Jamshīd al-Kāshī claimed to have discovered decimal fractions himself in the 15th century, J. Lennart Berggren notes that he was mistaken, as decimal fractions were first used five centuries before him by the Baghdadi mathematician Abu'l-Hasan al-Uqlidisi as early as the 10th century.

In Formal Education

Pedagogical Tools

In primary schools, fractions have been demonstrated through Cuisenaire rods, Fraction Bars, fraction strips, fraction circles, paper (for folding or cutting), pattern blocks, pie-shaped pieces, plastic rectangles, grid paper, dot paper, geoboards, counters and computer software.

Documents for Teachers

Several states in the United States have adopted learning trajectories from the Common Core State Standards Initiative's guidelines for mathematics education. Aside from sequencing the learning of fractions and operations with fractions, the document provides the following definition of a fraction: "A number expressible in the form $\frac{a}{b}$ where a is a whole number and b is a positive whole number. (The word *fraction* in the standards always refers to a non-negative number.)" The document itself also refers to negative fractions.

Unit Fraction

A unit fraction is a rational number written as a fraction where the numerator is one and the denominator is a positive integer. A unit fraction is therefore the reciprocal of a positive integer, $1/n$. Examples are 1/1, 1/2, 1/3, 1/4 ,1/5, etc.

Elementary Arithmetic

Multiplying any two unit fractions results in a product that is another unit fraction:

$$\frac{1}{x} \times \frac{1}{y} = \frac{1}{xy}.$$

However, adding, subtracting, or dividing two unit fractions produces a result that is generally not a unit fraction:

$$\frac{1}{x} + \frac{1}{y} = \frac{x+y}{xy}$$

$$\frac{1}{x} - \frac{1}{y} = \frac{y-x}{xy}$$

$$\frac{1}{x} \div \frac{1}{y} = \frac{y}{x}.$$

Modular Arithmetic

Unit fractions play an important role in modular arithmetic, as they may be used to reduce modular division to the calculation of greatest common divisors. Specifically, suppose that we wish to perform divisions by a value x, modulo y. In order for division by x to be well defined modulo y, x and y must be relatively prime. Then, by using the extended Euclidean algorithm for greatest common divisors we may find a and b such that

$$ax + by = 1,$$

from which it follows that

$$ax \equiv 1 \pmod{y},$$

or equivalently

$$a \equiv \frac{1}{x} \pmod{y}.$$

Thus, to divide by x (modulo y) we need merely instead multiply by a.

Finite Sums of Unit Fractions

Any positive rational number can be written as the sum of unit fractions, in multiple ways. For example,

$$\frac{4}{5} = \frac{1}{2} + \frac{1}{4} + \frac{1}{20} = \frac{1}{3} + \frac{1}{5} + \frac{1}{6} + \frac{1}{10}.$$

The ancient Egyptian civilisations used sums of distinct unit fractions in their notation for more general rational numbers, and so such sums are often called Egyptian fractions. There is still interest today in analyzing the methods used by the ancients to choose among the possible representations for a fractional number, and to calculate with such representations. The topic of Egyptian fractions has also seen interest in modern number theory; for instance, the Erdős–Graham conjecture and the Erdős–Straus conjecture concern sums of unit fractions, as does the definition of Ore's harmonic numbers.

In geometric group theory, triangle groups are classified into Euclidean, spherical, and hyperbolic cases according to whether an associated sum of unit fractions is equal to one, greater than one, or less than one respectively.

Series of Unit Fractions

Many well-known infinite series have terms that are unit fractions. These include:

- The harmonic series, the sum of all positive unit fractions. This sum diverges, and its partial sums

$$\frac{1}{1}+\frac{1}{2}+\frac{1}{3}+\cdots+\frac{1}{n}$$

  closely approximate $\ln n + \gamma$ as n increases.

- The Basel problem concerns the sum of the square unit fractions, which converges to $\pi^2/6$

- Apéry's constant is the sum of the cubed unit fractions.

- The binary geometric series, which adds to 2, and the reciprocal Fibonacci constant are additional examples of a series composed of unit fractions.

Matrices of Unit Fractions

The Hilbert matrix is the matrix with elements

$$B_{i,j}=\frac{1}{i+j-1}.$$

It has the unusual property that all elements in its inverse matrix are integers. Similarly, Richardson (2001) defined a matrix with elements

$$C_{i,j}=\frac{1}{F_{i+j-1}},$$

where F_i denotes the ith Fibonacci number. He calls this matrix the Filbert matrix and it has the same property of having an integer inverse.

Adjacent Fractions

Two fractions are called adjacent if their difference is a unit fraction.

Unit Fractions in Probability and Statistics

In a uniform distribution on a discrete space, all probabilities are equal unit fractions. Due to the principle of indifference, probabilities of this form arise frequently in statistical calculations. Additionally, Zipf's law states that, for many observed phenomena involving the selection of items from an ordered sequence, the probability that the nth item is selected is proportional to the unit fraction $1/n$.

Unit Fractions in Physics

The energy levels of photons that can be absorbed or emitted by a hydrogen atom are, according to the Rydberg formula, proportional to the differences of two unit fractions. An explanation for this phenomenon is provided by the Bohr model, according to which the energy levels of electron orbitals in a hydrogen atom are inversely proportional to square unit fractions, and the energy of a photon is quantized to the difference between two levels.

Arthur Eddington argued that the fine structure constant was a unit fraction, first 1/136 then 1/137. This contention has been falsified, given that current estimates of the fine structure constant are (to 6 significant digits) 1/137.036.

Dyadic Rational

In mathematics, a dyadic fraction or dyadic rational is a rational number whose denominator is a power of two, i.e., a number of the form $\frac{a}{2^b}$ where a is an integer and b is a natural number; for example, 1/2 or 3/8, but not 1/3. These are precisely the numbers whose binary expansion is finite.

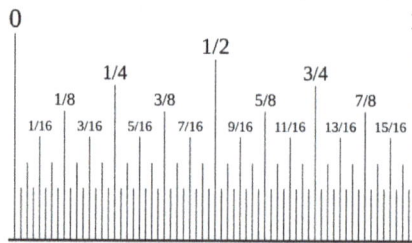

Dyadic rationals in the interval from 0 to 1.

Use in Measurement

The inch is customarily subdivided in dyadic rather than decimal fractions; similarly, the customary divisions of the gallon into half-gallons, quarts, and pints are dyadic. The ancient Egyptians also used dyadic fractions in measurement, with denominators up to 64.

Arithmetic

The sum, product, or difference of any two dyadic fractions is itself another dyadic fraction:

$$\frac{a}{2^b} + \frac{c}{2^d} = \frac{2^{d-b}a + c}{2^d} \quad (d \geq b)$$

$$\frac{a}{2^b} - \frac{c}{2^d} = \frac{2^{d-b}a - c}{2^d} \quad (d \geq b)$$

$$\frac{a}{2^b} - \frac{c}{2^d} = \frac{a - 2^{b-d}c}{2^b} \quad (d < b)$$

$$\frac{a}{2^b} \times \frac{c}{2^d} = \frac{a \times c}{2^{b+d}}.$$

However, the result of dividing one dyadic fraction by another is not necessarily a dyadic fraction.

Addition modulo 1 forms a group; this is the Prüfer 2-group.

Additional Properties

Because they are closed under addition, subtraction, and multiplication, but not division, the dyadic fractions form a subring of the rational numbers Q and an overring of the integers Z. Algebraically, this subring is the localization of the integers Z with respect to the set of powers of two.

The set of all dyadic fractions is dense in the real line: any real number x can be arbitrarily closely approximated by dyadic rationals of the form $\lfloor 2^i x \rfloor / 2^i$. Compared to other dense subsets of the real line, such as the rational numbers, the dyadic rationals are in some sense a relatively "small" dense set, which is why they sometimes occur in proofs.

Dual Group

Considering only the addition and subtraction operations of the dyadic rationals gives them the structure of an additive abelian group. The dual group of a group consists of its characters, group homomorphisms to the multiplicative group of the complex numbers, and in the spirit of Pontryagin duality the dual group of the additive dyadic rationals can also be viewed as a topological group. It is called the dyadic solenoid and is an example of a solenoid group and of a protorus.

The dyadic rationals are the direct limit of infinite cyclic subgroups of the rational numbers,

$$\varinjlim \left\{ 2^{-i}\mathbb{Z} \mid i = 0, 1, 2, \ldots \right\}$$

and their dual group can be constructed as the inverse limit of the unit circle group under the repeated squaring map

$$\zeta \mapsto \zeta^2.$$

An element of the dyadic solenoid can be represented as an infinite sequence of complex numbers $q_0, q_1, q_2, \ldots$, with the properties that each q_i lies on the unit circle and that, for all $i > 0$, $q_i^2 = q_{i-1}$. The group operation on these elements multiplies any two sequences componentwise. Each element of the dyadic solenoid corresponds to a character of the dyadic rationals that maps $a/2^b$ to the complex number q_b^a. Conversely, every character χ of the dyadic rationals corresponds to the element of the dyadic solenoid given by $q_i = \chi(1/2^i)$.

As a topological space the dyadic solenoid is a solenoid, and an indecomposable continuum.

Related Constructions

The surreal numbers are generated by an iterated construction principle which starts by generating all finite dyadic fractions, and then goes on to create new and strange kinds of infinite, infinitesimal and other numbers.

The binary van der Corput sequence is an equidistributed permutation of the positive dyadic rational numbers.

In Music

Time signatures in Western musical notation traditionally consist of dyadic fractions (for example: 2/2, 4/4, 6/8...), although non-dyadic time signatures have been introduced by composers in the twentieth century (for example: 2/♩., which would literally mean 2/⁸⁄₃, or 9/14). Non-dyadic time signatures are called *irrational* in musical terminology, but this usage does not correspond to the irrational numbers of mathematics, because they still consist of ratios of integers. Irrational time signatures in the mathematical sense are very rare, but one example ($\sqrt{42}/1$) appears in Conlon Nancarrow's *Studies for Player Piano*.

In Computing

As a data type used by computers, floating-point numbers are often defined as integers multiplied by positive or negative powers of two, and thus all numbers that can be represented for instance by IEEE floating-point datatypes are dyadic rationals. The same is true for the majority of fixed-point datatypes, which also uses powers of two implicitly in the majority of cases.

Repeating Decimal

A repeating or recurring decimal is decimal representation of a number whose decimal digits are periodic (repeating its values at regular intervals) and the infinitely-repeated portion is not zero. It can be shown that a number is rational if and only if its decimal representation is repeating or terminating (i.e. all except finitely many digits are zero). For example, the decimal representation of ▯ becomes periodic just after the decimal point, repeating the single digit "3" forever, i.e. 0.333.... A more complicated example is 3227/555, whose decimal becomes periodic after the *second* digit following the decimal point and then repeats the sequence "144" forever, i.e. 5.8144144144.... At present, there is no single universally accepted notation or phrasing for repeating decimals.

The infinitely-repeated digit sequence is called the repetend or reptend. If the repetend is a zero, this decimal representation is called a terminating decimal rather than a repeating decimal, since the zeros can be omitted and the decimal terminates before these zeros. Every terminating decimal representation can be written as a decimal fraction, a fraction whose divisor is a power of 10 (e.g. 1.585 = 1585/1000); it may also be written as a ratio of the form $k/2^n5^m$ (e.g. 1.585 = $317/2^35^2$). However, *every* number with a terminating decimal representation also trivially has a second, alternative representation as a repeating decimal whose repetend is the digit 9. This is obtained by decreasing the final non-zero digit by one and appending a repetend of 9, a fact that some find puzzling. 1.000... = 0.999... and 1.585000... = 1.584999... are two examples of this. (This type of repeating decimal can be obtained by long division if one uses a modified form of the usual division algorithm.)

Any number that cannot be expressed as a ratio of two integers is said to be irrational. Their decimal representation neither terminates nor infinitely repeats but extends for ever without regular repetition. Examples of such irrational numbers are the square root of 2 and pi.

Background

Notation

While there are several notational conventions for representing repeating decimals, none of them are accepted universally. In the United States, the convention is generally to indicate a repeating decimal by drawing a horizontal line (a vinculum) above the repetend ($\frac{1}{3} = 0.\overline{3}$). In the United Kingdom and mainland China, the convention is to place dots above the outermost numerals of the repetend ($\frac{1}{3} = 0.\dot{3}$). In some Latin American countries both vinculum and dots notation are used besides the arc notation over the repetend ($\frac{1}{3} = 0.\overset{\frown}{3}$). Another notation employed in parts of Europe is to enclose the repetend in parentheses ($\frac{1}{3} = 0.(3)$). Repeating decimals may also be rep resented by three periods (an ellipsis, e.g., 0.333...), although this method introduces uncertainty as to which digits should be repeated or even whether repetition is occurring at all, unless spaces are inserted between periods, since such ellipses are also employed for irrational decimals such as 3.14159...

In English, there are various ways to read repeating decimals aloud. Some common ones (for ⅓) include "zero point three repeating", "zero point three repeated", "zero point three recurring", and "zero point three into infinity". Mention of the initial zero may also be omitted.

Decimal Expansion and Recurrence Sequence

In order to convert a rational number represented as a fraction into decimal form, one may use long division. For example, consider the rational number 5/74:

```
           · ·
     0.0675
74 ) 5.00000
     4.44
       560
       518
        420
        370
         500
```

etc. Observe that at each step we have a remainder; the successive remainders displayed above are 56, 42, 50. When we arrive at 50 as the remainder, and bring down the "0", we find ourselves dividing 500 by 74, which is the same problem we began with. Therefore, the decimal repeats: 0.0675 675 675 ....

Every Rational Number is Either A Terminating or Repeating Decimal

For any given divisor, only finitely many different remainders can occur. In the example above, the 74 possible remainders are 0, 1, 2, ..., 73. If at any point in the division the remainder is 0, the expansion terminates at that point. If 0 never occurs as a remainder, then the division process continues for ever, and eventually a remainder must occur that has occurred before. The next step in the division will yield the same new digit in the quotient, and the same new remainder, as the previous time the remainder was the same. Therefore, the following division will repeat the same results.

Every Repeating or Terminating Decimal is A Rational Number

Each repeating decimal number satisfies a linear equation with integer coefficients, and its unique solution is a rational number. To illustrate the latter point, the number $\alpha = 5.8144144144...$ above satisfies the equation $10000\alpha - 10\alpha = 58144.144144... - 58.144144... = 58086$, whose solution is $\alpha = 58086/9990 = 3227/555$. The process of how to find these integer coefficients is described below.

The period length of $1/n$ are

> 0, 0, 1, 0, 0, 1, 6, 0, 1, 0, 2, 1, 6, 6, 1, 0, 16, 1, 18, 0, 6, 2, 22, 1, 0, 6, 3, 6, 28, 1, 15, 0, 2, 16, 6, 1, 3, 18, 6, 0, 5, 6, 21, 2, 1, 22, 46, 1, 42, 0, 16, 6, 13, 3, 2, 6, 18, 28, 58, 1, 60, 15, 6, 0, 6, 2, 33, 16, 22, 6, 35, 1, 8, 3, 1, ... (sequence A051626 in the OEIS).

The periodic part of $1/n$ are

> 0, 0, 3, 0, 0, 6, 142857, 0, 1, 0, 09, 3, 076923, 714285, 6, 0, 0588235294117647, 5, 052631578947368421, 0, 047619, 45, 0434782608695652173913, 6, 0, 384615, 037, 571428, 0344827586206896551724137931, 3, ... (sequence A036275 in the OEIS).

The period length of $1/(n$th prime) are

> 0, 1, 0, 6, 2, 6, 16, 18, 22, 28, 15, 3, 5, 21, 46, 13, 58, 60, 33, 35, 8, 13, 41, 44, 96, 4, 34, 53, 108, 112, 42, 130, 8, 46, 148, 75, 78, 81, 166, 43, 178, 180, 95, 192, 98, 99, 30, 222, 113, 228, 232, 7, 30, 50, 256, 262, 268, 5, 69, 28, ... (sequence A002371 in the OEIS).

The least prime p which $1/p$ with period length n are

> 3, 11, 37, 101, 41, 7, 239, 73, 333667, 9091, 21649, 9901, 53, 909091, 31, 17, 2071723, 19, 111111111111111111, 3541, 43, 23, 11111111111111111111111, 99990001, 21401, 859, 757, 29, 3191, 211, ... (sequence A007138 in the OEIS).

The least prime p which k/p has n different cycles ($1 \leq k \leq p-1$) are

> 7, 3, 103, 53, 11, 79, 211, 41, 73, 281, 353, 37, 2393, 449, 3061, 1889, 137, 2467, 16189, 641, 3109, 4973, 11087, 1321, 101, 7151, 7669, 757, 38629, 1231, ... (sequence A054471 in the OEIS).

Fractions with Prime Denominators

A fraction in lowest terms with a prime denominator other than 2 or 5 (i.e. coprime to 10) always produces a repeating decimal. The length of the repetend (period of the repeating decimal) of $1/p$

is equal to the order of 10 modulo p. If 10 is a primitive root modulo p, the repetend length is equal to $p - 1$; if not, the repetend length is a factor of $p - 1$. This result can be deduced from Fermat's little theorem, which states that $10^{p-1} \equiv 1 \pmod{p}$.

The base-10 repetend of the reciprocal of any prime number greater than 5 is divisible by 9.

If the repetend length of $1/p$ for prime p is equal to $p - 1$ then the repetend, expressed as an integer, is called a cyclic number.

Cyclic Numbers

Examples of fractions belonging to this group are:

- $1/7 = 0.\overline{142857}$, 6 repeating digits

- $1/17 = 0.\overline{05882352\ 94117647}$, 16 repeating digits

- $1/19 = 0.\overline{052631578\ 947368421}$, 18 repeating digits

- $1/23 = 0.\overline{04347826086\ 95652173913}$, 22 repeating digits

- $1/29 = 0.\overline{03448275862068\ 96551724137931}$, 28 repeating digits

- $1/47 = 0.\overline{0212765957446808510\overline{6382}\ 9787234042553191489\overline{3617}}$, 46 repeating digits

- $1/59 = 0.\overline{01694915254237288135593220338\ 98305084745762711864406779661}$, 58 repeating digits

- $1/61 = 0.\overline{016393442622950819672131147540\ 983606557377049180327868852459}$, 60 repeating digits

- $1/97 = 0.\overline{010309278350515463917525773195876288659793814432\ 9896907216494845\ 360824742268041237113402061855 67}$, 96 repeating digits

The list can go on to include the fractions 1/109, 1/113, 1/131, 1/149, 1/167, 1/179, 1/181, 1/193, etc. (sequence A001913 in the OEIS).

Every *proper* multiple of a cyclic number (that is, a multiple having the same number of digits) is a rotation.

- $1/7 = 1 \times 0.142857 \ldots = 0.142857 \ldots$

- $2/7 = 2 \times 0.142857 \ldots = 0.285714 \ldots$

- $3/7 = 3 \times 0.142857 \ldots = 0.428571 \ldots$

- $4/7 = 4 \times 0.142857 \ldots = 0.571428 \ldots$

- $5/7 = 5 \times 0.142857 \ldots = 0.714285 \ldots$

- $6/7 = 6 \times 0.142857 \ldots = 0.857142 \ldots$

The reason for the cyclic behavior is apparent from an arithmetic exercise of long division of $\frac{1}{7}$: the sequential remainders are the cyclic sequence {1, 3, 2, 6, 4, 5}. See also the article 142857 for more properties of this cyclic number.

A Fraction which is cyclic thus has a recurring decimal of even length that divides into two sequences in 9's complement form. For example 1/7 starts '142' and is followed by '857' while 6/7 (by rotation) starts '857' followed by *its* 9's complement '142'.

A *proper prime* is a prime p which ends in the digit 1 in base 10 and whose reciprocal in base 10 has a repetend with length p-1. In such primes, each digit 0, 1, …, 9 appears in the repeating sequence the same number of times as does each other digit (namely, $(p$-1$)/10$ times). They are:

> 61, 131, 181, 461, 491, 541, 571, 701, 811, 821, 941, 971, 1021, 1051, 1091, 1171, 1181, 1291, 1301, 1349, 1381, 1531, 1571, 1621, 1741, 1811, 1829, 1861, … (sequence A073761 in the OEIS).

A prime is a proper prime if and only if it is a full reptend prime and congruent to 1 mod 10.

If a prime p is both full reptend prime and safe prime, then $1/p$ will produce a stream of $p - 1$ pseudo-random digits. Those primes are

> 7, 23, 47, 59, 167, 179, 263, 383, 503, 863, 887, 983, 1019, 1367, 1487, 1619, 1823, … (sequence A000353 in the OEIS).

Other Reciprocals of Primes

Some reciprocals of primes that do not generate cyclic numbers are:

- $1/3 = 0.\overline{3}$, which has a period of 1.
- $1/11 = 0.\overline{09}$, which has a period of 2.
- $1/13 = 0.\overline{076923}$, which has a period of 6.
- $1/31 = 0.\overline{032258064516129}$, which has a period of 15.
- $1/37 = 0.\overline{027}$, which has a period of 3.
- $1/41 = 0.\overline{02439}$, which has a period of 5.
- $1/43 = 0.\overline{023255813953488372093}$, which has a period of 21.
- $1/53 = 0.\overline{0188679245283}$, which has a period of 13.
- $1/67 = 0.\overline{014925373134328358208955223880597}$, which has a period of 33.

(sequence A006559 in the OEIS)

The reason is that 3 is a divisor of 9, 11 is a divisor of 99, 41 is a divisor of 99999, etc. To find the period of $1/p$, we can check whether the prime p divides some number 999…999 in which the number of digits divides $p - 1$. Since the period is never greater than $p - 1$, we can obtain this by calculating

$\dfrac{10^{p-1}-1}{p}$. For example, for 11 we get

$$\frac{10^{11-1}-1}{11}=909090909$$

and then by inspection find the repetend 09 and period of 2.

Those reciprocals of primes can be associated with several sequences of repeating decimals. For example, the multiples of 1/13 can be divided into two sets, with different repetends. The first set is:

- 1/13 = 0.076923 …

- 10/13 = 0.769230 …

- 9/13 = 0.692307 …

- 12/13 = 0.923076 …

- 3/13 = 0.230769 …

- 4/13 = 0.307692 …,

where the repetend of each fraction is a cyclic re-arrangement of 076923. The second set is:

- 2/13 = 0.153846 …

- 7/13 = 0.538461 …

- 5/13 = 0.384615 …

- 11/13 = 0.846153 …

- 6/13 = 0.461538 …

- 8/13 = 0.615384 …,

where the repetend of each fraction is a cyclic re-arrangement of 153846.

In general, the set of proper multiples of reciprocals of a prime p consists of n subsets, each with repetend length k, where $nk = p - 1$.

Totient Rule

For an arbitrary integer n the length $\lambda(n)$ of the repetend of 1/n divides $\phi(n)$, where ϕ is the totient function. The length is equal to $\phi(n)$ if and only if 10 is a primitive root modulo n.

In particular, it follows that $\lambda(p) = p - 1$ if and only iff p is a prime and 10 is a primitive root modulo p. Then, the decimal expansions of n/p for $n = 1, 2, ..., p - 1$, all have periods of length $p - 1$ and differ only by a cyclic permutation. Such numbers p are called full repetend primes.

Reciprocals of Composite Integers Coprime to 10

If p is a prime other than 2 or 5, the decimal representation of the fraction $\dfrac{1}{p^2}$ repeats, e.g.:

1/49 = 0.020408163265306122448 979591836734693877551.

The period (repetend length) must be a factor of $\lambda(49) = 42$, where $\lambda(n)$ is known as the Carmichael function. This follows from Carmichael's theorem which states that if n is a positive integer then $\lambda(n)$ is the smallest integer m such that

$$a^m \equiv 1 \pmod{n}$$

for every integer a that is coprime to n.

The period of $\frac{1}{p^2}$ is usually pT_p, where T_p is the period of $\dfrac{1}{p}$. There are three known primes for which this is not true, and for those the period of $\dfrac{1}{p^2}$ is the same as the period of $\frac{1}{p}$ because p^2 divides $10^{p-1}-1$. These three primes are 3, 487 and 56598313 (sequence A045616 in the OEIS).

Similarly, the period of $\frac{1}{p^k}$ is usually $p^{k-1}T_p$

If p and q are primes other than 2 or 5, the decimal representation of the fraction $\frac{1}{pq}$ repeats. An example is 1/119:

119 = 7 × 17

$\lambda(7 \times 17) = \text{LCM}(\lambda(7), \lambda(17))$

= LCM(6, 16)

= 48,

where LCM denotes the least common multiple.

The period T of $\frac{1}{pq}$ is a factor of $\lambda(pq)$ and it happens to be 48 in this case:

1/119 = 0.0084033613445378151260050 42016806722689075630252521.

The period T of $\frac{1}{pq}$ is LCM(T_p, T_q), where T_p is the period of $\frac{1}{p}$ and T_q is the period of $\dfrac{1}{q}$.

If p, q, r etc. are primes other than 2 or 5, and k, ℓ, m etc. are positive integers, then $\dfrac{1}{p^k q^\ell r^m \cdots}$ is a repeating decimal with a period of $\text{LCM}(T_{p^k}, T_{q^\ell}, T_{r^m}, \ldots)$ where T_{p^k}, T_{q^ℓ}, T_{r^m}, etc. are respectively the period of the repeating decimals $\dfrac{1}{p^k}, \dfrac{1}{q^\ell}, \dfrac{1}{r^m}$, etc. as defined above.

Reciprocals of Integers Not Co-prime to 10

An integer that is not co-prime to 10 but has a prime factor other than 2 or 5 has a reciprocal that is eventually periodic, but with a non-repeating sequence of digits that precede the repeating part. The reciprocal can be expressed as:

$$\frac{1}{2^a 5^b\, p^k q^\ell \cdots},$$

where a and b are not both zero.

This fraction can also be expressed as:

$$\frac{5^{a-b}}{10^a\, p^k q^\ell \cdots},$$

if $a > b$, or as

$$\frac{2^{b-a}}{10^b\, p^k q^\ell \cdots},$$

if $b > a$, or as

$$\frac{1}{10^a\, p^k q^\ell \cdots},$$

if $a = b$.

The decimal has:

- An initial transient of $\max(a, b)$ digits after the decimal point. Some or all of the digits in the transient can be zeros.

- A subsequent repetend which is the same as that for the fraction $\dfrac{1}{p^k q^\ell \cdots}$.

For example $1/28 = 0.03571428571428\ldots$:

- the initial non-repeating digits are 03; and

- the subsequent repeating digits are 571428.

Converting Repeating Decimals to Fractions

Given a repeating decimal, it is possible to calculate the fraction that produced it. For example:

$2x = 0.333333\ldots$
$10x = 3.333333\ldots$ (multiplying each side of the above line by 10)
$9x = 3$ (subtracting the 1st line from the 2nd)
$x = 3/9 = 1/3$ (reducing to lowest terms)

A Shortcut

The procedure below can be applied in particular if the repetend has n digits, all of which are 0 except the final one which is 1. For instance for $n = 7$:

$$7.48181818\ldots = 7.3 + 0.18181818\ldots$$

$$= \frac{73}{10} + \frac{18}{99} = \frac{73}{10} + \frac{9 \times 2}{9 \times 11} = \frac{73}{10} + \frac{2}{11}$$

$$= \frac{11 \times 73 + 10 \times 2}{10 \times 11} = \frac{823}{110}$$

So this particular repeating decimal corresponds to the fraction $1/(10^n - 1)$, where the denominator is the number written as n digits 9. Knowing just that, a general repeating decimal can be expressed as a fraction without having to solve an equation. For example, one could reason:

It is possible to get a general formula expressing a repeating decimal with an n digit period (repetend length), beginning right after the decimal point, as a fraction:

$x = 0.(A_1 A_2 \ldots A_n)$

$10^n x = A_1 A_2 \ldots A_n.(A_1 A_2 \ldots A_n)$

$(10^n - 1)x = 99\ldots99x = A_1 A_2 \ldots A_n$

$x = A_1 A_2 \ldots A_n / (10^n - 1)$

$= A_1 A_2 \ldots A_n / 99\ldots99$

More explicitly one gets the following cases.

If the repeating decimal is between 0 and 1, and the repeating block is n digits long, first occurring right after the decimal point, then the fraction (not necessarily reduced) will be the integer number represented by the n-digit block divided by the one represented by n digits 9. For example,

- $0.444444\ldots = 4/9$ since the repeating block is 4 (a 1-digit block),

- $0.565656\ldots = 56/99$ since the repeating block is 56 (a 2-digit block),

- $0.012012\ldots = 12/999$ since the repeating block is 012 (a 3-digit block), and this further reduces to $4/333$.

- $0.9999999\ldots = 9/9 = 1$, since the repeating block is 9 (also a 1-digit block)

If the repeating decimal is as above, except that there are k (extra) digits 0 between the decimal point and the repeating n-digit block, then one can simply add k digits 0 after the n digits 9 of the denominator (and as before the fraction may subsequently be simplified). For example,

- $0.000444\ldots = 4/9000$ since the repeating block is 4 and this block is preceded by 3 zeros,

- $0.005656\ldots = 56/9900$ since the repeating block is 56 and it is preceded by 2 zeros,

- 0.00012012 … = 12/99900 = 2/16650 since the repeating block is 012 and it is preceded by 2 (!) zeros.

Any repeating decimal not of the form described above can be written as a sum of a terminating decimal and a repeating decimal of one of the two above types (actually the first type suffices, but that could require the terminating decimal to be negative). For example,

- 1.23444 … = 1.23 + 0.00444 … = 123/100 + 4/900 = 1107/900 + 4/900 = 1111/900 or alternatively 1.23444 … = 0.79 + 0.44444 … = 79/100 + 4/9 = 711/900 + 400/900 = 1111/900

- 0.3789789 … = 0.3 + 0.0789789 … = 3/10 + 789/9990 = 2997/9990 + 789/9990 = 3786/9990 = 631/1665 or alternatively 0.3789789 … = −0.6 + 0.9789789 … = −6/10 + 978/999 = −5994/9990 + 9780/9990 = 3786/9990 = 631/1665

It follows that any repeating decimal with period n, and k digits after the decimal point that do not belong to the repeating part, can be written as a (not necessarily reduced) fraction whose denominator is $(10^n - 1)10^k$.

Conversely the period of the repeating decimal of a fraction c/d will be (at most) the smallest number n such that $10^n - 1$ is divisible by d.

For example, the fraction 2/7 has $d = 7$, and the smallest k that makes $10^k - 1$ divisible by 7 is $k = 6$, because $999999 = 7 \times 142857$. The period of the fraction 2/7 is therefore 6.

Repeating Decimals as Infinite Series

A repeating decimal can also be expressed as an infinite series. That is, a repeating decimal can be regarded as the sum of an infinite number of rational numbers. To take the simplest example,

$$\sum_{n=1}^{\infty} \frac{1}{10^n} = \frac{1}{10} + \frac{1}{100} + \frac{1}{1000} + \cdots = 0.\overline{1}$$

The above series is a geometric series with the first term as 1/10 and the common factor 1/10. Because the absolute value of the common factor is less than 1, we can say that the geometric series converges and find the exact value in the form of a fraction by using the following formula where a is the first term of the series and r is the common factor.

$$\frac{a}{1-r} = \frac{\frac{1}{10}}{1-\frac{1}{10}} = \frac{1}{9} = 0.\overline{1}$$

Multiplication and Cyclic Permutation

The cyclic behavior of repeating decimals in multiplication also leads to the construction of integers which are cyclically permuted when multiplied by certain numbers. For example, 102564 x 4 = 410256. Note that 102564 is the repetend of 4/39 and 410256 the repetend of 16/39.

Other Properties of Repetend Lengths

Various properties of repetend lengths (periods) are given by Mitchell and Dickson.

The period of $1/k$ for integer k is always $\leq k - 1$.

If p is prime, the period of $1/p$ divides evenly into $p - 1$.

If k is composite, the period of $1/k$ is strictly less than $k - 1$.

The period of c/k, for c coprime to k, equals the period of $1/k$.

If $k = 2^a 5^b n$ where $n > 1$ and n is not divisible by 2 or 5, then the length of the transient of $1/k$ is $\max(a, b)$, and the period equals r, where r is the smallest integer such that $10^r \equiv 1 \pmod{n}$.

If $p, p', p'', \ldots$ are distinct primes, then the period of $1/(pp'p''\ldots)$ equals the lowest common multiple of the periods of $1/p$, $1/p'$, $1/p''$, $\ldots$.

If k and k' have no common prime factors other than 2 and/or 5, then the period of $\dfrac{1}{kk'}$ equals the least common multiple of the periods of $\dfrac{1}{k}$ and $\dfrac{1}{k'}$.

For prime p, if $\operatorname{period}(\dfrac{1}{p}) = \operatorname{period}(\dfrac{1}{p^2}) = \cdots = \operatorname{period}(\dfrac{1}{p^m})$ but $\operatorname{period}(\dfrac{1}{p^m}) \neq \operatorname{period}(\dfrac{1}{p^{m+1}})$, then

for $c \geq 0$ we have $.\operatorname{period}(\dfrac{1}{p^{m+c}}) = p^c \cdot \operatorname{period}(\dfrac{1}{p})$

If p is a proper prime ending in a 1 – that is, if the repetend of $1/p$ is a cyclic number of length $p - 1$ and $p = 10h + 1$ for some h – then each digit 0, 1, ..., 9 appears in the repetend exactly $h = (p - 1)/10$ times.

Extension to Other Bases

Various features of repeating decimals extend to the representation of numbers in all other integer bases, not just base 10:

- Any number can be represented as an integer component followed by a radix point (the generalization of a decimal point to non-decimal systems) followed by a finite or infinite number of digits.

- A rational number has a terminating sequence after the radix point if all the prime factors of the denominator of the fully reduced fractional form are also factors of the base. This terminating representation is equivalent to a representation with a repeating sequence that can be constructed from the terminating form by decreasing the last digit by 1 and appending an infinite sequence of a digit representing a number that is one less than the base.

- A rational number has an infinitely repeating sequence of finite length less than the value of the fully reduced fraction's denominator if the reduced fraction's denominator contains a prime factor that is not a factor of the base. The repeating sequence is preceded after the radix point by a transient of finite length if the reduced fraction also shares a prime factor with the base.

- An irrational number has a representation of infinite length that never repeats itself.

For example, in duodecimal, 1/2 = 0.6, 1/3 = 0.4, 1/4 = 0.3 and 1/6 = 0.2 all terminate; 1/5 = 0.2497 repeats with period 4, in contrast with the equivalent decimal expansion of 0.2; 1/7 = 0.186Ϩ35 has period 6 in duodecimal, just as it does in decimal.

If b is an integer base and k is an integer,

$$\frac{1}{k} = \frac{1}{b} + \frac{(b-k)^1}{b^2} + \frac{(b-k)^2}{b^3} + \frac{(b-k)^3}{b^4} + \frac{(b-k)^4}{b^5} + \cdots + \frac{(b-k)^{N-1}}{b^N} + \cdots [6pt]$$

For example 1/7 in duodecimal:

$$\frac{1}{7} = \frac{1}{10} + \frac{5}{10^2} + \frac{21}{10^3} + \frac{\tau5}{10^4} + \frac{441}{10^5} + \frac{1985}{10^6} + \cdots [6pt]$$

Which is 0.186Ϩ35 (base 12). Note that 10 (base 12) is 12 (base 10), 10^2 (base 12) is 144 (base 10), 21 (base 12) is 25 (base 10), Ϩ5 (base 12) is 125 (base 10), ...

Applications to Cryptography

Repeating decimals (also called decimal sequences) have found cryptographic and error-correction coding applications. In these applications repeating decimals to base 2 are generally used which gives rise to binary sequences. The maximum length binary sequence for $1/p$ (when 2 is a primitive root of p) is given by:

$$a(i) = 2^i \bmod p \bmod 2$$

These sequences of period p-1 have an autocorrelation function that has a negative peak of -1 for shift of $(p-1)/2$. The randomness of these sequences has been examined by diehard tests.

Cyclic Number

A cyclic number is an integer in which cyclic permutations of the digits are successive multiples of the number. The most widely known is 142857:

142857 × 1 = 142857

142857 × 2 = 285714

142857 × 3 = 428571

142857 × 4 = 571428

142857 × 5 = 714285

142857 × 6 = 857142

Details

To qualify as a cyclic number, it is required that successive multiples be cyclic permutations. Thus, the number 076923 would not be considered a cyclic number, because even though all cyclic permutations are multiples, they are not successive multiples:

$$076923 \times 1 = 076923$$

$$076923 \times 3 = 230769$$

$$076923 \times 4 = 307692$$

$$076923 \times 9 = 692307$$

$$076923 \times 10 = 769230$$

$$076923 \times 12 = 923076$$

The following trivial cases are typically excluded:

1. single digits, e.g.: 5

2. repeated digits, e.g.: 555

3. repeated cyclic numbers, e.g.: 142857142857

If leading zeros are not permitted on numerals, then 142857 is the only cyclic number in decimal, due to the necessary structure given in the next section. Allowing leading zeros, the sequence of cyclic numbers begins:

$$(10^6-1) / 7 = 142857 \text{ (6 digits)}$$

$$(10^{16}-1) / 17 = 0588235294117647 \text{ (16 digits)}$$

$$(10^{18}-1) / 19 = 052631578947368421 \text{ (18 digits)}$$

$$(10^{22}-1) / 23 = 0434782608695652173913 \text{ (22 digits)}$$

$$(10^{28}-1) / 29 = 0344827586206896551724137931 \text{ (28 digits)}$$

$$(10^{46}-1) / 47 = 0212765957446808510638297872340425531914893617 \text{ (46 digits)}$$

$$(10^{58}-1) / 59 = 0169491525423728813559322033898305084745762711864406779661$$
(58 digits)

$$(10^{60}-1) / 61 = 016393442622950819672131147540983606557377049180327868852459$$
(60 digits)

Relation to Repeating Decimals

Cyclic numbers are related to the recurring digital representations of unit fractions. A cyclic number of length L is the digital representation of

$$1/(L + 1).$$

Conversely, if the digital period of $1/p$ (where p is prime) is

$$p - 1,$$

then the digits represent a cyclic number.

For example:

$$1/7 = 0.142857\ 142857....$$

Multiples of these fractions exhibit cyclic permutation:

$$1/7 = 0.142857\ 142857...$$

$$2/7 = 0.285714\ 285714...$$

$$3/7 = 0.428571\ 428571...$$

$$4/7 = 0.571428\ 571428...$$

$$5/7 = 0.714285\ 714285...$$

$$6/7 = 0.857142\ 857142....$$

Form of Cyclic Numbers

From the relation to unit fractions, it can be shown that cyclic numbers are of the form of the Fermat quotient

$$\frac{b^{p-1}-1}{p}$$

where b is the number base (10 for decimal), and p is a prime that does not divide b. (Primes p that give cyclic numbers in base b are called full reptend primes or long primes in base b).

For example, the case $b = 10$, $p = 7$ gives the cyclic number 142857, and the case $b = 12$, $p = 5$ gives the cyclic number 2497.

Not all values of p will yield a cyclic number using this formula; for example, the case $b = 10$, $p = 13$ gives 076923076923, and the case $b = 12$, $p = 19$ gives 076B45076B45076B45. These failed cases will always contain a repetition of digits (possibly several).

The first values of p for which this formula produces cyclic numbers in decimal ($b = 10$) are (sequence A001913 in the OEIS)

7, 17, 19, 23, 29, 47, 59, 61, 97, 109, 113, 131, 149, 167, 179, 181, 193, 223, 229, 233, 257, 263, 269, 313, 337, 367, 379, 383, 389, 419, 433, 461, 487, 491, 499, 503, 509, 541, 571, 577, 593, 619, 647, 659, 701, 709, 727, 743, 811, 821, 823, 857, 863, 887, 937, 941, 953, 971, 977, 983, ...

For $b = 12$ (duodecimal), these ps are (sequence A019340 in the OEIS)

5, 7, 17, 31, 41, 43, 53, 67, 101, 103, 113, 127, 137, 139, 149, 151, 163, 173, 197, 223, 257, 269, 281, 283, 293, 317, 353, 367, 379, 389, 401, 449, 461, 509, 523, 547, 557, 569, 571, 593, 607, 617, 619, 631, 641, 653, 691, 701, 739, 751, 761, 773, 787, 797, 809, 821, 857, 881, 929, 953, 967, 977, 991, ...

For $b = 2$ (binary), these ps are (sequence A001122 in the OEIS)

3, 5, 11, 13, 19, 29, 37, 53, 59, 61, 67, 83, 101, 107, 131, 139, 149, 163, 173, 179, 181, 197, 211, 227, 269, 293, 317, 347, 349, 373, 379, 389, 419, 421, 443, 461, 467, 491, 509, 523, 541, 547, 557, 563, 587, 613, 619, 653, 659, 661, 677, 701, 709, 757, 773, 787, 797, 821, 827, 829, 853, 859, 877, 883, 907, 941, 947, ...

For $b = 3$ (ternary), these ps are (sequence A019334 in the OEIS)

2, 5, 7, 17, 19, 29, 31, 43, 53, 79, 89, 101, 113, 127, 137, 139, 149, 163, 173, 197, 199, 211, 223, 233, 257, 269, 281, 283, 293, 317, 331, 353, 379, 389, 401, 449, 461, 463, 487, 509, 521, 557, 569, 571, 593, 607, 617, 631, 641, 653, 677, 691, 701, 739, 751, 773, 797, 809, 811, 821, 823, 857, 859, 881, 907, 929, 941, 953, 977, ...

There are no such ps in the hexadecimal system.

The known pattern to this sequence comes from algebraic number theory, specifically, this sequence is the set of primes p such that b is a primitive root modulo p. A conjecture of Emil Artin is that this sequence contains 37.395..% of the primes (for b in A085397).

Construction of Cyclic Numbers

Cyclic numbers can be constructed by the following procedure:

Let b be the number base (10 for decimal)
Let p be a prime that does not divide b.
Let $t = 0$.
Let $r = 1$.
Let $n = 0$.
loop:

Let $t = t + 1$

Let $x = r \cdot b$

Let $d = \text{int}(x / p)$

Let $r = x \bmod p$

Let $n = n \cdot b + d$

If $r \neq 1$ then repeat the loop.

if $t = p - 1$ then n is a cyclic number.

This procedure works by computing the digits of $1/p$ in base b, by long division. r is the remainder at each step, and d is the digit produced.

The step

$$n = n \cdot b + d$$

serves simply to collect the digits. For computers not capable of expressing very large integers, the digits may be output or collected in another way.

Note that if *t* ever exceeds *p*/2, then the number must be cyclic, without the need to compute the remaining digits.

Properties of Cyclic Numbers

- When multiplied by their generating prime, results in a sequence of '*base*–1' digits (9 in decimal). *Decimal 142857 × 7 = 999999.*

- When split in two,three four etc...regarding base 10,100,1000 etc.. by its digits and added the result is a sequence of 9's. *14 + 28 + 57 = 99, 142 + 857 = 999, 1428 + 5714+ 2857 = 9999 etc. ...* (This is a special case of Midy's Theorem.)

- All cyclic numbers are divisible by '*base*–1' (9 in decimal) and the sum of the remainder is the a multiple of the divisor. (This follows from the previous point.)

Other Numeric Bases

Using the above technique, cyclic numbers can be found in other numeric bases. (Note that not all of these follow the second rule (all successive multiples being cyclic permutations) listed in the Special Cases section above) In each of these cases the digits across half the period add up to the base minus one. Thus for binary the sum of the bits across half the period is 1; for ternary it is 2, and so on.

In binary, the sequence of cyclic numbers begins: (sequence A001122 in the OEIS)

11 (3) → 01

101 (5) → 0011

1011 (11) → 0001011101

1101 (13) → 000100111011

10011 (19) → 00001101011110101

11101 (29) → 0000100011010011110111001011

100101 (37) → 00000110111010110011111001000101001 1

In ternary: (sequence A019334 in the OEIS)

2 (2) → 1

12 (5) → 0121

21 (7) → 010212

122 (17) → 0011202122110201

201 (19) → 001102100221120122

1002 (29) → 000221010201112220012120211

1011 (31) → 000212111221020222010111001202

In quaternary:

(none)

In quinary: (sequence A019335 in the OEIS)

2 (2) → 2

3 (3) → 13

12 (7) → 032412

32 (17) → 0121340243231042

43 (23) → 01020413321434240311123

122 (37) → 00314212204011334244130232240433102

133 (43) → 0024231412234340431114420213032210104013333

In senary: (sequence A167794 in the OEIS)

15 (11) → 0313452421

21 (13) → 024340531215

25 (17) → 0204122453514331

105 (41) → 0051335412440330234455042201431152253211

135 (59) → 003354440223510413432425030145522011533204514212313052541

141 (61) → 0033125040441544530143423202205522430515114011025412132353335

211 (79) → 002422325434441304033512354102140052450553133230121114251522043201453415503105

In base 7: (sequence A019337 in the OEIS)

2 (2) → 3

5 (5) → 1254

14 (11) → 0431162355

16 (13) → 035245631421

23 (17) → 0261143464055232

32 (23) → 0206251134364604155323

56 (41) → 0112363262135202250565543034045314644161

In octal: (sequence A019338 in the OEIS)

3 (3) → 25

5 (5) → 1463

13 (11) → 0564272135

35 (29) → 0215173454106475626043236713

65 (53) → 0115220717545336140465103476625570602324416373126743

73 (59) → 0105330745756511606404255436276724470320212661713735223415

123 (83) → 0061262710366576352321570224030531344173277165150674112014254562
075537472464336045

In nonary:

2 (2) → 4

(no others)

In base 11: (sequence A019339 in the OEIS)

2 (2) → 5

3 (3) → 37

12 (13) → 093425A17685

16 (17) → 07132651A3978459

21 (23) → 05296243390A581486771A

27 (29) → 04199534608387A69115764A2723

29 (31) → 039A32146818574A71078964292536

In duodecimal: (sequence A019340 in the OEIS)

5 (5) → 2497

7 (7) → 186A35

15 (17) → 08579214B36429A7

27 (31) → 0478AA093598166B74311B28623A55

35 (41) → 036190A653277397A9B4B85A2B15689448241207

37 (43) → 0342295A3AA730A068456B879926181148B1B53765

45 (53) → 02872B3A23205525A784640AA4B9349081989B6696143757B117

In base 13: (sequence A019341 in the OEIS)

2 (2) → 6

5 (5) → 27A5

B (11) → 12495BA837

16 (19) → 08B82976AC414A3562

25 (31) → 055B42692C21347C7718A63A0AB985

2B (37) → 0474BC3B3215368A25C85810919AB79642A7

32 (41) → 04177C08322B13645926C8B550C49AA1B96873A6

In base 14: (sequence A019342 in the OEIS)

3 (3) → 49

13 (17) → 0B75A9C4D2683419

15 (19) → 0A45C7522D398168BB

19 (23) → 0874391B7CAD569A4C2613

21 (29) → 06A89925B163C0D73544B82C7A1D

3B (53) → 039AB8A075793610B146C21828DA43253D6864A7CD2C971BC5B5

43 (59) → 03471937B8ACB5659A2BC15D09D74DA96C4A62531287843B21C80D4069

In base 15: (sequence A019343 in the OEIS)

2 (2) → 7

D (13) → 124936DCA5B8

14 (19) → 0BC9718A3E3257D64B

18 (23) → 09BB1487291E533DA67C5D

1E (29) → 07B5A528BD6ACDE73949C6318421

27 (37) → 061339AE2C87A8194CE8DBB540C26746D5A2

2B (41) → 0574B51C68BA922DD80AE97A39D286345CC116E4

In hexadecimal:

(none)

In base 17: (sequence A019344 in the OEIS)

2 (2) → 8

3 (3) → 5B

5 (5) → 36DA

7 (7) → 274E9C

B (11) → 194ADF7C63

16 (23) → 0C9A5F8ED52G476B1823BE

1E (31) → 09583E469EDC11AG7B8D2CA7234FF6

In base 18: (sequence A019345 in the OEIS)

5 (5) → 3AE7

B (11) → 1B834H69ED

1B (29) → 0B31F95A9GDAE4H6EG28C781463D

21 (37) → 08DB37565F184FA3G0H946EACBC2G9D27E1H

27 (43) → 079B57H2GD721C293DEBCHA86CA0F14AFG5F8E4365

2H (53) → 0620C41682CG57EAFB3D4788EGHBFH5DGB9F51CA3726E4DA9931

35 (59) → 058F4A6CEBAC3BG30G89DD227GE0AHC92D7B53675E61EH19844FFA13H7

In base 19: (sequence A019346 in the OEIS)

2 (2) → 9

7 (7) → 2DAG58

B (11) → 1DFA6H538C

D (13) → 18EBD2HA475G

14 (23) → 0FD4291C784I35EG9H6BAE

1A (29) → 0C89FDE7G73HD1I6A9354B2BF15H

1I (37) → 09E73B5C631A52AEGHI94BF7D6CFH8DG8421

In base 20: (sequence A019347 in the OEIS)

3 (3) → 6D

D (13) → 1AF7DGI94C63

H (17) → 13ABF5HCIG984E27

13 (23) → 0H7GA8DI546J2C39B61EFD

1H (37) → 0AG469EBHGF2E11C8CJ93FDA58234H5II7B7

23 (43) → 0960IC1H43E878GEHD9F6JADJ17I2FG5BCB3526A4D

27 (47) → 08A4522B15ACF67D3GBI5J2JB9FEHH8IE974DC6G381E0H

In base 21: (sequence A019348 in the OEIS)

2 (2) → A

J (19) → 1248HE7F9JIGC36D5B

12 (23) → 0J3DECG92FAK1H7684BI5A

18 (29) → 0F475198EA2IH7K5GDFJBC6AI23D

1A (31) → 0E4FC4179A382EIK6G58GJDBAHCI62

2B (53) → 086F9AEDI4FHH927J8F13K47B1KCE5BA672G533BID1C5JH0GD9J

38 (71) → 06493BB50C8I721A13HFE42K27EA785J4F7KEGBH99FK8C2DIJAJH356GI-0ID6ADCF1G5D

In base 22: (sequence A019349 in the OEIS)

5 (5) → 48HD

H (17) → 16A7GI2CKFBE53J9

J (19) → 13A95H826KIBCG4DJF

19 (31) → 0FDAE45EJJ3C194L68B7HG722I9KCH

1F (37) → 0D1H57G143CAFA2872L8K4GE5KHI9B6BJDEJ

1J (41) → 0BHFC7B5JIH3GDKK8CJ6LA469EAG234I5811D92F

23 (47) → 0A6C3G897L18JEB5361J44ELBF9I5DCE0KD27AGIFK2HH7

In base 23: (sequence A019350 in the OEIS)

2 (2) → B

3 (3) → 7F

5 (5) → 4DI9

H (17) → 182G59AILEK6HDC4

21 (47) → 0B5K1AHE496JD4KCGEFF3L0MBH2LC58IDG39I2A6877J1M

2D (59) → 08M51CJK65AC1LJ27I79846E9H3BFME0HLA32GHCAL13KF4FDEIG8D-5JB7

3K (89) → 05LG6ADG0BK9CL4910HJ2J8I21CF5FHD4327B8C3864EMH16GC96MB-2DA1IDLM53K3E4KLA7H759IJKFBEAJEGI8

In base 24: (sequence A019351 in the OEIS)

7 (7) → 3A6KDH

B (11) → 248HALJF6D

D (13) → 1L795CM3GEIB

H (17) → 19L45FCGME2JI8B7

17 (31) → 0IDMAK327HJ8C96N5A1D3KLG64FBEH

1D (37) → 0FDEM1735K2E6BG54CN8A91MGKI3L9HC7IJB

1H (41) → 0E14284G98IHDB2M5KBGN9MJLFJ7EF56ACL1I3C7

In base 25:

2 (2) → C

(no others)

Note that in ternary ($b = 3$), the case $p = 2$ yields 1 as a cyclic number. While single digits may be considered trivial cases, it may be useful for completeness of the theory to consider them only when they are generated in this way.

It can be shown that no cyclic numbers (other than trivial single digits) exist in any numeric base which is a perfect square, that is, base 4, 9, 16, 25, etc.

Egyptian Fraction

An Egyptian fraction is a finite sum of distinct unit fractions, such as $\frac{1}{2} + \frac{1}{3} + \frac{1}{16}$. That is, each fraction in the expression has a numerator equal to 1 and a denominator that is a positive integer, and all the denominators differ from each other. The value of an expression of this type is a positive rational number a/b; for instance the Egyptian fraction above sums to 43/48. Every positive rational number can be represented by an Egyptian fraction. Sums of this type, and similar sums also including 2/3 and 3/4 as summands, were used as a serious notation for rational numbers by the ancient Egyptians, and continued to be used by other civilizations into medieval times. In modern mathematical notation, Egyptian fractions have been superseded by vulgar fractions and decimal notation. However, Egyptian fractions continue to be an object of study in modern number theory and recreational mathematics, as well as in modern historical studies of ancient mathematics.

Motivating Applications

Beyond their historical use, Egyptian fractions have some practical advantages over other representations of fractional numbers. For instance, Egyptian fractions can help in dividing a number of objects into equal shares (Knott). For example, if one wants to divide 5 pizzas equally among 8 diners, the Egyptian fraction

- $5/8 = 1/2 + 1/8$

means that each diner gets half a pizza plus another eighth of a pizza, e.g. by splitting 4 pizzas into 8 halves, and the remaining pizza into 8 eighths.

Similarly, although one could divide 13 pizzas among 12 diners by giving each diner one pizza and splitting the remaining pizza into 12 parts (perhaps destroying it), one could note that

- $13/12 = 1/2 + 1/3 + 1/4$

and split 6 pizzas into halves, 4 into thirds and the remaining 3 into quarters, and then give each diner one half, one third and one quarter.

Early History

Egyptian fraction notation was developed in the Middle Kingdom of Egypt, altering the Old Kingdom's Eye of Horus numeration system. Five early texts in which Egyptian fractions appear were the Egyptian Mathematical Leather Roll, the Moscow Mathematical Papyrus, the Reisner Papyrus, the Kahun Papyrus and the Akhmim Wooden Tablet. A later text, the Rhind Mathematical Papyrus, introduced improved ways of writing Egyptian fractions. The Rhind papyrus was written by Ahmes and dates from the Second Intermediate Period; it includes a table of Egyptian fraction expansions for rational numbers $2/n$, as well as 84 word problems. Solutions to each problem were written out in scribal shorthand, with the final answers of all 84 problems being expressed in Egyptian fraction notation. $2/n$ tables similar to the one on the Rhind papyrus also appear on some of the other texts. However, as the Kahun Papyrus shows, vulgar fractions were also used by scribes within their calculations.

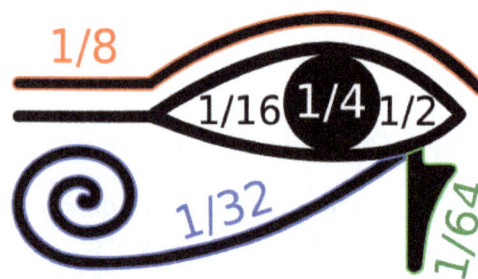

Eye of Horus

Notation

(The Egyptians also used an alternative notation modified from the Old Kingdom to denote a special set of fractions of the form $1/2^k$ (for $k = 1, 2, ..., 6$) and sums of these numbers, which are necessarily dyadic rational numbers. These have been called "Horus-Eye fractions" after a theory (now discredited) that they were based on the parts of the Eye of Horus symbol. They were used in the Middle Kingdom in conjunction with the later notation for Egyptian fractions to subdivide a hekat, the primary ancient Egyptian volume measure for grain, bread, and other small quantities of volume, as described in the Akhmim Wooden Tablet. If any remainder was left after expressing a quantity in Eye of Horus fractions of a hekat, the remainder was written using the usual Egyptian fraction notation as multiples of a *ro*, a unit equal to 1/320 of a hekat.

Calculation Methods

Modern historians of mathematics have studied the Rhind papyrus and other ancient sources in an attempt to discover the methods the Egyptians used in calculating with Egyptian fractions. In particular, study in this area has concentrated on understanding the tables of expansions for numbers of the form $2/n$ in the Rhind papyrus. Although these expansions can generally be described as algebraic identities, the methods used by the Egyptians may not correspond directly to these identities. Additionally, the expansions in the table do not match any single identity; rather, different identities match the expansions for prime and for composite denominators, and more than one identity fits the numbers of each type:

- For small odd prime denominators p, the expansion $2/p = 2/(p + 1) + 2/p(p + 1)$ was used.

- For larger prime denominators, an expansion of the form $2/p = 1/A + (2A − p)/Ap$ was used, where A is a number with many divisors (such as a practical number) between $p/2$ and p. The remaining term $(2A − p)/Ap$ was expanded by representing the number $(2A − p)/Ap$ as a sum of divisors of A and forming a fraction d/Ap for each such divisor d in this sum. As an example, Ahmes' expansion $1/24 + 1/111 + 1/296$ for $2/37$ fits this pattern with $A = 24$ and $(2A − p)/Ap = 11 = 3 + 8$, as $1/24 + 1/111 + 1/296 = 1/24 + 3/(24 \times 37) + 8/(24 \times 37)$. There may be many different expansions of this type for a given p; however, as K. S. Brown observed, the expansion chosen by the Egyptians was often the one that caused the largest denominator to be as small as possible, among all expansions fitting this pattern.

- For composite denominators, factored as $p \times q$, one can expand $2/pq$ using the identity $2/pq = 1/aq + 1/apq$, where $a = (p+1)/2$. For instance, applying this method for $pq = 21$ gives $p = 3$, $q = 7$, and $a = (3+1)/2 = 2$, producing the expansion $2/21 = 1/14 + 1/42$ from the Rhind papyrus. Some authors have preferred to write this expansion as $2/A \times A/pq$, where $A = p+1$; replacing the second term of this product by $p/pq + 1/pq$, applying the distributive law to the product, and simplifying leads to an expression equivalent to the first expansion described here. This method appears to have been used for many of the composite numbers in the Rhind papyrus, but there are exceptions, notably $2/35$, $2/91$, and $2/95$.

- One can also expand $2/pq$ as $1/pr + 1/qr$, where $r = (p+q)/2$. For instance, Ahmes expands $2/35 = 1/30 + 1/42$, where $p = 5$, $q = 7$, and $r = (5+7)/2 = 6$. Later scribes used a more general form of this expansion, $n/pq = 1/pr + 1/qr$, where $r = (p + q)/n$, which works when $p + q$ is a multiple of n (Eves 1953).

- 100r some other composite denominators, the expansion for $2/pq$ has the form of an expansion for $2/q$ with each denominator multiplied by p. For instance, $95 = 5 \times 19$, and $2/19 = 1/12 + 1/76 + 1/114$ (as can be found using the method for primes with $A = 12$), so $2/95 =$

1/(5×12) + 1/(5×76) + 1/(5×114) = 1/60 + 1/380 + 1/570. This expression can be simplified as 1/380 + 1/570 = 1/228 but the Rhind papyrus uses the unsimplified form.

- The final (prime) expansion in the Rhind papyrus, 2/101, does not fit any of these forms, but instead uses an expansion $2/p = 1/p + 1/2p + 1/3p + 1/6p$ that may be applied regardless of the value of p. That is, 2/101 = 1/101 + 1/202 + 1/303 + 1/606. A related expansion was also used in the Egyptian Mathematical Leather Roll for several cases.

Later Usage

Egyptian fraction notation continued to be used in Greek times and into the Middle Ages, despite complaints as early as Ptolemy's Almagest about the clumsiness of the notation compared to alternatives such as the Babylonian base-60 notation. An important text of medieval mathematics, the *Liber Abaci* (1202) of Leonardo of Pisa (more commonly known as Fibonacci), provides some insight into the uses of Egyptian fractions in the Middle Ages, and introduces topics that continue to be important in modern mathematical study of these series.

The primary subject of the *Liber Abaci* is calculations involving decimal and vulgar fraction notation, which eventually replaced Egyptian fractions. Fibonacci himself used a complex notation for fractions involving a combination of a mixed radix notation with sums of fractions. Many of the calculations throughout Fibonacci's book involve numbers represented as Egyptian fractions, and one section of this book provides a list of methods for conversion of vulgar fractions to Egyptian fractions. If the number is not already a unit fraction, the first method in this list is to attempt to split the numerator into a sum of divisors of the denominator; this is possible whenever the denominator is a practical number, and *Liber Abaci* includes tables of expansions of this type for the practical numbers 6, 8, 12, 20, 24, 60, and 100.

The next several methods involve algebraic identities such as $\frac{a}{ab-1} = \frac{1}{b} + \frac{1}{b(ab-1)}$. For instance, Fibonacci represents the fraction $\frac{8}{11}$ by splitting the numerator into a sum of two numbers, each of which divides one plus the denominator: $\frac{8}{11} = \frac{6}{11} + \frac{2}{11}$. Fibonacci applies the algebraic identity above to each these two parts, producing the expansion $\frac{8}{11} = \frac{1}{2} + \frac{1}{22} + \frac{1}{6} + \frac{1}{66}$. Fibonacci describes similar methods for denominators that are two or three less than a number with many factors.

In the rare case that these other methods all fail, Fibonacci suggests a greedy algorithm for computing Egyptian fractions, in which one repeatedly chooses the unit fraction with the smallest denominator that is no larger than the remaining fraction to be expanded: that is, in more modern notation, we replace a fraction x/y by the expansion

$$\frac{x}{y} = \frac{1}{\lceil y/x \rceil} + \frac{(-y) \bmod x}{y \lceil y/x \rceil},$$

where $\lceil \ldots \rceil$ represents the ceiling function.

Fibonacci suggests switching to another method after the first such expansion, but he also gives examples in which this greedy expansion was iterated until a complete Egyptian fraction expansion was constructed: $\dfrac{4}{13} = \dfrac{1}{4} + \dfrac{1}{18} + \dfrac{1}{468}$ and $\dfrac{17}{29} = \dfrac{1}{2} + \dfrac{1}{12} + \dfrac{1}{348}$.

As later mathematicians showed, each greedy expansion reduces the numerator of the remaining fraction to be expanded, so this method always terminates with a finite expansion. However, compared to ancient Egyptian expansions or to more modern methods, this method may produce expansions that are quite long, with large denominators, and Fibonacci himself noted the awkwardness of the expansions produced by this method. For instance, the greedy method expands

$$\frac{5}{121} = \frac{1}{25} + \frac{1}{757} + \frac{1}{763309} + \frac{1}{873960180913} + \frac{1}{1527612795642093418846225},$$

while other methods lead to the much better expansion

$$\frac{5}{121} = \frac{1}{33} + \frac{1}{121} + \frac{1}{363}.$$

Sylvester's sequence 2, 3, 7, 43, 1807, ... can be viewed as generated by an infinite greedy expansion of this type for the number one, where at each step we choose the denominator $\lfloor y/x \rfloor + 1$ instead of $\lceil y/x \rceil$, and sometimes Fibonacci's greedy algorithm is attributed to Sylvester.

After his description of the greedy algorithm, Fibonacci suggests yet another method, expanding a fraction a/b by searching for a number c having many divisors, with $b/2 < c < b$, replacing a/b by ac/bc, and expanding as a sum of divisors of bc, similar to the method proposed by ac Hultsch and Bruins to explain some of the expansions in the Rhind papyrus.

Modern Number Theory

Although Egyptian fractions are no longer used in most practical applications of mathematics, modern number theorists have continued to study many different problems related to them. These include problems of bounding the length or maximum denominator in Egyptian fraction representations, finding expansions of certain special forms or in which the denominators are all of some special type, the termination of various methods for Egyptian fraction expansion, and showing that expansions exist for any sufficiently dense set of sufficiently smooth numbers.

- One of the earliest publications of Paul Erdős proved that it is not possible for a harmonic progression to form an Egyptian fraction representation of an integer. The reason is that, necessarily, at least one denominator of the progression will be divisible by a prime number that does not divide any other denominator.

- The Erdős–Graham conjecture in combinatorial number theory states that, if the unit fractions are partitioned into finitely many subsets, then one of the subsets has a subset of itself whose reciprocals sum to one. That is, for every $r > 0$, and every r-coloring of the integers greater than one, there is a finite monochromatic subset S of these integers such that

$$\sum_{n \in S} 1/n = 1.$$

The conjecture was proven in 2003 by Ernest S. Croot, III.

- Znám's problem and primary pseudoperfect numbers are closely related to the existence of Egyptian fractions of the form

$$\sum \frac{1}{x_i} + \prod \frac{1}{x_i} = 1.$$

For instance, the primary pseudoperfect number 1806 is the product of the prime numbers 2, 3, 7, and 43, and gives rise to the Egyptian fraction $1 = 1/2 + 1/3 + 1/7 + 1/43 + 1/1806$.

- Egyptian fractions are normally defined as requiring all denominators to be distinct, but this requirement can be relaxed to allow repeated denominators. However, this relaxed form of Egyptian fractions does not allow for any number to be represented using fewer fractions, as any expansion with repeated fractions can be converted to an Egyptian fraction of equal or smaller length by repeated application of the replacement

$$\frac{1}{k} + \frac{1}{k} = \frac{2}{k+1} + \frac{2}{k(k+1)}$$

if k is odd, or simply by replacing $1/k + 1/k$ by $2/k$ if k is even. This result was first proven by Takenouchi (1921).

- Graham and Jewett proved that it is similarly possible to convert expansions with repeated denominators to (longer) Egyptian fractions, via the replacement

$$\frac{1}{k} + \frac{1}{k} = \frac{1}{k} + \frac{1}{k+1} + \frac{1}{k(k+1)}.$$

This method can lead to long expansions with large denominators, such as

$$\frac{4}{5} = \frac{1}{5} + \frac{1}{6} + \frac{1}{7} + \frac{1}{8} + \frac{1}{30} + \frac{1}{31} + \frac{1}{32} + \frac{1}{42} + \frac{1}{43} + \frac{1}{56} + \frac{1}{930} + \frac{1}{931} + \frac{1}{992} + \frac{1}{1806} + \frac{1}{865830}.$$

Botts (1967) had originally used this replacement technique to show that any rational number has Egyptian fraction representations with arbitrarily large minimum denominators.

- Any fraction x/y has an Egyptian fraction representation in which the maximum denominator is bounded by

$$O\left(y \log y (\log \log y)^4 (\log \log \log y)^2\right)$$

and a representation with at most

$$O\left(\sqrt{\log y}\right)$$

terms. The number of terms must sometimes be at least proportional to $\log \log y$; for instance this is true for the fractions in the sequence 1/2, 2/3, 6/7, 42/43, 1806/1807, ... whose denominators form Sylvester's sequence. It has been conjectured that $O(\log \log y)$ terms are always enough. It is also possible to find representations in which both the maximum denominator and the number of terms are small.

- Graham (1964) characterized the numbers that can be represented by Egyptian fractions in which all denominators are nth powers. In particular, a rational number q can be represented as an Egyptian fraction with square denominators if and only if q lies in one of the two half-open intervals

$$\left[0, \frac{\pi^2}{6} - 1\right) \cup \left[1, \frac{\pi^2}{6}\right).$$

- Martin (1999) showed that any rational number has very dense expansions, using a constant fraction of the denominators up to N for any sufficiently large N.

- Engel expansion, sometimes called an *Egyptian product*, is a form of Egyptian fraction expansion in which each denominator is a multiple of the previous one:

$$x = \frac{1}{a_1} + \frac{1}{a_1 a_2} + \frac{1}{a_1 a_2 a_3} + \cdots.$$

In addition, the sequence of multipliers a_i is required to be nondecreasing. Every rational number has a finite Engel expansion, while irrational numbers have an infinite Engel expansion.

- Anshel & Goldfeld (1991) study numbers that have multiple distinct Egyptian fraction representations with the same number of terms and the same product of denominators; for instance, one of the examples they supply is

$$\frac{5}{12} = \frac{1}{4} + \frac{1}{10} + \frac{1}{15} = \frac{1}{5} + \frac{1}{6} + \frac{1}{20}.$$

Unlike the ancient Egyptians, they allow denominators to be repeated in these expansions. They apply their results for this problem to the characterization of free products of Abelian groups by a small number of numerical parameters: the rank of the commutator subgroup, the number of terms in the free product, and the product of the orders of the factors.

- The number of different n-term Egyptian fraction representations of the number one is bounded above and below by double exponential functions of n.

Open Problems

Some notable problems remain unsolved with regard to Egyptian fractions, despite considerable effort by mathematicians.

- The Erdős–Straus conjecture concerns the length of the shortest expansion for a fraction of the form $4/n$. Does an expansion

$$\frac{4}{n} = \frac{1}{x} + \frac{1}{y} + \frac{1}{z}$$

  exist for every n? It is known to be true for all $n < 10^{14}$, and for all but a vanishingly small fraction of possible values of n, but the general truth of the conjecture remains unknown.

- It is unknown whether an odd greedy expansion exists for every fraction with an odd denominator. If Fibonacci's greedy method is modified so that it always chooses the smallest possible *odd* denominator, under what conditions does this modified algorithm produce a finite expansion? An obvious necessary condition is that the starting fraction x/y have an odd denominator y, and it is conjectured but not known that this is also a sufficient condition. It is known that every x/y with odd y has an expansion into distinct odd unit fractions, constructed using a different method than the greedy algorithm.

- It is possible to use brute-force search algorithms to find the Egyptian fraction representation of a given number with the fewest possible terms or minimizing the largest denominator; however, such algorithms can be quite inefficient. The existence of polynomial time algorithms for these problems, or more generally the computational complexity of such problems, remains unknown.

Guy (2004) describes these problems in more detail and lists numerous additional open problems.

References

- Ambrose, Gavin; et al. (2006). The Fundamentals of Typography, (2nd ed.). Lausanne: AVA Publishing. p. 74. ISBN 978-2-940411-76-4..

- Greer, A. (1986). New comprehensive mathematics for 'O' level (2nd ed., reprinted. ed.). Cheltenham: Thornes. p. 5. ISBN 9780859501590. Retrieved 2014-07-29.

- Eves, Howard (1990). An introduction to the history of mathematics (6th ed.). Philadelphia: Saunders College Pub. ISBN 0-03-029558-0.

- Berggren, J. Lennart (2007). "Mathematics in Medieval Islam". The Mathematics of Egypt, Mesopotamia, China, India, and Islam: A Sourcebook. Princeton University Press. p. 518. ISBN 978-0-691-11485-9.

- Guy, Richard K. (2004), "D11. Egyptian Fractions", Unsolved problems in number theory (3rd ed.), Springer-Verlag, pp. 252–262, ISBN 978-0-387-20860-2.

- Welsh, Alan H. (1996), Aspects of statistical inference, Wiley Series in Probability and Statistics, 246, John Wiley and Sons, p. 66, ISBN 978-0-471-11591-5.

- Saichev, Alexander; Malevergne, Yannick; Sornette, Didier (2009), Theory of Zipf's Law and Beyond, Lecture Notes in Economics and Mathematical Systems, 632, Springer-Verlag, ISBN 978-3-642-02945-5.

- Yang, Fujia; Hamilton, Joseph H. (2009), Modern Atomic and Nuclear Physics, World Scientific, pp. 81–86,

ISBN 978-981-283-678-6.

- Kilmister, Clive William (1994), Eddington's search for a fundamental theory: a key to the universe, Cambridge University Press, ISBN 978-0-521-37165-0.

- Eves, Howard (1953), An Introduction to the History of Mathematics, Holt, Reinhard, and Winston, ISBN 0-03-029558-0.

- Gardner, Milo (2002), "The Egyptian Mathematical Leather Roll, attested short term and long term", in Gratton-Guinness, Ivor, History of the Mathematical Sciences, Hindustan Book Co, pp. 119–134, ISBN 81-85931-45-3

- Guy, Richard K. (2004), "D11. Egyptian Fractions", Unsolved problems in number theory (3rd ed.), Springer-Verlag, pp. 252–262, ISBN 978-0-387-20860-2.

- Laurel (31 March 2004). "Math Forum – Ask Dr. Math:Can Negative Fractions Also Be Proper or Improper?". Retrieved 2014-10-30.

- "Common Core State Standards for Mathematics" (PDF). Common Core State Standards Initiative. 2010. p. 85. Retrieved 2013-10-10.

Arithmetic Operations: An Integrated Study

Some of the arithmetic operations explained in the section are algebraic operation, addition, subtraction, method of complements, multiplication and division. These operations are performed on numbers and variables. The topics elucidated in this chapter are of vital importance and provide a better understanding of arithmetic operations.

Algebraic Operation

Algebraic operations in the solution to the quadratic equation. The radical sign, √ denoting a square root, is equivalent to exponentiation to the power of ½. The ± sign represents the equation written with either a + and with a - sign.

$$x = \frac{-b \pm \sqrt{b^2 - 4ac}}{2a}$$

In mathematics, a basic algebraic operation is any one of the traditional operations of arithmetic, which are addition, subtraction, multiplication, division, raising to an integer power, and taking roots (fractional power). These operations may be performed on numbers, in which case they are often called arithmetic operations. They may also be performed, in a similar way, on variables, algebraic expressions, and, more generally on elements of algebraic structures, such as groups and fields.

The term *algebraic operation* may also be used for operations that may be defined by compounding basic algebraic operations, such as the dot product. In calculus and mathematical analysis, *algebraic operation* is also used for the operations that may be defined by purely algebraic methods. For example, exponentiation with an integer or rational exponent is an algebraic operation, but not the general exponentiation with a real or complex exponent. Also, the derivative is an operation that is not algebraic.

Notation

Multiplication symbols are usually omitted, and implied when there is no operator between two variables or terms, or when a coefficient is used. For example, $3 \times x^2$ is written as $3x^2$, and $2 \times x \times y$ is written as $2xy$. Sometimes multiplication symbols are replaced with either a dot, or center-dot, so that $x \times y$ is written as either $x \cdot y$ or $x \cdot y$. Plain text, programming languages, and calculators also use a single asterisk to represent the multiplication symbol, and it must be explicitly used, for example, $3x$ is written as 3 * x.

Rather than using the obelus symbol, ÷, division is usual represented with a vinculum, a horizontal line, e.g. $3/x + 1$. In plain text and programming languages a slash (also called a solidus) is used, e.g. 3 / (x + 1).

Exponents are usually formatted using superscripts, e.g. x^2. In plain text, and in the TeX mark-up language, the caret symbol, ^, represents exponents, so x^2 is written as x 2. In programming languages such as Ada, Fortran, Perl, Python and Ruby, a double asterisk is used, so x^2 is written as x ** 2.

The plus-minus sign, ±, is used as a shorthand notation for two expressions written as one, representing one expression with a plus sign, the other with a minus sign. For example $y = x \pm 1$ represents the two equations $y = x + 1$ and $y = x - 1$. Sometimes it is used for denoting positive-or-negative term such as $\pm x$.

Addition

Addition (often signified by the plus symbol "+") is one of the four basic operations of arithmetic, with the others being subtraction, multiplication and division. The addition of two whole numbers is the total amount of those quantities combined. For example, in the picture on the right, there is a combination of three apples and two apples together, making a total of five apples. This observation is equivalent to the mathematical expression "3 + 2 = 5" i.e., "3 *add* 2 is equal to 5".

3 + 2 = 5 with apples, a popular choice in textbooks

Besides counting fruits, addition can also represent combining other physical objects. Using systematic generalizations, addition can also be defined on more abstract quantities, such as integers, rational numbers, real numbers and complex numbers and other abstract objects such as vectors and matrices.

In arithmetic, rules for addition involving fractions and negative numbers have been devised amongst others. In algebra, addition is studied more abstractly.

Addition has several important properties. It is commutative, meaning that order does not matter, and it is associative, meaning that when one adds more than two numbers, the order in which addition is performed does not matter. Repeated addition of 1 is the same as counting; addition of 0 does not change a number. Addition also obeys predictable rules concerning related operations such as subtraction and multiplication.

Performing addition is one of the simplest numerical tasks. Addition of very small numbers is accessible to toddlers; the most basic task, 1 + 1, can be performed by infants as young as five months and even some non-human animals. In primary education, students are taught to add numbers in the decimal system, starting with single digits and progressively tackling more difficult problems. Mechanical aids range from the ancient abacus to the modern computer, where research on the most efficient implementations of addition continues to this day.

Notation and Terminology

The plus sign

Addition is written using the plus sign "+" between the terms; that is, in infix notation. The result is expressed with an equals sign. For example,

$1+1 = 2$ ("one plus one equals two")

$2 + 2 = 4$ ("two plus two equals four")

$3 + 3 = 6$ ("three plus three equals six")

$5 + 4 + 2 = 11$ ("associativity" below)

$3 + 3 + 3 + 3 = 12$ ("multiplication" below)

There are also situations where addition is "understood" even though no symbol appears:

$$\begin{array}{r} 5 \\ \underline{12} \\ \mathbf{\mathit{17}} \end{array}$$

Columnar addition: 5 + 12 = 17

- A column of numbers, with the last number in the column underlined, usually indicates that the numbers in the column are to be added, with the sum written below the underlined number.

- A whole number followed immediately by a fraction indicates the sum of the two, called a *mixed number*. For example,

$3\frac{1}{2} = 3 + \frac{1}{2} = 3.5.$

This notation can cause confusion since in most other contexts juxtaposition denotes multiplication instead.

The sum of a series of related numbers can be expressed through capital sigma notation, which compactly denotes iteration. For example,

$$\sum_{k=1}^{5} k^2 = 1^2 + 2^2 + 3^2 + 4^2 + 5^2 = 55.$$

The numbers or the objects to be added in general addition are collectively referred to as the terms, the addends or the summands; this terminology carries over to the summation of multiple terms. This is to be distinguished from *factors*, which are multiplied. Some authors call the first addend the *augend*. In fact, during the Renaissance, many authors did not consider the first addend an "addend" at all. Today, due to the commutative property of addition, "augend" is rarely used, and both terms are generally called addends.

All of the above terminology derives from Latin. "Addition" and "add" are English words derived from the Latin verb *addere*, which is in turn a compound of *ad* "to" and *dare* "to give", from the Proto-Indo-European root *deh_3*- "to give"; thus to *add* is to *give to*. Using the gerundive suffix *-nd* results in "addend", "thing to be added". Likewise from *augere* "to increase", one gets "augend", "thing to be increased".

The resultant	12
To whom it shall be addede	8
The nombre to be addede	4

Redrawn illustration from *The Art of Nombryng*, one of the first English arithmetic texts, in the 15th century

"Sum" and "summand" derive from the Latin noun *summa* "the highest, the top" and associated verb *summare*. This is appropriate not only because the sum of two positive numbers is greater than either, but because it was common for the ancient Greeks and Romans to add upward, contrary to the modern practice of adding downward, so that a sum was literally higher than the addends. *Addere* and *summare* date back at least to Boethius, if not to earlier Roman writers such as Vitruvius and Frontinus; Boethius also used several other terms for the addition operation. The later Middle English terms "adden" and "adding" were popularized by Chaucer.

The plus sign "+" (Unicode:U+002B; ASCII: +) is an abbreviation of the Latin word *et*, meaning "and". It appears in mathematical works dating back to at least 1489.

Interpretations

Addition is used to model countless physical processes. Even for the simple case of adding natural numbers, there are many possible interpretations and even more visual representations.

Combining Sets

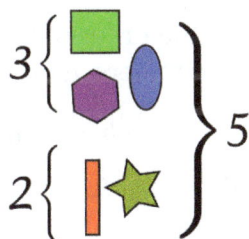

Possibly the most fundamental interpretation of addition lies in combining sets:

- When two or more disjoint collections are combined into a single collection, the number of objects in the single collection is the sum of the number of objects in the original collections.

This interpretation is easy to visualize, with little danger of ambiguity. It is also useful in higher mathematics; for the rigorous definition it inspires. However, it is not obvious how one should extend this version of addition to include fractional numbers or negative numbers.

One possible fix is to consider collections of objects that can be easily divided, such as pies or, still better, segmented rods. Rather than just combining collections of segments, rods can be joined end-to-end, which illustrates another conception of addition: adding not the rods but the lengths of the rods.

Extending A Length

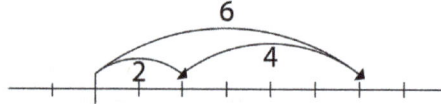

A number-line visualization of the algebraic addition 2 + 4 = 6. A translation by 2 followed by a translation by 4 is the same as a translation by 6.

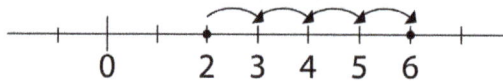

A number-line visualization of the unary addition 2 + 4 = 6. A translation by 4 is equivalent to four translations by 1.

A second interpretation of addition comes from extending an initial length by a given length:

- When an original length is extended by a given amount, the final length is the sum of the original length and the length of the extension.

The sum $a + b$ can be interpreted as a binary operation that combines a and b, in an algebraic sense, or it can be interpreted as the addition of b more units to a. Under the latter interpretation, the parts of a sum $a + b$ play asymmetric roles, and the operation $a + b$ is viewed as applying the unary operation $+b$ to a. Instead of calling both a and b addends, it is more appropriate to call a the augend in this case, since a plays a passive role. The unary view is also useful when discussing subtraction, because each unary addition operation has an inverse unary subtraction operation, and *vice versa*.

Properties

Commutativity

Addition is commutative: one can change the order of the terms in a sum, and the result is the same. Symbolically, if a and b are any two numbers, then

$$a + b = b + a.$$

4 + 2 = 2 + 4 with blocks

The fact that addition is commutative is known as the "commutative law of addition". This phrase suggests that there are other commutative laws: for example, there is a commutative law of multiplication. However, many binary operations are not commutative, such as subtraction and division, so it is misleading to speak of an unqualified "commutative law".

Associativity

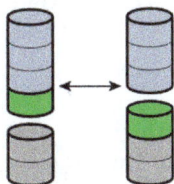

2+(1+3) = (2+1)+3 with segmented rods

Addition is associative: when adding three or more numbers, the order of operations does not matter .

As an example, should the expression $a + b + c$ be defined to mean $(a + b) + c$ or $a + (b + c)$? That addition is associative tells us that the choice of definition is irrelevant. For any three numbers a, b, and c, it is true that $(a + b) + c = a + (b + c)$. For example, $(1 + 2) + 3 = 3 + 3 = 6 = 1 + 5 = 1 + (2 + 3)$.

When addition is used together with other operations, the order of operations becomes important. In the standard order of operations, addition is a lower priority than exponentiation, nth roots, multiplication and division, but is given equal priority to subtraction.

Identity Element

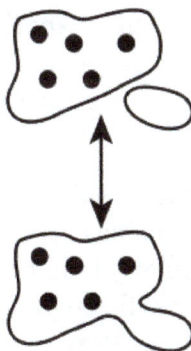

5 + 0 = 5 with bags of dots

When adding zero to any number, the quantity does not change; zero is the identity element for addition, also known as the additive identity. In symbols, for any a,

$$a + 0 = 0 + a = a.$$

This law was first identified in Brahmagupta's *Brahmasphutasiddhanta* in 628 AD, although he wrote it as three separate laws, depending on whether a is negative, positive, or zero itself, and he used words rather than algebraic symbols. Later Indian mathematicians refined the concept; around the year 830, Mahavira wrote, "zero becomes the same as what is added to it", corresponding to the unary statement $0 + a = a$. In the 12th century, Bhaskara wrote, "In the addition of cipher, or subtraction of it, the quantity, positive or negative, remains the same", corresponding to the unary statement $a + 0 = a$.

Successor

In the context of integers, addition of one also plays a special role: for any integer a, the integer (a + 1) is the least integer greater than a, also known as the successor of a. For instance, 3 is the successor of 2 and 7 is the successor of 6. Because of this succession, the value of $a + b$ can also be seen as the bth successor of a, making addition iterated succession. For examples, 6 + 2 is 8, because 8 is the successor of 7, which is the successor of 6, making 8 the 2nd successor of 6.

Units

To numerically add physical quantities with units, they must be expressed with common units. For example, adding 50 millilitres to 150 millilitres gives 200 millilitres. However, if a measure of 5 feet is extended by 2 inches, the sum is 62 inches, since 60 inches is synonymous with 5 feet. On the other hand, it is usually meaningless to try to add 3 meters and 4 square meters, since those units are incomparable; this sort of consideration is fundamental in dimensional analysis.

Performing Addition

Innate Ability

Studies on mathematical development starting around the 1980s have exploited the phenomenon of habituation: infants look longer at situations that are unexpected. A seminal experiment by Karen Wynn in 1992 involving Mickey Mouse dolls manipulated behind a screen demonstrated that five-month-old infants *expect* 1 + 1 to be 2, and they are comparatively surprised when a physical situation seems to imply that 1 + 1 is either 1 or 3. This finding has since been affirmed by a variety of laboratories using different methodologies. Another 1992 experiment with older toddlers, between 18 and 35 months, exploited their development of motor control by allowing them to retrieve ping-pong balls from a box; the youngest responded well for small numbers, while older subjects were able to compute sums up to 5.

Even some nonhuman animals show a limited ability to add, particularly primates. In a 1995 experiment imitating Wynn's 1992 result (but using eggplants instead of dolls), rhesus macaque and cottontop tamarin monkeys performed similarly to human infants. More dramatically, after being taught the meanings of the Arabic numerals 0 through 4, one chimpanzee was able to compute the sum of two numerals without further training. More recently, Asian elephants have demonstrated an ability to perform basic arithmetic.

Learning Addition As Children

Typically, children first master counting. When given a problem that requires that two items and three items be combined, young children model the situation with physical objects, often fingers or a drawing, and then count the total. As they gain experience, they learn or discover the strategy of "counting-on": asked to find two plus three, children count three past two, saying "three, four, *five*" (usually ticking off fingers), and arriving at five. This strategy seems almost universal; children can easily pick it up from peers or teachers. Most discover it independently. With additional experience, children learn to add more quickly by exploiting the commutativity of addition by counting up from the larger number, in this case starting with three and counting "four, *five*." Eventually children begin to recall certain addition facts ("number bonds"), either through experience or rote memorization. Once some facts are committed to memory, children begin to derive unknown facts from known ones. For example, a child asked to add six and seven may know that 6 + 6 = 12 and then reason that 6 + 7 is one more, or 13. Such derived facts can be found very quickly and most elementary school students eventually rely on a mixture of memorized and derived facts to add fluently.

Different nations introduce whole numbers and arithmetic at different ages, with many countries teaching addition in pre-school. However, throughout the world, addition is taught by the end of the first year of elementary school.

Addition Table

Children are often presented with the addition table of pairs of numbers from 1 to 10 to memorize. Knowing this, one can perform any addition.

Addition table				
Addition table of 1	**Addition table of 2**	**Addition table of 3**	**Addition table of 4**	**Addition table of 5**
1 + 0 = 1	2 + 0 = 2	3 + 0 = 3	4 + 0 = 4	5 + 0 = 5
1 + 1 = 2	2 + 1 = 3	3 + 1 = 4	4 + 1 = 5	5 + 1 = 6
1 + 2 = 3	2 + 2 = 4	3 + 2 = 5	4 + 2 = 6	5 + 2 = 7
1 + 3 = 4	2 + 3 = 5	3 + 3 = 6	4 + 3 = 7	5 + 3 = 8
1 + 4 = 5	2 + 4 = 6	3 + 4 = 7	4 + 4 = 8	5 + 4 = 9
1 + 5 = 6	2 + 5 = 7	3 + 5 = 8	4 + 5 = 9	5 + 5 = 10
1 + 6 = 7	2 + 6 = 8	3 + 6 = 9	4 + 6 = 10	5 + 6 = 11
1 + 7 = 8	2 + 7 = 9	3 + 7 = 10	4 + 7 = 11	5 + 7 = 12
1 + 8 = 9	2 + 8 = 10	3 + 8 = 11	4 + 8 = 12	5 + 8 = 13
1 + 9 = 10	2 + 9 = 11	3 + 9 = 12	4 + 9 = 13	5 + 9 = 14
1 + 10 = 11	2 + 10 = 12	3 + 10 = 13	4 + 10 = 14	5 + 10 = 15
Addition table of 6	**Addition table of 7**	**Addition table of 8**	**Addition table of 9**	**Addition table of 10**
6 + 0 = 6	7 + 0 = 7	8 + 0 = 8	9 + 0 = 9	10 + 0 = 10
6 + 1 = 7	7 + 1 = 8	8 + 1 = 9	9 + 1 = 10	10 + 1 = 11
6 + 2 = 8	7 + 2 = 9	8 + 2 = 10	9 + 2 = 11	10 + 2 = 12
6 + 3 = 9	7 + 3 = 10	8 + 3 = 11	9 + 3 = 12	10 + 3 = 13
6 + 4 = 10	7 + 4 = 11	8 + 4 = 12	9 + 4 = 13	10 + 4 = 14
6 + 5 = 11	7 + 5 = 12	8 + 5 = 13	9 + 5 = 14	10 + 5 = 15
6 + 6 = 12	7 + 6 = 13	8 + 6 = 14	9 + 6 = 15	10 + 6 = 16
6 + 7 = 13	7 + 7 = 14	8 + 7 = 15	9 + 7 = 16	10 + 7 = 17
6 + 8 = 14	7 + 8 = 15	8 + 8 = 16	9 + 8 = 17	10 + 8 = 18
6 + 9 = 15	7 + 9 = 16	8 + 9 = 17	9 + 9 = 18	10 + 9 = 19
6 + 10 = 16	7 + 10 = 17	8 + 10 = 18	9 + 10 = 19	10 + 10 = 20

Decimal System

The prerequisite to addition in the decimal system is the fluent recall or derivation of the 100 single-digit "addition facts". One could memorize all the facts by rote, but pattern-based strategies are more enlightening and, for most people, more efficient:

- *Commutative property*: Mentioned above, using the pattern $a + b = b + a$ reduces the number of "addition facts" from 100 to 55.

- *One or two more*: Adding 1 or 2 is a basic task, and it can be accomplished through counting on or, ultimately, intuition.

- *Zero*: Since zero is the additive identity, adding zero is trivial. Nonetheless, in the teaching of arithmetic, some students are introduced to addition as a process that always increases the addends; word problems may help rationalize the "exception" of zero.

- *Doubles*: Adding a number to itself is related to counting by two and to multiplication. Doubles facts form a backbone for many related facts, and students find them relatively easy to grasp.

- *Near-doubles*: Sums such as $6 + 7 = 13$ can be quickly derived from the doubles fact $6 + 6 = 12$ by adding one more, or from $7 + 7 = 14$ but subtracting one.

- *Five and ten*: Sums of the form $5 + x$ and $10 + x$ are usually memorized early and can be used for deriving other facts. For example, $6 + 7 = 13$ can be derived from $5 + 7 = 12$ by adding one more.

- *Making ten*: An advanced strategy uses 10 as an intermediate for sums involving 8 or 9; for example, $8 + 6 = 8 + 2 + 4 = 10 + 4 = 14$.

As students grow older, they commit more facts to memory, and learn to derive other facts rapidly and fluently. Many students never commit all the facts to memory, but can still find any basic fact quickly.

Carry

The standard algorithm for adding multidigit numbers is to align the addends vertically and add the columns, starting from the ones column on the right. If a column exceeds ten, the extra digit is "carried" into the next column. For example, in the addition $27 + 59$

$$
\begin{array}{r}
1 \\
27 \\
+59 \\
\hline
86
\end{array}
$$

$7 + 9 = 16$, and the digit 1 is the carry. An alternate strategy starts adding from the most significant digit on the left; this route makes carrying a little clumsier, but it is faster at getting a rough estimate of the sum. There are many alternative methods.

Addition of Decimal Fractions

Decimal fractions can be added by a simple modification of the above process. One aligns two decimal fractions above each other, with the decimal point in the same location. If necessary, one can add trailing zeros to a shorter decimal to make it the same length as the longer decimal. Finally, one performs the same addition process as above, except the decimal point is placed in the answer, exactly where it was placed in the summands.

As an example, 45.1 + 4.34 can be solved as follows:

$$
\begin{array}{r}
45.10 \\
+\,04.34 \\
\hline
49.44
\end{array}
$$

Scientific Notation

In scientific notation, numbers are written in the form $x = a \times 10^{b}$, where a is the significand and 10^{b} is the exponential part. Addition requires two numbers in scientific notation to be represented using the same exponential part, so that the significand can be simply added or subtracted.

For example:

$$2.34 \times 10^{-5} + 5.67 \times 10^{-6} = 2.34 \times 10^{-5} + 0.567 \times 10^{-5} = 2.907 \times 10^{-5}$$

Addition in Other Bases

Addition in other bases is very similar to decimal addition. As an example, one can consider addition in binary. Adding two single-digit binary numbers is relatively simple, using a form of carrying:

$0 + 0 \rightarrow 0$

$0 + 1 \rightarrow 1$

$1 + 0 \rightarrow 1$

$1 + 1 \rightarrow 0$, carry 1 (since $1 + 1 = 2 = 0 + (1 \times 2^{1})$)

Adding two "1" digits produces a digit "0", while 1 must be added to the next column. This is similar to what happens in decimal when certain single-digit numbers are added together; if the result equals or exceeds the value of the radix (10), the digit to the left is incremented:

$5 + 5 \rightarrow 0$, carry 1 (since $5 + 5 = 10 = 0 + (1 \times 10^{1})$)

$7 + 9 \rightarrow 6$, carry 1 (since $7 + 9 = 16 = 6 + (1 \times 10^{1})$)

This is known as *carrying*. When the result of an addition exceeds the value of a digit, the procedure is to "carry" the excess amount divided by the radix (that is, 10/10) to the left, adding it to the next positional value. This is correct since the next position has a weight that is higher by a factor equal to the radix. Carrying works the same way in binary:

$$
\begin{array}{r}
1\,1\,1\,1\,1 \quad \text{(carried digits)} \\
0\,1\,1\,0\,1 \\
+\quad 1\,0\,1\,1\,1 \\
\hline
1\,0\,0\,1\,0\,0 = 36
\end{array}
$$

In this example, two numerals are being added together: 01101_2 (13_{10}) and 10111_2 (23_{10}). The top row shows the carry bits used. Starting in the rightmost column, $1 + 1 = 10_2$. The 1 is carried to the left, and the 0 is written at the bottom of the rightmost column. The second column from the right is added: $1 + 0 + 1 = 10_2$ again; the 1 is carried, and 0 is written at the bottom. The third column: $1 + 1 + 1 = 11_2$. This time, a 1 is carried, and a 1 is written in the bottom row. Proceeding like this gives the final answer 100100_2 (36 decimal).

Computers

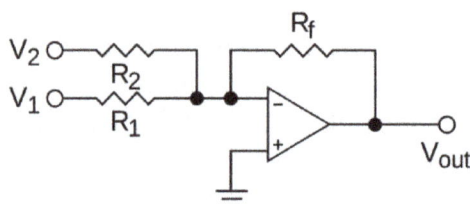

Addition with an op-amp. See Summing amplifier for details.

Analog computers work directly with physical quantities, so their addition mechanisms depend on the form of the addends. A mechanical adder might represent two addends as the positions of sliding blocks, in which case they can be added with an averaging lever. If the addends are the rotation speeds of two shafts, they can be added with a differential. A hydraulic adder can add the pressures in two chambers by exploiting Newton's second law to balance forces on an assembly of pistons. The most common situation for a general-purpose analog computer is to add two voltages (referenced to ground); this can be accomplished roughly with a resistor network, but a better design exploits an operational amplifier.

Addition is also fundamental to the operation of digital computers, where the efficiency of addition, in particular the carry mechanism, is an important limitation to overall performance.

Part of Charles Babbage's Difference Engine including the addition and carry mechanisms

The abacus, also called a counting frame, is a calculating tool that was in use centuries before the adoption of the written modern numeral system and is still widely used by merchants, traders and clerks in Asia, Africa, and elsewhere; it dates back to at least 2700–2300 BC, when it was used in Sumer.

Blaise Pascal invented the mechanical calculator in 1642; it was the first operational adding machine. It made use of a gravity-assisted carry mechanism. It was the only operational mechanical calculator in the 17th century and the earliest automatic, digital computers. Pascal's calculator was limited by its carry mechanism, which forced its wheels to only turn one way so it could add. To subtract, the operator had to use the Pascal's calculator's complement, which required as many steps as an addition. Giovanni Poleni followed Pascal, building the second functional mechanical calculator in 1709, a calculating clock made of wood that, once setup, could multiply two numbers automatically.

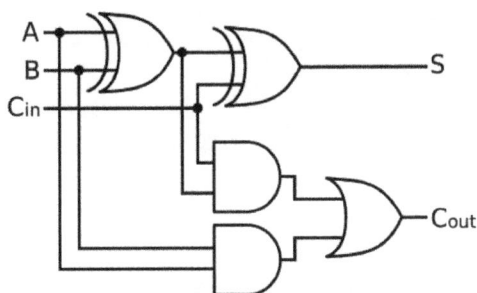

"Full adder" logic circuit that adds two binary digits, A and B, along with a carry input C_{in}, producing the sum bit, S, and a carry output, C_{out}.

Adders execute integer addition in electronic digital computers, usually using binary arithmetic. The simplest architecture is the ripple carry adder, which follows the standard multi-digit algorithm. One slight improvement is the carry skip design, again following human intuition; one does not perform all the carries in computing 999 + 1, but one bypasses the group of 9s and skips to the answer.

In practice, comutational addition may achieved via XOR and AND bitwise logical operations in conjunction with bitshift operations as shown in the pseudocode below. Both XOR and AND gates are straightforward to realize in digital logic allowing the realization of full adder circuits which in turn may be combined into more complex logical operations. In modern digital computers, integer addition is typically the fastest arithmetic instruction, yet it has the largest impact on performance, since it underlies all floating-point operations as well as such basic tasks as address generation during memory access and fetching instructions during branching. To increase speed, modern designs calculate digits in parallel; these schemes go by such names as carry select, carry lookahead, and the Ling pseudocarry. Many implementations are, in fact, hybrids of these last three designs. Unlike addition on paper, addition on a computer often changes the addends. On the ancient abacus and adding board, both addends are destroyed, leaving only the sum. The influence of the abacus on mathematical thinking was strong enough that early Latin texts often claimed that in the process of adding "a number to a number", both numbers vanish. In modern times, the ADD instruction of a microprocessor replaces the augend with the sum but preserves the addend. In a high-level programming language, evaluating $a + b$ does not change either a or b; if the goal is to replace a with the sum this must be explicitly requested, typically with the statement $a = a + b$. Some languages such as C or C++ allow this to be abbreviated as a += b.

```
// Iterative Algorithm

int add(int x, int y){

    int carry = 0;
```

```
   while (y != 0){

      carry = AND(x, y);    // Logical AND

      x     = XOR(x, y);    // Logical XOR

      y     = carry << 1;  // left bitshift carry by one

   }

   return x;

}

// Recursive Algorithm

int add(int x, int y){

   return x if (y == 0) else add(XOR(x, y) , AND(x, y) << 1);

}
```

On a computer, if the result of an addition is too large to store, an arithmetic overflow occurs, resulting in an incorrect answer. Unanticipated arithmetic overflow is a fairly common cause of program errors. Such overflow bugs may be hard to discover and diagnose because they may manifest themselves only for very large input data sets, which are less likely to be used in validation tests. One especially notable such error was the Y2K bug, where overflow errors due to using a 2-digit format for years caused significant computer problems in 2000.

Addition of Numbers

To prove the usual properties of addition, one must first define addition for the context in question. Addition is first defined on the natural numbers. In set theory, addition is then extended to progressively larger sets that include the natural numbers: the integers, the rational numbers, and the real numbers. (In mathematics education, positive fractions are added before negative numbers are even considered; this is also the historical route.)

Natural Numbers

There are two popular ways to define the sum of two natural numbers a and b. If one defines natural numbers to be the cardinalities of finite sets, (the cardinality of a set is the number of elements in the set), then it is appropriate to define their sum as follows:

- Let N(S) be the cardinality of a set S. Take two disjoint sets A and B, with N(A) = a and N(B) = b. Then $a + b$ is defined as $N(A \cup B)$.

Here, A U B is the union of A and B. An alternate version of this definition allows A and B to possibly overlap and then takes their disjoint union, a mechanism that allows common elements to be separated out and therefore counted twice.

The other popular definition is recursive:

- Let n^+ be the successor of n, that is the number following n in the natural numbers, so $0^+=1$, $1^+=2$. Define $a + 0 = a$. Define the general sum recursively by $a + (b^+) = (a + b)^+$. Hence $1 + 1 = 1 + 0^+ = (1 + 0)^+ = 1^+ = 2$.

Again, there are minor variations upon this definition in the literature. Taken literally, the above definition is an application of the Recursion Theorem on the partially ordered set $\mathbb{N}^2$. On the other hand, some sources prefer to use a restricted Recursion Theorem that applies only to the set of natural numbers. One then considers a to be temporarily "fixed", applies recursion on b to define a function "$a +$ ", and pastes these unary operations for all a together to form the full binary operation.

This recursive formulation of addition was developed by Dedekind as early as 1854, and he would expand upon it in the following decades. He proved the associative and commutative properties, among others, through mathematical induction.

Integers

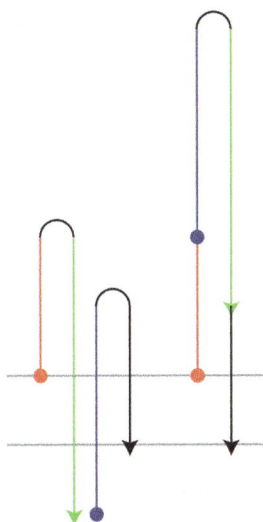

Defining $(-2) + 1$ using only addition of positive numbers: $(2 - 4) + (3 - 2) = 5 - 6$.

The simplest conception of an integer is that it consists of an absolute value (which is a natural number) and a sign (generally either positive or negative). The integer zero is a special third case, being neither positive nor negative. The corresponding definition of addition must proceed by cases:

- For an integer n, let $|n|$ be its absolute value. Let a and b be integers. If either a or b is zero, treat it as an identity. If a and b are both positive, define $a + b = |a| + |b|$. If a and b are both negative, define $a + b = -(|a|+|b|)$. If a and b have different signs, define $a + b$ to be the difference between $|a|$ and $|b|$, with the sign of the term whose absolute value is larger. As an example, $-6 + 4 = -2$; because -6 and 4 have different signs, their absolute values are subtracted, and since the negative term is larger, the answer is negative.

Although this definition can be useful for concrete problems, it is far too complicated to produce elegant general proofs; there are too many cases to consider.

A much more convenient conception of the integers is the Grothendieck group construction. The essential observation is that every integer can be expressed (not uniquely) as the difference of two natural numbers, so we may as well *define* an integer as the difference of two natural numbers. Addition is then defined to be compatible with subtraction:

- Given two integers $a - b$ and $c - d$, where a, b, c, and d are natural numbers, define $(a - b) + (c - d) = (a + c) - (b + d)$.

Rational Numbers (Fractions)

Addition of rational numbers can be computed using the least common denominator, but a conceptually simpler definition involves only integer addition and multiplication:

- Define $\dfrac{a}{b} + \dfrac{c}{d} = \dfrac{ad + bc}{bd}$.

As an example, the sum $\dfrac{3}{4} + \dfrac{1}{8} = \dfrac{3 \times 8 + 4 \times 1}{4 \times 8} = \dfrac{24 + 4}{32} = \dfrac{28}{32} = \dfrac{7}{8}$..

Addition of fractions is much simpler when the denominators are the same; in this case, one can simply add the numerators while leaving the denominator the same: $\dfrac{a}{c} + \dfrac{b}{c} = \dfrac{a + b}{c}$, so

$\dfrac{1}{4} + \dfrac{2}{4} = \dfrac{1 + 2}{4} = \dfrac{3}{4}$.

The commutativity and associativity of rational addition is an easy consequence of the laws of integer arithmetic. For a more rigorous and general discussion.

Real Numbers

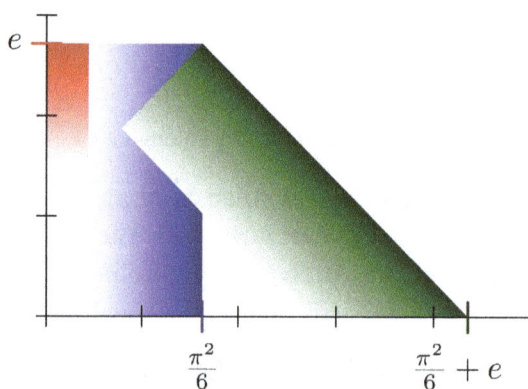

Adding $\pi^2/6$ and e using Dedekind cuts of rationals

A common construction of the set of real numbers is the Dedekind completion of the set of rational numbers. A real number is defined to be a Dedekind cut of rationals: a non-empty set of rationals that is closed downward and has no greatest element. The sum of real numbers a and b is defined element by element:

- Define $a + b = \{q + r \mid q \in a, r \in b\}$.

This definition was first published, in a slightly modified form, by Richard Dedekind in 1872. The commutativity and associativity of real addition are immediate; defining the real number 0 to be the set of negative rationals, it is easily seen to be the additive identity. Probably the trickiest part of this construction pertaining to addition is the definition of additive inverses.

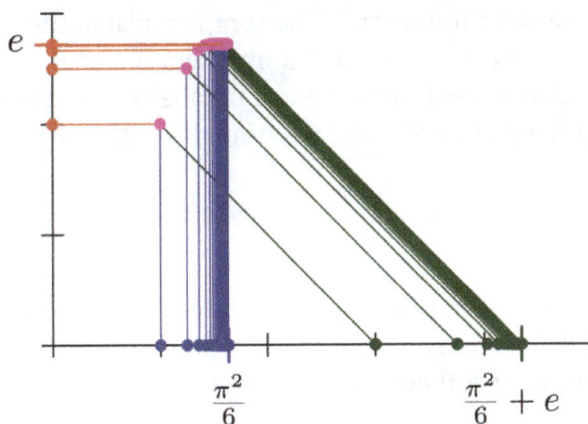

Adding $\pi^2/6$ and e using Cauchy sequences of rationals

Unfortunately, dealing with multiplication of Dedekind cuts is a time-consuming case-by-case process similar to the addition of signed integers. Another approach is the metric completion of the rational numbers. A real number is essentially defined to be the a limit of a Cauchy sequence of rationals, $\lim a_n$. Addition is defined term by term:

- Define $\lim_n a_n + \lim_n b_n = \lim_n (a_n + b_n)$.

This definition was first published by Georg Cantor, also in 1872, although his formalism was slightly different. One must prove that this operation is well-defined, dealing with co-Cauchy sequences. Once that task is done, all the properties of real addition follow immediately from the properties of rational numbers. Furthermore, the other arithmetic operations, including multiplication, have straightforward, analogous definitions.

Complex Numbers

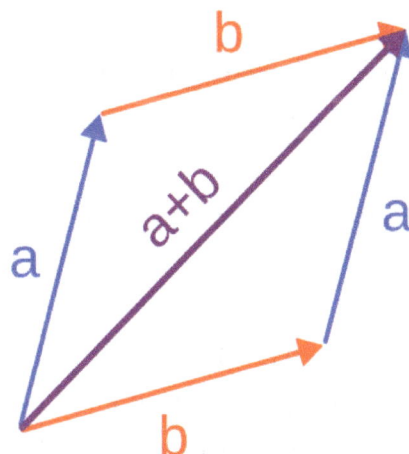

Addition of two complex numbers can be done geometrically by constructing a parallelogram.

Complex numbers are added by adding the real and imaginary parts of the summands. That is to say:

$$(a+bi)+(c+di)=(a+c)+(b+d)i.$$

Using the visualization of complex numbers in the complex plane, the addition has the following geometric interpretation: the sum of two complex numbers A and B, interpreted as points of the complex plane, is the point X obtained by building a parallelogram three of whose vertices are O, A and B. Equivalently, X is the point such that the triangles with vertices O, A, B, and X, B, A, are congruent.

Generalizations

There are many binary operations that can be viewed as generalizations of the addition operation on the real numbers. The field of abstract algebra is centrally concerned with such generalized operations, and they also appear in set theory and category theory.

Addition in Abstract Algebra

Vector Addition

In linear algebra, a vector space is an algebraic structure that allows for adding any two vectors and for scaling vectors. A familiar vector space is the set of all ordered pairs of real numbers; the ordered pair (a,b) is interpreted as a vector from the origin in the Euclidean plane to the point (a,b) in the plane. The sum of two vectors is obtained by adding their individual coordinates:

$$(a,b) + (c,d) = (a+c,b+d).$$

This addition operation is central to classical mechanics, in which vectors are interpreted as forces.

Matrix Addition

Matrix addition is defined for two matrices of the same dimensions. The sum of two $m \times n$ (pronounced "m by n") matrices A and B, denoted by A + B, is again an $m \times n$ matrix computed by adding corresponding elements:

$$\mathbf{A+B} = \begin{bmatrix} a_{11} & a_{12} & \cdots & a_{1n} \\ a_{21} & a_{22} & \cdots & a_{2n} \\ \vdots & \vdots & \ddots & \vdots \\ a_{m1} & a_{m2} & \cdots & a_{mn} \end{bmatrix} + \begin{bmatrix} b_{11} & b_{12} & \cdots & b_{1n} \\ b_{21} & b_{22} & \cdots & b_{2n} \\ \vdots & \vdots & \ddots & \vdots \\ b_{m1} & b_{m2} & \cdots & b_{mn} \end{bmatrix} = \begin{bmatrix} a_{11}+b_{11} & a_{12}+b_{12} & \cdots & a_{1n}+b_{1n} \\ a_{21}+b_{21} & a_{22}+b_{22} & \cdots & a_{2n}+b_{2n} \\ \vdots & \vdots & \ddots & \vdots \\ a_{m1}+b_{m1} & a_{m2}+b_{m2} & \cdots & a_{mn}+b_{mn} \end{bmatrix}$$

For example:

$$\begin{bmatrix} 1 & 3 \\ 1 & 0 \\ 1 & 2 \end{bmatrix} + \begin{bmatrix} 0 & 0 \\ 7 & 5 \\ 2 & 1 \end{bmatrix} = \begin{bmatrix} 1+0 & 3+0 \\ 1+7 & 0+5 \\ 1+2 & 2+1 \end{bmatrix} = \begin{bmatrix} 1 & 3 \\ 8 & 5 \\ 3 & 3 \end{bmatrix}$$

Modular Arithmetic

In modular arithmetic, the set of integers modulo 12 has twelve elements; it inherits an addition operation from the integers that is central to musical set theory. The set of integers modulo 2 has just two elements; the addition operation it inherits is known in Boolean logic as the "exclusive or" function. In geometry, the sum of two angle measures is often taken to be their sum as real numbers modulo 2π. This amounts to an addition operation on the circle, which in turn generalizes to addition operations on many-dimensional tori.

General Addition

The general theory of abstract algebra allows an "addition" operation to be any associative and commutative operation on a set. Basic algebraic structures with such an addition operation include commutative monoids and abelian groups.

Addition in Set Theory and Category Theory

A far-reaching generalization of addition of natural numbers is the addition of ordinal numbers and cardinal numbers in set theory. These give two different generalizations of addition of natural numbers to the transfinite. Unlike most addition operations, addition of ordinal numbers is not commutative. Addition of cardinal numbers, however, is a commutative operation closely related to the disjoint union operation.

In category theory, disjoint union is seen as a particular case of the coproduct operation, and general coproducts are perhaps the most abstract of all the generalizations of addition. Some coproducts, such as *Direct sum* and *Wedge sum*, are named to evoke their connection with addition.

Related Operations

Addition, along with subtraction, multiplication and division, is considered one of the basic operations and is used in elementary arithmetic.

Arithmetic

Subtraction can be thought of as a kind of addition—that is, the addition of an additive inverse. Subtraction is itself a sort of inverse to addition, in that adding x and subtracting x are inverse functions.

Given a set with an addition operation, one cannot always define a corresponding subtraction operation on that set; the set of natural numbers is a simple example. On the other hand, a subtraction operation uniquely determines an addition operation, an additive inverse operation, and an additive identity; for this reason, an additive group can be described as a set that is closed under subtraction.

Multiplication can be thought of as repeated addition. If a single term x appears in a sum n times, then the sum is the product of n and x. If n is not a natural number, the product may still make sense; for example, multiplication by -1 yields the additive inverse of a number.

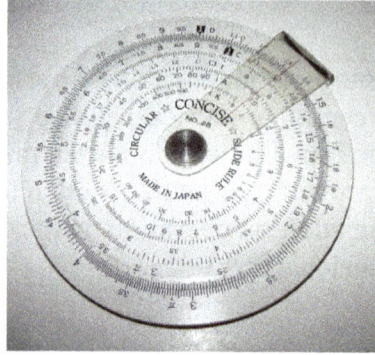

A circular slide rule

In the real and complex numbers, addition and multiplication can be interchanged by the exponential function:

$$e^{a+b} = e^a \, e^b.$$

This identity allows multiplication to be carried out by consulting a table of logarithms and computing addition by hand; it also enables multiplication on a slide rule. The formula is still a good first-order approximation in the broad context of Lie groups, where it relates multiplication of infinitesimal group elements with addition of vectors in the associated Lie algebra.

There are even more generalizations of multiplication than addition. In general, multiplication operations always distribute over addition; this requirement is formalized in the definition of a ring. In some contexts, such as the integers, distributivity over addition and the existence of a multiplicative identity is enough to uniquely determine the multiplication operation. The distributive property also provides information about addition; by expanding the product $(1 + 1)(a + b)$ in both ways, one concludes that addition is forced to be commutative. For this reason, ring addition is commutative in general.

Division is an arithmetic operation remotely related to addition. Since $a/b = a(b^{-1})$, division is right distributive over addition: $(a + b) / c = a / c + b / c$. However, division is not left distributive over addition; $1/ (2 + 2)$ is not the same as $1/2 + 1/2$.

Ordering

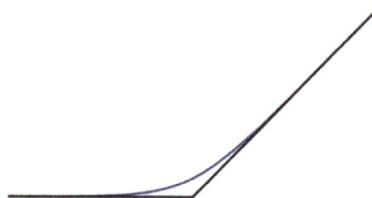

Log-log plot of $x + 1$ and max $(x, 1)$ from $x = 0.001$ to 1000

The maximum operation "max (a, b)" is a binary operation similar to addition. In fact, if two non-negative numbers a and b are of different orders of magnitude, then their sum is approximately equal to their maximum. This approximation is extremely useful in the applications of mathematics, for example in truncating Taylor series. However, it presents a perpetual difficulty in numerical analysis, essentially since "max" is not invertible. If b is much greater than a, then a straightforward calculation of $(a + b) - b$ can accumulate an unacceptable round-off error, perhaps even returning zero.

The approximation becomes exact in a kind of infinite limit; if either a or b is an infinite cardinal number, their cardinal sum is exactly equal to the greater of the two. Accordingly, there is no subtraction operation for infinite cardinals.

Maximization is commutative and associative, like addition. Furthermore, since addition preserves the ordering of real numbers, addition distributes over "max" in the same way that multiplication distributes over addition:

$$a + \max (b, c) = \max (a + b, a + c).$$

For these reasons, in tropical geometry one replaces multiplication with addition and addition with maximization. In this context, addition is called "tropical multiplication", maximization is called "tropical addition", and the tropical "additive identity" is negative infinity. Some authors prefer to replace addition with minimization; then the additive identity is positive infinity.

Tying these observations together, tropical addition is approximately related to regular addition through the logarithm:

$$\log (a + b) \approx \max (\log a, \log b),$$

which becomes more accurate as the base of the logarithm increases. The approximation can be made exact by extracting a constant h, named by analogy with Planck's constant from quantum mechanics, and taking the "classical limit" as h tends to zero:

$$\max(a,b) = \lim_{h \to 0} h \log(e^{a/h} + e^{b/h}).$$

In this sense, the maximum operation is a *dequantized* version of addition.

Other Ways to Add

Incrementation, also known as the successor operation, is the addition of 1 to a number.

Summation describes the addition of arbitrarily many numbers, usually more than just two. It includes the idea of the sum of a single number, which is itself, and the empty sum, which is zero. An infinite summation is a delicate procedure known as a series.

Counting a finite set is equivalent to summing 1 over the set.

Integration is a kind of "summation" over a continuum, or more precisely and generally, over a differentiable manifold. Integration over a zero-dimensional manifold reduces to summation.

Linear combinations combine multiplication and summation; they are sums in which each term has a multiplier, usually a real or complex number. Linear combinations are especially useful in contexts where straightforward addition would violate some normalization rule, such as mixing of strategies in game theory or superposition of states in quantum mechanics.

Convolution is used to add two independent random variables defined by distribution functions.

Its usual definition combines integration, subtraction, and multiplication. In general, convolution is useful as a kind of domain-side addition; by contrast, vector addition is a kind of range-side addition.

Subtraction

Subtraction is a mathematical operation that represents the operation of removing objects from a collection. It is signified by the minus sign (−). For example, in the picture on the right, there are 5 − 2 apples—meaning 5 apples with 2 taken away, which is a total of 3 apples. Therefore, 5 − 2 = 3. Besides counting fruits, subtraction can also represent combining other physical and abstract quantities using different kinds of objects including negative numbers, fractions, irrational numbers, vectors, decimals, functions, and matrices.

"5 − 2 = 3" (verbally, "five minus two equals three")

$$^{6}7_{1}28$$
$$- 51$$
$$\overline{677}$$

An example problem

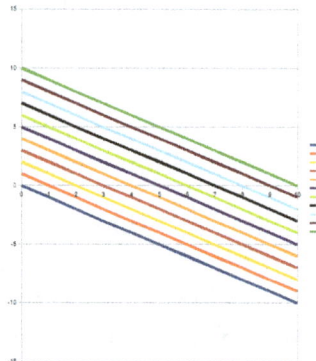

Subtraction of numbers 0–10. Line labels = minuend. X axis = subtrahend. Y axis = difference.

Placard outside shop in Bordeaux advertising subtraction of 20% from the price of a second perfume

Subtraction follows several important patterns. It is anticommutative, meaning that changing the order changes the sign of the answer. It is not associative, meaning that when one subtracts more than two numbers, the order in which subtraction is performed matters. Subtraction of 0 does not change a number. Subtraction also obeys predictable rules concerning related operations such as addition and multiplication. All of these rules can be proven, starting with the subtraction of integers and generalizing up through the real numbers and beyond. General binary operations that continue these patterns are studied in abstract algebra.

Performing subtraction is one of the simplest numerical tasks. Subtraction of very small numbers is accessible to young children. In primary education, students are taught to subtract numbers in the decimal system, starting with single digits and progressively tackling more difficult problems.

Notation and Terminology

Subtraction is written using the minus sign "−" between the terms; that is, in infix notation. The result is expressed with an equals sign. For example,

$2 - 1 = 1$ (verbally, "two minus one equals one")

$4 - 2 = 2$ (verbally, "four minus two equals two")

$6 - 3 = 3$ (verbally, "six minus three equals three")

$4 - 6 = -2$ (verbally, "four minus six equals negative two")

There are also situations where subtraction is "understood" even though no symbol appears:

- A column of two numbers, with the lower number in red, usually indicates that the lower number in the column is to be subtracted, with the difference written below, under a line. This is most common in accounting.

Formally, the number being subtracted is known as the *subtrahend*, while the number it is subtracted from is the *minuend*. The result is the *difference*.

All of this terminology derives from Latin. "Subtraction" is an English word derived from the Latin verb *subtrahere*, which is in turn a compound of *sub* "from under" and *trahere* "to pull"; thus to *subtract* is to *draw from below, take away.* Using the gerundive suffix *-nd* results in "subtra-

hend", "thing to be subtracted". Likewise from *minuere* "to reduce or diminish", one gets "minuend", "thing to be diminished".

Of Integers and Real Numbers

Integers

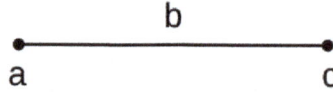

Imagine a line segment of length b with the left end labeled a and the right end labeled c. Starting
from a, it takes b steps to the right to reach c. This movement to the right is modeled mathematically by addition:

$$a + b = c.$$

From c, it takes b steps to the *left* to get back to a. This movement to the left is modeled by subtraction:

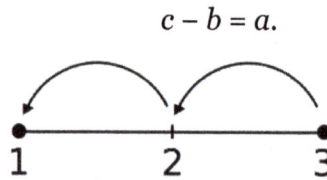

$$c - b = a.$$

Now, a line segment labeled with the numbers 1, 2, and 3. From position 3, it takes no steps to the
left to stay at 3, so $3 - 0 = 3$. It takes 2 steps to the left to get to position 1, so $3 - 2 = 1$. This picture
is inadequate to describe what would happen after going 3 steps to the left of position 3. To represent such an operation, the line must be extended.

To subtract arbitrary natural numbers, one begins with a line containing every natural number (0,
1, 2, 3, 4, 5, 6, ...). From 3, it takes 3 steps to the left to get to 0, so $3 - 3 = 0$. But $3 - 4$ is still invalid
since it again leaves the line. The natural numbers are not a useful context for subtraction.

The solution is to consider the integer number line (..., −3, −2, −1, 0, 1, 2, 3, ...). From 3, it takes 4
steps to the left to get to −1:

$$3 - 4 = -1.$$

Natural Numbers

Subtraction of natural numbers is not closed. The difference is not a natural number unless the
minuend is greater than or equal to the subtrahend. For example, 26 cannot be subtracted from 11
to give a natural number. Such a case uses one of two approaches:

1. Say that 26 cannot be subtracted from 11; subtraction becomes a partial function.

2. Give the answer as an integer representing a negative number, so the result of subtracting
    26 from 11 is −15.

Real Numbers

Subtraction of real numbers is defined as addition of signed numbers. Specifically, a number is subtracted by adding its additive inverse. Then we have $3 - \pi = 3 + (-\pi)$. This helps to keep the ring of real numbers "simple" by avoiding the introduction of "new" operators such as subtraction. Ordinarily a ring only has two operations defined on it; in the case of the integers, these are addition and multiplication. A ring already has the concept of additive inverses, but it does not have any notion of a separate subtraction operation, so the use of signed addition as subtraction allows us to apply the ring axioms to subtraction without needing to prove anything.

Properties

Anticommutativity

Subtraction is anti-commutative, meaning that if one reverses the terms in a difference left-to-right, the result is the negative of the original result. Symbolically, if a and b are any two numbers, then

$$a - b = -(b - a).$$

Non-associativity

Subtraction is non-associative, which comes up when one tries to define repeated subtraction. Should the expression

$$\text{``}a - b - c\text{''}$$

be defined to mean $(a - b) - c$ or $a - (b - c)$? These two possibilities give different answers. To resolve this issue, one must establish an order of operations, with different orders giving different results.

Predecessor

In the context of integers, subtraction of one also plays a special role: for any integer a, the integer $(a - 1)$ is the largest integer less than a, also known as the predecessor of a.

Units of Measurement

When subtracting two numbers with units of measurement such as kilograms or pounds, they must have the same unit. In most cases the difference will have the same unit as the original numbers.

Percentages

Changes in percentages can be reported in at least two forms, percentage change and percentage point change. Percentage change represents the relative change between the two quantities as a percentage, while percentage point change is simply the number obtained by subtracting the two percentages.

As an example, suppose that 30% of widgets made in a factory are defective. Six months later, 20% of widgets are defective. The percentage change is -33 1/3%, while the percentage point change is -10 percentage points.

In Computing

The method of complements is a technique used to subtract one number from another using only addition of positive numbers. This method was commonly used in mechanical calculators and is still used in modern computers.

Binary digit	Ones' complement
0	1
1	0

To subtract a binary number y (the subtrahend) from another number x (the minuend), the ones' complement of y is added to x and one is added to the sum. The leading digit '1' of the result is then discarded.

The method of complements is especially useful in binary (radix 2) since the ones' complement is very easily obtained by inverting each bit (changing '0' to '1' and vice versa). And adding 1 to get the two's complement can be done by simulating a carry into the least significant bit. For example:

```
  01100100  (x, equals decimal 100)

- 00010110  (y, equals decimal 22)
```

becomes the sum:

```
  01100100  (x)

+ 11101001  (ones' complement of y)

+_____1  (to get the two's complement)

 101001110
```

Dropping the initial "1" gives the answer: 01001110 (equals decimal 78)

The Teaching of Subtraction in Schools

Methods used to teach subtraction to elementary school vary from country to country, and within a country, different methods are in fashion at different times. In what is, in the United States, called traditional mathematics, a specific process is taught to students at the end of the 1st year or during the 2nd year for use with multi-digit whole numbers, and is extended in either the fourth or fifth grade to include decimal representations of fractional numbers.

In America

Almost all American schools currently teach a method of subtraction using borrowing or regrouping (the decomposition algorithm) and a system of markings called crutches. Although a method of borrowing had been known and published in textbooks previously, the use of crutches in American schools spread after William A. Brownell published a study claiming that crutches were beneficial to students using this method. This system caught on rapidly, displacing the other methods of subtraction in use in America at that time.

In Europe

Some European schools employ a method of subtraction called the Austrian method, also known as the additions method. There is no borrowing in this method. There are also crutches (markings to aid memory), which vary by country.

Comparing The Two Main Methods

Both these methods break up the subtraction as a process of one digit subtractions by place value. Starting with a least significant digit, a subtraction of subtrahend:

$$s_j \, s_{j-1} \, \dots \, s_1$$

from minuend

$$m_k \, m_{k-1} \, \dots \, m_1,$$

where each s_i and m_i is a digit, proceeds by writing down $m_1 - s_1$, $m_2 - s_2$, and so forth, as long as s_i does not exceed m_i. Otherwise, m_i is increased by 10 and some other digit is modified to correct for this increase. The American method corrects by attempting to decrease the minuend digit m_{i+1} by one (or continuing the borrow leftwards until there is a non-zero digit from which to borrow). The European method corrects by increasing the subtrahend digit s_{i+1} by one.

The Austrian method does not reduce the 7 to 6. Rather it increases the subtrahend hundred's digit by one. A small mark is made near or below this digit (depending on the school). Then the subtraction proceeds by asking what number when increased by 1, and 5 is added to it, makes 7. The answer is 1, and is written down in the result's hundred's place.

There is an additional subtlety in that the student always employs a mental subtraction table in the American method. The Austrian method often encourages the student to mentally use the addition table in reverse. In the example above, rather than adding 1 to 5, getting 6, and subtracting that from 7, the student is asked to consider what number, when increased by 1, and 5 is added to it, makes 7.

Subtraction By Hand

Austrian Method

Example:

$$
\begin{array}{r}
753 \\
-491 \\
\hline
\end{array}
\qquad
\begin{array}{r}
753 \\
-491 \\
\hline
2
\end{array}
$$

1 + ... = 3   The difference is written under the line.

$$\begin{array}{r} 7\textcolor{orange}{53} \\ -4\textcolor{orange}{91} \\ \hline 2 \end{array}$$

$$\begin{array}{r} 7\textcolor{orange}{53} \\ -4\textcolor{orange}{91} \\ {\scriptstyle 1} \\ \hline 2 \end{array}$$

$9 + \ldots = 5$
The required sum (5) is too small!

So, we add 10 to it and put a 1 under the next higher place in the subtrahend.

$$\begin{array}{r} 7\textcolor{orange}{5}3 \\ -4\textcolor{orange}{9}1 \\ {\scriptstyle 1} \\ \hline 62 \end{array}$$

$$\begin{array}{r} \textcolor{orange}{7}53 \\ -\textcolor{orange}{4}91 \\ {\scriptstyle 1} \\ \hline 62 \end{array}$$

$9 + \ldots = 15$
Now we can find the difference like before.

$(4 + 1) + \ldots = 7$

$$\begin{array}{r} \textcolor{orange}{7}53 \\ -\textcolor{orange}{4}91 \\ {\scriptstyle 1} \\ \hline 262 \end{array}$$

$$\begin{array}{r} 753 \\ -491 \\ {\scriptstyle 1} \\ \hline 262 \end{array}$$

The difference is written under the line.

The total difference.

Subtraction from Left to Right

Example:

$$\begin{array}{r} \textcolor{orange}{7}53 \\ -\textcolor{orange}{4}91 \\ \hline 3 \end{array}$$

$$\begin{array}{r} 7\textcolor{orange}{53} \\ -4\textcolor{orange}{91} \\ \hline 2 \end{array}$$

$7 - 4 = 3$
This result is only penciled in.

Because the next digit of the minuend is smaller than the subtrahend, we subtract one from our penciled-in-number and mentally add ten to the next.

$$753$$
$$-491$$
$$26$$

$15 - 9 = 6$

$$753$$
$$-491$$
$$26$$

Because the next digit in the minuend
is not smaller than the subtrahend,
We keep this number.

$$753$$
$$-491$$
$$262$$

$3 - 1 = 2$

American Method

In this method, each digit of the subtrahend is subtracted from the digit above it starting from right to left. If the top number is too small to subtract the bottom number from it, we add 10 to it; this 10 is 'borrowed' from the top digit to the left, which we subtract 1 from. Then we move on to subtracting the next digit and borrowing as needed, until every digit has been subtracted. Example:

$$753$$
$$-491$$

$3 - 1 = ...$

$$753$$
$$-491$$
$$2$$

We write the difference under the line.

$$753$$
$$-491$$
$$2$$

$5 - 9 = ...$
The minuend (5) is too small!

$$\overset{6\ 15}{7\!\!\!/5\,3}$$
$$-491$$
$$2$$

So, we add 10 to it. The 10 is
'borrowed' from the digit on the left,
which goes down by 1.

$$
\begin{array}{r}
{\scriptstyle 6\ 15} \\
7\cancel{5}3 \\
-49\textcolor{orange}{1} \\
\hline
62
\end{array}
$$

$15 - 9 = \ldots$
Now the subtraction works, and we write the difference under the line.

$$
\begin{array}{r}
{\scriptstyle 6\ 15} \\
7\cancel{5}3 \\
-49\textcolor{orange}{1} \\
\hline
62
\end{array}
$$

We write the difference under the line.

$$
\begin{array}{r}
{\scriptstyle 6\ 15} \\
7\cancel{5}3 \\
-\textcolor{orange}{4}91 \\
\hline
62
\end{array}
$$

$6 - 4 = \ldots$

$$
\begin{array}{r}
{\scriptstyle 6\ 15} \\
7\cancel{5}3 \\
-491 \\
\hline
262
\end{array}
$$

The Total Difference.

Trade First

A variant of the American method where all borrowing is done before all subtraction.

Example:

$$
\begin{array}{r}
{\scriptstyle 4\ 11} \\
75\cancel{1} \\
-49\textcolor{orange}{3} \\
\hline
\end{array}
$$

$1 - 3 =$ not possible.
We add a 10 to the 1. Because the 10 is 'borrowed' from the nearby 5, the 5 is lowered by 1.

$$
\begin{array}{r}
{\scriptstyle 6\ 14} \\
{\scriptstyle \cancel{4}\ 11} \\
7\cancel{5}\cancel{1} \\
-4\textcolor{orange}{9}3 \\
\hline
\end{array}
$$

$4 - 9 =$ not possible.
So we proceed as in step 1.

$$
\begin{array}{r}
{\scriptstyle 6\ 14} \\
{\scriptstyle \cancel{4}\ 11} \\
7\cancel{5}\cancel{1} \\
-49\textcolor{orange}{3} \\
\hline
8
\end{array}
$$

Working from right to left:
$11 - 3 = 8$

$$
\begin{array}{r}
{\scriptstyle 6\ 14} \\
{\scriptstyle \cancel{4}\ 11} \\
7\cancel{5}\cancel{1} \\
-4\textcolor{orange}{9}3 \\
\hline
58
\end{array}
$$

$14 - 9 = 5$

$$
\begin{array}{r}
{\scriptstyle 6\ 14} \\
{\scriptstyle \cancel{4}\ 11} \\
7\cancel{5}\cancel{1} \\
-\textcolor{orange}{4}93 \\
\hline
258
\end{array}
$$

$6 - 4 = 2$

Partial Differences

The partial differences method is different from other vertical subtraction methods because no borrowing or carrying takes place. In their place, one places plus or minus signs depending on whether the minuend is greater or smaller than the subtrahend. The sum of the partial differences is the total difference.

Example:

```
  753
 -491
 + 3 0 0
```
•

The smaller number is subtracted from the greater:
$700 - 400 = 300$
Because the minuend is greater than the subtrahend, this difference has a plus sign.

```
  753
 -491
 + 3 0 0
 -   4 0
```
•

The smaller number is subtracted from the greater:
$90 - 50 = 40$
Because the minuend is smaller than the subtrahend, this difference has a minus sign.

```
  753
 -491
 + 3 0 0
 -   4 0
 +     2
```
•

The smaller number is subtracted from the greater:
$3 - 1 = 2$
Because the minuend is greater than the subtrahend, this difference has a plus sign.

```
  753
 -491
 + 3 0 0
 -   4 0
 +     2
   2 6 2
```
•

$+ 300 - 40 + 2 = 262$

Nonvertical Methods

Counting Up

Instead of finding the difference digit by digit, one can count up the numbers between the subtrahend and the minuend.

Example:

1234 – 567 = can be found by the following steps:

- 567 + 3 = 570
- 570 + 30 = 600
- 600 + 400 = 1000
- 1000 + 234 = 1234

Add up the value from each step to get the total difference: 3 + 30 + 400 + 234 = 667.

Breaking Up The Subtraction

Another method that is useful for mental arithmetic is to split up the subtraction into small steps.

Example:

1234 – 567 = can be solved in the following way:

- 1234 – 500 = 734
- 734 – 60 = 674
- 674 – 7 = 667

Same Change

The same change method uses the fact that adding or subtracting the same number from the minuend and subtrahend does not change the answer. One adds the amount needed to get zeros in the subtrahend.

Example:

"1234 – 567 =" can be solved as follows:

- 1234 – 567 = 1237 – 570 = 1267 – 600 = 667

Method of Complements

In mathematics and computing, the method of complements is a technique used to subtract one number from another using only addition of positive numbers. This method was commonly used in mechanical calculators and is still used in modern computers.

Complement numbers on an adding machine c. 1910. The smaller numbers, for use when subtracting, are the nines' complement of the larger numbers, which are used when adding.

The *nines' complement* of a number is formed by replacing each digit with nine minus that digit. To subtract a decimal number y (the subtrahend) from another number x (the minuend) two methods may be used:

In the first method the nines' complement of x is added to y. Then the nines' complement of the result obtained is formed to produce the desired result.

In the second method the nines' complement of y is added to x and one is added to the sum. The leading digit '1' of the result is then discarded. Discarding the initial '1' is especially convenient on calculators or computers that use a fixed number of digits: there is nowhere for it to go so it is simply lost during the calculation. The nines' complement plus one is known as the *tens' complement*.

The method of complements can be extended to other number bases (radices); in particular, it is used on most digital computers to perform subtraction, represent negative numbers in base 2 or binary arithmetic and test underflow and overflow in calculation.

Numeric Complements

The radix complement of an n digit number y in radix b is, by definition, $b^n - y$. The radix complement is most easily obtained by adding 1 to the diminished radix complement, which is $(b^n - 1) - y$. Since $(b^n - 1)$ is the digit $b - 1$ repeated n times (because $b^n - 1 = b^n - 1^n = (b - 1)(b^{n-1} + b^{n-2} + ... + b + 1) = (b - 1)b^{n-1} + ... + (b - 1)$. The diminished radix complement of a number is found by complementing each digit with respect to $b - 1$ (that is, subtracting each digit in y from $b - 1$).

The subtraction of y from x may be performed as follows. Adding the diminished radix complement of x to y results in the value $b^n - 1 - x + y$ or $b^n - 1 - (x - y)$ which is the diminished radix complement of $x - y$, except for possible padding digits $b - 1$.. The diminished radix complement of this is the value $x - y$. Alternatively, adding the radix complement of y to x results in the value $x + b^n - y$ or $x - y + b^n$. Assuming $y \le x$, the result will always be greater or equal to b^n and dropping the initial '1' is the same as subtracting b^n, making the result $x - y + b^n - b^n$ or just $x - y$, the desired result.

In the decimal numbering system, the radix complement is called the *ten's complement* and the diminished radix complement the *nines' complement*. In binary, the radix complement is called the *two's complement* and the diminished radix complement the *ones' complement*. The naming of complements in other bases is similar. Some people, notably Donald Knuth, recommend using the placement of the apostrophe to distinguish between the radix complement and the diminished radix complement. In this usage, the *four's complement* refers to the radix complement of a number in base four while *fours' complement* is the diminished radix complement of a number in base 5. However, the distinction is not important when the radix is apparent (nearly always), and the subtle difference in apostrophe placement is not common practice. Most writers use *one's* and *nine's complement*, and many style manuals leave out the apostrophe, recommending *ones* and *nines complement*.

Decimal Example

Digit	Nines' complement
0	9
1	8
2	7
3	6
4	5
5	4
6	3
7	2
8	1
9	0

The nines' complement of a decimal digit is the number that must be added to it to produce 9; the complement of 3 is 6, the complement of 7 is 2, and so on, see table. To form the nines' complement of a larger number, each digit is replaced by its nines' complement.

Consider the following subtraction problem:

```
  873   (x, the minuend)
- 218   (y, the subtrahend)
```

First Method

We compute the nines' complement of the minuend, 873. Add that to the subtrahend 218, then calculate the nines' complement of the result.

```
  126   (nines' complement of x)
+ 218   (y, the subtrahend)
=
  344
```

Now calculate the nines' complement of the result

```
344 (result)
655 (nine's complement of result, the correct answer)
```

Second Method

We compute the nines' complement of 218, which is 781. Because 218 is three digits long, this is the same as subtracting 218 from 999.

Next, the sum of x and the nines' complement of y is taken:

```
   873   (x)
+  781   (nines' complement of y = 999-y)
=
  1654   (999 + x-y)
```

The leading "1" digit is then dropped, giving 654.

```
  1654
 -1000   -(999 + 1)
=
   654   (x-y-1)
```

This is not yet correct. We have essentially added 999 to the equation in the first step. Then we removed 1000 when we dropped the leading 1 in the result 1654 above. This will thus make the answer we get (654) one less than the correct answer $(x - y)$. To fix this, we must add 1 to our answer:

```
   654
    +1
=
   655   (x-y)
```

Adding a 1 gives 655, the correct answer to our original subtraction problem.

Magnitude of Numbers

In the following example the result of the subtraction has fewer digits than x:

```
   123410 (x, the minuend)
 - 123401 (y, the subtrahend)
```

Using the first method the sum of the nines' complement of x and y is

```
   876589 (nines' complement of x)
+  123401 (y)
=
   999990
```

The nines' complement of 999990 is 000009. Removing the leading zeros gives 9 the desired result.

If the subtrahend, y, has fewer digits than the minuend, x, leading zeros must be added in the second method. These zeros become leading nines when the complement is taken. For example:

```
  48032  (x)
-   391  (y)
```

can be rewritten

```
  48032  (x)
- 00391  (y with leading zeros)
```

Replacing 00391 with its nines' complement and adding 1 produces the sum:

```
  48032  (x)
+ 99608  (nines' complement of y)
+     1
=
 147641
```

Dropping the leading "1" gives the correct answer: 47641.

Binary Example

Binary digit	Ones'complement
0	1
1	0

The method of complements is especially useful in binary (radix 2) since the ones' complement is very easily obtained by inverting each bit (changing '0' to '1' and vice versa). And adding 1 to get the two's complement can be done by simulating a carry into the least significant bit. For example:

```
  01100100  (x, equals decimal 100)
- 00010110  (y, equals decimal 22)
```

becomes the sum:

```
  01100100  (x)
+ 11101001  (ones' complement of y)
+        1  (to get the two's complement)
=
 101001110
```

Dropping the initial "1" gives the answer: 01001110 (equals decimal 78)

Negative Number Representations

The method of complements normally assumes that the operands are positive and that $y \leq x$, logical constraints given that adding and subtracting arbitrary integers is normally done by comparing signs, adding the two or subtracting the smaller from the larger, and giving the result the correct sign.

Let's see what happens if $x < y$. In that case, there will not be a "1" digit to cross out after the addition since $x - y + b^n$ will be less than b^n. For example, (in decimal):

```
  185   (x)
- 329   (y)
```

Complementing y and adding gives:

```
  185   (x)
+ 670   (nines' complement of y)
+   1

  856
```

This is obviously the wrong answer; the expected answer is -144. But it isn't as far off as it seems; 856 happens to be the ten's complement of 144. This issue can be addressed in three ways:

- Ignore the issue. This is reasonable if a person is operating a calculating device that doesn't support negative numbers since comparing the two operands before the calculation so they can be entered in the proper order, and verifying that the result is reasonable, is easy for humans to do.

- Represent negative numbers as radix complements of their positive counterparts. Numbers less than $b^n / 2$ are considered positive; the rest are considered negative (and their magnitude can be obtained by taking the radix complement). This works best for even radices since the sign can be determined by looking at the first digit. For example, numbers in ten's complement notation are positive if the first digit is 0, 1, 2, 3, or 4, and negative if 5, 6, 7, 8, or 9. And it works very well in binary since the first bit can be considered a sign bit: the number is positive if the sign bit is 0 and negative if it is 1. Indeed, two's complement is used in most modern computers to represent signed numbers.

- Complement the result if there is no carry out of the most significant digit (an indication that x was less than y). This is easier to implement with digital circuits than comparing and swapping the operands. But since taking the radix complement requires adding 1, it is difficult to do directly. Fortunately, a trick can be used to get around this addition: Instead of always setting a carry into the least significant digit when subtracting, the carry out of the most significant digit is used as the carry input into the least significant digit (an operation called an *end-around carry*). So if $y \leq x$, the carry from the most significant digit that would normally be ignored is added, producing the correct result. And if not, the 1 is not added and the result is one less than the radix complement of the answer, or the diminished radix complement, which does not require an addition to obtain. This method is used by computers that use sign-and-magnitude to represent signed numbers.....

Practical Uses

The method of complements was used in many mechanical calculators as an alternative to running the gears backwards. For example:

- Pascal's calculator had two sets of result digits, a black set displaying the normal result and a red set displaying the nines' complement of this. A horizontal slat was used to cover up one of these sets, exposing the other. To subtract, the red digits were exposed and set to 0. Then the nines' complement of the minuend was entered. On some machine this could be done by dialing in the minuhend using inner wheels of complements (i.e. without having to mentally determine the nines' complement of the minuhend). In displaying that data in the complement window (red set), the operator could see the nines' complement of the nines' complement of the minuhend, that is the minuhend. The slat was then moved to expose the black digits (which now displayed the nines' complement of the minuhend) and the subtrahend was added by dialing it in. Finally, the operator had to move the slat again to read the correct answer.

• The Comptometer had nines' complement digits printed in smaller type along with the normal digits on each key. To subtract, the operator was expected to mentally subtract 1 from the subtrahend and enter the result using the smaller digits. Since subtracting 1 before complementing is equivalent to adding 1 afterwards, the operator would thus effectively add the ten's complement of the subtrahend. The operator also needed to hold down the "subtraction cutoff tab" corresponding to the leftmost digit of the answer. This tab prevented the carry from being propagated past it, the Comptometer's method of dropping the initial 1 from the result.

• The Curta calculator used the method of complements for subtraction, and managed to hide this from the user. Numbers were entered using digit input slides along the side of the device. The number on each slide was added to a result counter by a gearing mechanism which engaged cams on a rotating "echelon drum" (a.k.a. "step drum"). The drum was turned by use of a crank on the top of the instrument. The number of cams encountered by each digit as the crank turned was determined by the value of that digit. For example, if a slide is set to its "6" position, a row of 6 cams would be encountered around the drum corresponding to that position. For subtraction, the drum was shifted slightly before it was turned, which moved a different row of cams into position. This alternate row contained the nines' complement of the digits. Thus, the row of 6 cams that had been in position for addition now had a row with 3 cams. The shifted drum also engaged one extra cam which added 1 to the result (as required for the method of complements). The always present tens' complement "overflow 1" which carried out beyond the most significant digit of the results register was, in effect, discarded.

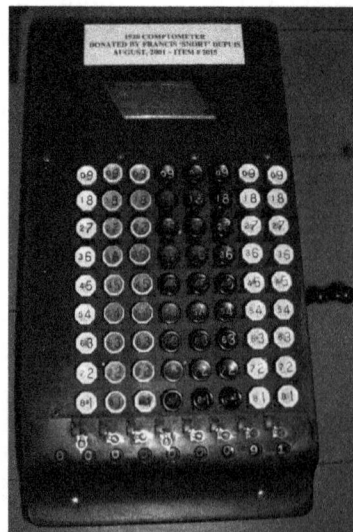

Comptometer from the 1920s, with nines' complements marked on each key

In Computers

Use of the method of complements is ubiquitous in digital computers, regardless of the representation used for signed numbers. However, the circuitry required depends on the representation:

- If two's complement representation is used, subtraction requires only inverting the bits of the subtrahend and setting a carry into the rightmost bit.

- Using ones' complement representation requires inverting the bits of the subtrahend and connecting the carry out of the most significant bit to the carry in of the least significant bit (end-around carry).

- Using sign-magnitude representation requires only complementing the sign bit of the subtrahend and adding, but the addition/subtraction logic needs to compare the sign bits, complement one of the inputs if they are different, implement an end-around carry, and complement the result if there was no carry from the most significant bit.

Manual Uses

The method of complements was used to correct errors when accounting books were written by hand. To remove an entry from a column of numbers, the accountant could add a new entry with the ten's complement of the number to subtract. A bar was added over the digits of this entry to denote its special status. It was then possible to add the whole column of figures to obtain the corrected result.

Complementing the sum is handy for cashiers making change for a purchase from currency in a single denomination of 1 raised to an integer power of the currency's base. For decimal currencies that would be 10, 100, 1,000, etc., e.g. a $10.00 bill.

In Grade School Education

In grade schools, students are sometimes taught the method of complements as a shortcut useful in mental arithmetic. Subtraction is done by adding the ten's complement of the subtrahend, which is the nines' complement plus 1. The result of this addition used when it is clear that the difference will be positive, otherwise the ten's complement of the addition's result is used with it marked as negative. The same technique works for subtracting on an adding machine.

Multiplication

Four bags with three marbles per bag gives twelve marbles ($4 \times 3 = 12$).

$$3 \times 2 = 6$$

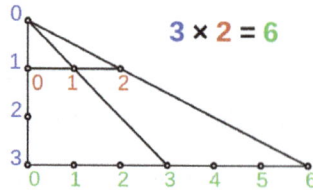

Multiplication can also be thought of as scaling. Here we see 2 being multiplied by 3 using scaling, giving 6 as a result

Animation for the multiplication $2 \times 3 = 6$.

$4 \times 5 = 20$, the rectangle is composed of 20 squares, having dimensions of 4 by 5.

Area of a cloth 4.5m × 2.5m = 11.25m²; $4\frac{1}{2} \times 2\frac{1}{2} = 11\frac{1}{4}$

Multiplication (often denoted by the cross symbol "×", by a point "·", by juxtaposition, or, on computers, by an asterisk "*") is one of the four elementary, mathematical operations of arithmetic; with the others being addition, subtraction and division.

The multiplication of whole numbers may be thought as a repeated addition; that is, the multiplication of two numbers is equivalent to adding as many copies of one of them, the *multiplicand*, as the value of the other one, the *multiplier*. Normally, the multiplier is written first and multiplicand second, though this can vary, as the distinction is not very meaningful:

$$a \times b = \underbrace{b + \cdots + b}_{a} = \underbrace{a + \cdots + a}_{b}$$

For example, 4 multiplied by 3 (often written as 3×4 and said as "3 times 4") can be calculated by adding 3 copies of 4 together:

$$3 \times 4 = 4 + 4 + 4 = 12$$

Here 3 and 4 are the "factors" and 12 is the "product".

One of the main properties of multiplication is the commutative property, adding 3 copies of 4 gives the same result as adding 4 copies of 3:

$$4 \times 3 = 3 + 3 + 3 + 3 = 12$$

The multiplication of integers (including negative numbers), rational numbers (fractions) and real numbers is defined by a systematic generalization of this basic definition.

Multiplication can also be visualized as counting objects arranged in a rectangle (for whole numbers) or as finding the area of a rectangle whose sides have given lengths. The area of a rectangle does not depend on which side is measured first, which illustrates the commutative property. The product of two measurements is a new type of measurement, for instance multiplying the lengths of the two sides of a rectangle gives its area, this is the subject of dimensional analysis.

The inverse operation of multiplication is division. For example, since 4 multiplied by 3 equals 12, then 12 divided by 3 equals 4. Multiplication by 3, followed by division by 3, yields the original number (since the division of a number other than 0 by itself equals 1).

Multiplication is also defined for other types of numbers, such as complex numbers, and more abstract constructs, like matrices. For these more abstract constructs, the order that the operands are multiplied sometimes does matter. A listing of the many different kinds of products that are used in mathematics is given in the product (mathematics) page.

Notation and Terminology

The multiplication sign ×

In arithmetic, multiplication is often written using the sign "×" between the terms; that is, in infix notation. For example,

$$2 \times 3 = 6 \text{ (verbally, "two times three equals six")}$$

$$2 \times 3 \times 5 = 6 \times 5 = 30$$

$$2 \times 3 \times 5 = 6 \times 5 = 30$$

$$2 \times 2 \times 2 \times 2 \times 2 = 32$$

The sign is encoded in Unicode at U+00D7 × MULTIPLICATION SIGN (HTML × · ×).

There are other mathematical notations for multiplication:

- Multiplication is also denoted by dot signs, usually a middle-position dot (rarely period):

$$5 \cdot 2 \quad \text{or} \quad 5.2$$

The middle dot notation, encoded in Unicode as U+22C5. DOT OPERATOR, is standard in the United States, the United Kingdom, and other countries where the period is used as a decimal point. When the dot operator character is not accessible, the interpunct ($\cdot$) is used. In other countries that use a comma as a decimal mark, either the period or a middle dot is used for multiplication.

- In algebra, multiplication involving variables is often written as a juxtaposition (e.g., xy for x times y or $5x$ for five times x). The notation can also be used for quantities that are surrounded by parentheses (e.g., 5(2) or (5)(2) for five times two). This implicit usage of multiplication can cause ambiguity when the concatenated variables happen to match the name of another variable, when a variable name in front of a parenthesis can be confused with a function name, or in the correct determination of the order of operations.

- In matrix multiplication, there is a distinction between the cross and the dot symbols. The cross symbol generally denotes the taking a cross product of two vectors, yielding a vector as the result, while the dot denotes taking the dot product of two vectors, resulting in a scalar.

In computer programming, the asterisk (as in 5*2) is still the most common notation. This is due to the fact that most computers historically were limited to small character sets (such as ASCII and EBCDIC) that lacked a multiplication sign (such as • or ×), while the asterisk appeared on every keyboard. This usage originated in the FORTRAN programming language.

The numbers to be multiplied are generally called the "factors". The number to be multiplied is called the "multiplicand", while the number of times the multiplicand is to be multiplied comes from the "multiplier". Usually the multiplier is placed first and the multiplicand is placed second, however sometimes the first factor is the multiplicand and the second the multiplier. Additionally, there are some sources in which the term "multiplicand" is regarded as a synonym for "factor". In algebra, a number that is the multiplier of a variable or expression (e.g., the 3 in $3xy^2$) is called a coefficient.

The result of a multiplication is called a product. A product of integers is a multiple of each factor. For example, 15 is the product of 3 and 5, and is both a multiple of 3 and a multiple of 5.

Computation

The common methods for multiplying numbers using pencil and paper require a multiplication table of memorized or consulted products of small numbers (typically any two numbers from 0 to 9), however one method, the peasant multiplication algorithm, does not.

Multiplying numbers to more than a couple of decimal places by hand is tedious and error prone. Common logarithms were invented to simplify such calculations. The slide rule allowed numbers to be quickly multiplied to about three places of accuracy. Beginning in the early twentieth century, mechanical calculators, such as the Marchant, automated multiplication of up to 10 digit numbers. Modern electronic computers and calculators have greatly reduced the need for multiplication by hand.

Historical Algorithms

Methods of multiplication were documented in the Egyptian, Greek, Indian and Chinese civilizations.

The Ishango bone, dated to about 18,000 to 20,000 BC, hints at a knowledge of multiplication in the Upper Paleolithic era in Central Africa.

Egyptians

The Egyptian method of multiplication of integers and fractions, documented in the Ahmes Papyrus, was by successive additions and doubling. For instance, to find the product of 13 and 21 one had to double 21 three times, obtaining $2 \times 21 = 42$, $4 \times 21 = 2 \times 42 = 84$, $8 \times 21 = 2 \times 84 = 168$. The full product could then be found by adding the appropriate terms found in the doubling sequence:

$$13 \times 21 = (1 + 4 + 8) \times 21 = (1 \times 21) + (4 \times 21) + (8 \times 21) = 21 + 84 + 168 = 273.$$

Babylonians

The Babylonians used a sexagesimal positional number system, analogous to the modern day decimal system. Thus, Babylonian multiplication was very similar to modern decimal multiplication. Because of the relative difficulty of remembering 60×60 different products, Babylonian mathematicians employed multiplication tables. These tables consisted of a list of the first twenty multiples of a certain *principal number n*: $n, 2n, ..., 20n$; followed by the multiples of $10n$: $30n$ $40n$, and $50n$. Then to compute any sexagesimal product, say $53n$, one only needed to add $50n$ and $3n$ computed from the table.

Chinese

$38 \times 76 = 2888$

In the mathematical text *Zhou Bi Suan Jing*, dated prior to 300 BC, and the *Nine Chapters on the Mathematical Art*, multiplication calculations were written out in words, although the early Chinese mathematicians employed Rod calculus involving place value addition, subtraction, multiplication and division. These place value decimal arithmetic algorithms were introduced by Al Khwarizmi to Arab countries in the early 9th century.

Modern Methods

Product of 45 and 256. Note the order of the numerals in 45 is reversed down the left column. The carry step of the multiplication can be performed at the final stage of the calculation (in bold), returning the final product of $45 \times 256 = 11520$. This is a variant of Lattice multiplication.

The modern method of multiplication based on the Hindu–Arabic numeral system was first described by Brahmagupta. Brahmagupta gave rules for addition, subtraction, multiplication and division. Henry Burchard Fine, then professor of Mathematics at Princeton University, wrote the following:

> *The Indians are the inventors not only of the positional decimal system itself, but of most of the processes involved in elementary reckoning with the system. Addition and subtraction they performed quite as they are performed nowadays; multiplication they effected in many ways, ours among them, but division they did cumbrously.*

Grid Method

Grid method multiplication or the box method, is used in primary schools in England and Wales to help teach and understanding of how multiple digit multiplication works. An example of multiplying 34 by 13 would be to lay the numbers out in a grid like:

	30	**4**
10	300	40
3	90	12

and then add the entries.

Computer Algorithms

The classical method of multiplying two n-digit numbers requires n^2 simple multiplications. Multiplication algorithms have been designed that reduce the computation time considerably when multiplying large numbers. In particular for very large numbers methods based on the Discrete Fourier Transform can reduce the number of simple multiplications to the order of $n \log_2(n) \log_2 \log_2(n)$.

Products of Measurements

One can only meaningfully add or subtract quantities of the same type but can multiply or divide quantities of different types. Four bags with three marbles each can be though of as:

$$[4 \text{ bags}] \times [3 \text{ marbles per bag}] = 12 \text{ marbles}.$$

When two measurements are multiplied together the product is of a type depending on the types of the measurements. The general theory is given by dimensional analysis. This analysis is routinely applied in physics but has also found applications in finance.

A common example is multiplying speed by time gives distance, so

$$50 \text{ kilometers per hour} \times 3 \text{ hours} = 150 \text{ kilometers}.$$

Other examples:

$$2.5 \text{ meters} \times 4.5 \text{ meters} = 11.25 \text{ square meters}$$

$$11 \text{ meters/second} \times 9 \text{ seconds} = 99 \text{ meters}$$

Products of Sequences

Capital Pi Notation

The product of a sequence of terms can be written with the product symbol, which derives from the capital letter Π (Pi) in the Greek alphabet. Unicode position U+220F ($\prod$) contains a glyph for denoting such a product, distinct from U+03A0 (Π), the letter. The meaning of this notation is given by:

$$\prod_{i=1}^{4} i = 1 \cdot 2 \cdot 3 \cdot 4,$$

that is

$$\prod_{i=1}^{4} i = 24.$$

The subscript gives the symbol for a dummy variable (i in this case), called the "index of multiplication" together with its lower bound (1), whereas the superscript (here 4) gives its upper bound. The lower and upper bound are expressions denoting integers. The factors of the product are obtained by taking the expression following the product operator, with successive integer values substituted for the index of multiplication, starting from the lower bound and incremented by 1 up to and including the upper bound. So, for example:

$$\prod_{i=1}^{6} i = 1 \cdot 2 \cdot 3 \cdot 4 \cdot 5 \cdot 6 = 720$$

More generally, the notation is defined as

$$\prod_{i=m}^{n} x_i = x_m \cdot x_{m+1} \cdot x_{m+2} \cdots x_{n-1} \cdot x_n,$$

where m and n are integers or expressions that evaluate to integers. In case $m = n$, the value of the product is the same as that of the single factor x_m. If $m > n$, the product is the empty product, with the value 1.

Infinite Products

One may also consider products of infinitely many terms; these are called infinite products. Notationally, we would replace n above by the lemniscate ∞. The product of such a series is defined as the limit of the product of the first n terms, as n grows without bound. That is, by definition,

$$\prod_{i=m}^{\infty} x_i = \lim_{n \to \infty} \prod_{i=m}^{n} x_i.$$

One can similarly replace m with negative infinity, and define:

$$\prod_{i=-\infty}^{\infty} x_i = \left(\lim_{m \to -\infty} \prod_{i=m}^{0} x_i \right) \cdot \left(\lim_{n \to \infty} \prod_{i=1}^{n} x_i \right),$$

provided both limits exist.

Multiplication with Set Theory

The product of non-negative integers can be defined with set theory using cardinal numbers or the Peano axioms. How to extend this to multiplying arbitrary integers, and then arbitrary rational numbers. The product of real numbers is defined in terms of products of rational numbers, see construction of the real numbers.

Multiplication in Group Theory

There are many sets that, under the operation of multiplication, satisfy the axioms that define group structure. These axioms are closure, associativity, and the inclusion of an identity element and inverses.

A simple example is the set of non-zero rational numbers. Here we have identity 1, as opposed to groups under addition where the identity is typically 0. Note that with the rationals, we must exclude zero because, under multiplication, it does not have an inverse: there is no rational number that can be multiplied by zero to result in 1. In this example we have an abelian group, but that is not always the case.

To see this, look at the set of invertible square matrices of a given dimension, over a given field. Now it is straightforward to verify closure, associativity, and inclusion of identity (the identity matrix) and inverses. However, matrix multiplication is not commutative, therefore this group is nonabelian.

Another fact of note is that the integers under multiplication is not a group, even if we exclude zero. This is easily seen by the nonexistence of an inverse for all elements other than 1 and −1.

Multiplication in group theory is typically notated either by a dot, or by juxtaposition (the omission of an operation symbol between elements). So multiplying element a by element b could be notated a · b or ab. When referring to a group via the indication of the set and operation, the dot is used, e.g., our first example could be indicated by $\left(\mathbb{Q} \setminus \{0\}, \cdot \right)$

Division (Mathematics)

Division is one of the four basic operations of arithmetic, the others being addition, subtraction, and multiplication. The division of two natural numbers is the process of calculating the number of times one number is contained within one another. For example, in the picture on the right, the 20 apples are divided into groups of five apples, and there exist four groups, meaning that five can be contained within 20 four times, or 20 ÷ 5 = 4. Division can also be thought of as the process of evaluating a fraction, and fractional notation (a/b and $\frac{a}{b}$) is commonly used to represent division.

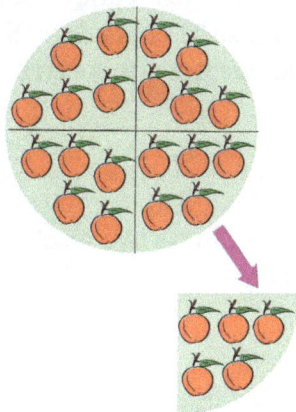

20 ÷ 5 = 4, because 20 apples contain five apples four times.

Division is the inverse of multiplication; if $a \times b = c$, then $a = c \div b$, as long as b is not zero. Division by zero is undefined for the real numbers and most other contexts, because if $b = 0$, then a cannot be deduced from b and c, as then c will always equal zero regardless of a. In some contexts, division by zero can be defined although to a limited extent, and limits involving division of a real number as it approaches zero are defined.

In division, the *dividend* is divided by the *divisor* to get a *quotient*. In the above example, 20 is the dividend, five is the divisor, and the quotient is four. In some cases, the divisor may not be contained fully by the dividend; for example, 10 ÷ 3 leaves a remainder of 1/3 as 10 is not a multiple of three. Normally, this remainder is added to the quotient so 10 ÷ 3 would equal 31/3 or 3.33 ..., but in the context of integer division, where numbers have no fractional part, the remainder is discarded.

Besides dividing apples, division can be applied to other physical and abstract objects. Division has been defined in several contexts, such as for the real and complex numbers and for more abstract contexts such as for vector spaces and fields.

Teaching division usually leads to the concept of fractions being introduced to school pupils. Unlike addition, subtraction, and multiplication, the set of all integers is not closed under division. Dividing two integers may result in a remainder. To complete the division of the remainder, the number system is extended to include fractions or rational numbers as they are more generally called.

Notation

Division is often shown in algebra and science by placing the *dividend* over the *divisor* with a horizontal line, also called a fraction bar, between them. For example, a divided by b is written

$$\frac{a}{b}$$

This can be read out loud as "a divided by b", "a by b" or "a over b". A way to express division all on one line is to write the *dividend* (or numerator), then a slash, then the *divisor* (or denominator), like this:

$$a >$$

This is the usual way to specify division in most computer programming languages since it can easily be typed as a simple sequence of ASCII characters. Some mathematical software, such as MATLAB and GNU Octave, allows the operands to be written in the reverse order by using the backslash as the division operator:

$$b \backslash a$$

A typographical variation halfway between these two forms uses a solidus (fraction slash) but elevates the dividend, and lowers the divisor:

$$^a\!/_b$$

Any of these forms can be used to display a fraction. A fraction is a division expression where both dividend and divisor are integers (typically called the *numerator* and *denominator*), and there is no implication that the division must be evaluated further. A second way to show division is to use the obelus (or division sign), common in arithmetic, in this manner:

$$a \div b$$

This form is infrequent except in elementary arithmetic. ISO 80000-2-9.6 states it should not be used. The obelus is also used alone to represent the division operation itself, as for instance as a label on a key of a calculator.

In some non-English-speaking cultures, "a divided by b" is written $a : b$. This notation was introduced in 1631 by William Oughtred in his *Clavis Mathematicae* and later popularized by Gottfried Wilhelm Leibniz. However, in English usage the colon is restricted to expressing the related concept of ratios (then "a is to b").

In elementary classes of some countries, the notation $b)\,a$ or $b\overline{)a}$ is used to denote a divided by b, especially when discussing long division; similarly, commonly used in Latin America, $b)a$ for short division (as shown in an example on that page). This notation was first introduced $\overline{\text{by}}$ Michael Stifel in *Arithmetica integra*, published in 1544.

Computing

Manual Methods

Division is often introduced through the notion of "sharing out" a set of objects, for example a pile of sweets, into a number of equal portions. Distributing the objects several at a time in each round of sharing to each portion leads to the idea of "chunking", i.e., division by repeated subtraction.

More systematic and more efficient (but also more formalised and more rule-based, and more removed from an overall holistic picture of what division is achieving), a person who knows the multiplication tables can divide two integers using pencil and paper using the method of short division, if the divisor is simple. Long division is used for larger integer divisors. If the dividend has a fractional part (expressed as a decimal fraction), one can continue the algorithm past the ones place as far as desired. If the divisor has a fractional part, we can restate the problem by moving the decimal to the right in both numbers until the divisor has no fraction.

A person can calculate division with an abacus by repeatedly placing the dividend on the abacus, and then subtracting the divisor the offset of each digit in the result, counting the number of divisions possible at each offset.

A person can use logarithm tables to divide two numbers, by subtracting the two numbers' logarithms, then looking up the antilogarithm of the result.

A person can calculate division with a slide rule by aligning the divisor on the C scale with the dividend on the D scale. The quotient can be found on the D scale where it is aligned with the left index on the C scale. The user is responsible, however, for mentally keeping track of the decimal point.

By Computer or With Computer Assistance

Modern computers compute division by methods that are faster than long division: see Division algorithm.

In modular arithmetic, some numbers have a multiplicative inverse with respect to the modulus. We can calculate division by multiplication in such a case. This approach is useful in computers that do not have a fast division instruction.

Properties

Division is right-distributive over addition and subtraction. That means:

$$\frac{a+b}{c} = (a+b) \div c = \frac{a}{c} + \frac{b}{c}$$

in the same way as in multiplication $(a+b) \times c = a \times c + b \times c$. But division is not left-distributive, i.e. we have

$$\frac{a}{b+c} = a \div (b+c) \neq \frac{a}{b} + \frac{a}{c}$$

unlike multiplication.

Euclidean Division

The Euclidean division is the mathematical formulation of the outcome of the usual process of division of integers. It asserts that, given two integers, a, the *dividend*, and b, the *divisor*, such that $b \neq 0$, there are unique integers q, the *quotient*, and r, the *remainder*, such that $a = bq + r$ and $0 \leq r < |b|$, where $|b|$ denotes the absolute value of b.

Of Integers

Division of integers is not closed. Apart from division by zero being undefined, the quotient is not an integer unless the dividend is an integer multiple of the divisor. For example, 26 cannot be divided by 11 to give an integer. Such a case uses one of five approaches:

1. Say that 26 cannot be divided by 11; division becomes a partial function.

2. Give an approximate answer as a decimal fraction or a mixed number, so $\frac{26}{11} \simeq 2.36$ or $\frac{26}{11} \simeq 2\frac{36}{100}$. This is the approach usually taken in numerical computation.

3. Give the answer as a fraction representing a rational number, so the result of the division of 26 by 11 is $\frac{26}{11}$. But, usually, the resulting fraction should be simplified: the result of the division of 52 by 22 is also $\frac{26}{11}$. This simplification may be done by factoring out the greatest common divisor.

4. Give the answer as an integer *quotient* and a *remainder*, so $\frac{26}{11} = 2$ remainder 4. To make the distinction with the previous case, this division, with two integers as result, is sometimes called *Euclidean division*, because it is the basis of the Euclidean algorithm.

5. Give the integer quotient as the answer, so $\frac{26}{11} = 2$. This is sometimes called *integer division*.

Dividing integers in a computer program requires special care. Some programming languages, such as C, treat integer division as in case 5 above, so the answer is an integer. Other languages, such as MATLAB and every computer algebra system return a rational number as the answer, as in case 3 above. These languages also provide functions to get the results of the other cases, either directly or from the result of case 3.

Names and symbols used for integer division include div, /, \, and %. Definitions vary regarding integer division when the dividend or the divisor is negative: rounding may be toward zero (so called T-division) or toward −∞ (F-division); rarer styles can occur – Modulo operation for the details.

Divisibility rules can sometimes be used to quickly determine whether one integer divides exactly into another.

Of Rational Numbers

The result of dividing two rational numbers is another rational number when the divisor is not 0. The division of two rational numbers p/q and r/s can be computed as

$$\frac{p/q}{r/s} = \frac{p}{q} \times \frac{s}{r} = \frac{ps}{qr}.$$

All four quantities are integers, and only p may be 0. This definition ensures that division is the inverse operation of multiplication.

Of Real Numbers

Division of two real numbers results in another real number when the divisor is not 0. It is defined such $a/b = c$ if and only if $a = cb$ and $b \neq 0$.

By zero

Division of any number by zero (where the divisor is zero) is undefined. This is because zero multiplied by any finite number always results in a product of zero. Entry of such an expression into most calculators produces an error message.

Of Complex Numbers

Dividing two complex numbers results in another complex number when the divisor is not 0, which is found using the conjugate of the denominator:

$$\frac{p+iq}{r+is} = \frac{(p+iq)(r-is)}{(r+is)(r-is)} = \frac{pr+qs+i(qr-ps)}{r^2+s^2} = \frac{pr+qs}{r^2+s^2} + i\frac{qr-ps}{r^2+s^2}.$$

This process of multiplying and dividing by $r-is$ is called 'realisation' or (by analogy) rationalisation. All four quantities p, q, r, s are real numbers, and r and s may not both be 0.

Division for complex numbers expressed in polar form is simpler than the definition above:

$$\frac{pe^{iq}}{re^{is}} = \frac{pe^{iq}e^{-is}}{re^{is}e^{-is}} = \frac{p}{r}e^{i(q-s)}.$$

Again all four quantities p, q, r, s are real numbers, and r may not be 0.

Of Polynomials

One can define the division operation for polynomials in one variable over a field. Then, as in the case of integers, one has a remainder. Euclidean division of polynomials, and, for hand-written computation, polynomial long division or synthetic division.

Of Matrices

One can define a division operation for matrices. The usual way to do this is to define $A / B = AB^{-1}$, where B^{-1} denotes the inverse of B, but it is far more common to write out AB^{-1} explicitly to avoid confusion. An elementwise division can also be defined in terms of the Hadamard product.

Left and Right Division

Because matrix multiplication is not commutative, one can also define a left division or so-called *backslash-division* as $A \setminus B = A^{-1}B$. For this to be well defined, B^{-1} need not exist, however A^{-1} does need to exist. To avoid confusion, division as defined by $A / B = AB^{-1}$ is sometimes called *right division* or *slash-division* in this context.

Note that with left and right division defined this way, $A/(BC)$ is in general not the same as $(A/B)/C$ and nor is $(AB)\backslash C$ the same as $A\backslash(B\backslash C)$, but $A/(BC) = (A/C)/B$ and $(AB)\backslash C = B\backslash(A\backslash C)$.

Pseudoinverse

To avoid problems when A^{-1} and/or B^{-1} do not exist, division can also be defined as multiplication with the pseudoinverse, i.e., $A / B = AB^+$ and $A \backslash B = A^+B$, where A^+ and B^+ denote the pseudoinverse of A and B.

Euclidean Division

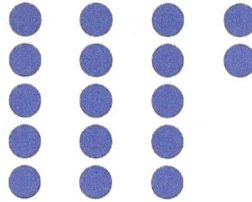

17 is divided into 3 groups of 5 with 2 left over. Here the dividend is 17, the divisor is 5, the quotient is 3, and the remainder is 2.

$$17 = 5 \times 3 + 2$$

In arithmetic, Euclidean division is the process of division of two integers, which produces a quotient and a remainder. This is a theorem that states that the quotient and remainder exist and are unique, under some conditions. Because of this uniqueness, *Euclidean division* is often considered without referring to any method of computation, and without explicitly computing the quotient and the remainder. The methods of computation are called integer division algorithms, the most well-known being long division.

Euclidean division, and algorithms to compute it, are fundamental for many questions concerning integers, such as the Euclidean algorithm for finding the greatest common divisor of two integers, and modular arithmetic, for which only remainders are considered. The operation consisting in computing only the remainder is called the *modulo operation*.

The pie has 9 slices, so each of the 4 people receive 2 slices and 1 is left over.

Statement of The Theorem

Given two integers a and b, with $b \neq 0$, there exist unique integers q and r such that

$$a = bq + r$$

and

$$0 \leq r < |b|,$$

where $|b|$ denotes the absolute value of b.

The four integers that appear in this theorem have been given names: a is called the dividend, b is called the divisor, q is called the quotient and r is called the remainder.

The computation of the quotient and the remainder from the dividend and the divisor is called division or, in case of ambiguity, Euclidean division. The theorem is frequently referred to as the *division algorithm*, although it is a theorem and not an algorithm, because its proof as given below also provides a simple division algorithm for computing q and r.

Division is not defined in the case where $b = 0$.

For the remainder and the modulo operation, there are other conventions than $0 \leq r < |b|$, see § Generalized division algorithms.

History

Although "Euclidean division" is named after Euclid, it seems that he did not know the existence and uniqueness theorem, and that the only computation method that he knew was the division by repeated subtraction.

Before the discovery of Hindu–Arabic numeral system, which has been introduced during 12th century in Europe by Fibonacci, division was extremely difficult, and only the best mathematicians were able to do it. In fact, the long division algorithm requires this notation.

The term "Euclidean division" was introduced during the 20th century as a shorthand for "division of Euclidean rings". It has been rapidly adopted by mathematicians for distinguishing this division from the other kinds of division of numbers.

Intuitive Example

Suppose that a pie has 9 slices and they are to be divided evenly among 4 people. Using Euclidean division, 9 divided by 4 is 2 with remainder 1. In other words, each person receives 2 slices of pie, and there is 1 slice left over.

This can be confirmed using multiplication, the inverse of division: if each of the 4 people received 2 slices, then $4 \times 2 = 8$ slices were given out in all. Adding the 1 slice remaining, the result is 9 slices. In summary: $9 = 4 \times 2 + 1$.

In general, if the number of slices is denoted a and the number of people is b, one can divide the pie evenly among the people such that each person receives q slices (the quotient) and some number of slices $r < b$ are left over (the remainder). Regardless, the equation $a = bq + r$ holds.

If 9 slices were divided among 3 people instead of 4, each would receive 3 and no slices would be left over. In this case the remainder is zero, and it is said that 3 *evenly divides* 9, or that 3 *divides* 9.

Euclidean division can also be extended to negative integers using the same formula; for example $-9 = 4 \times (-3) + 3$, so -9 divided by 4 is -3 with remainder 3.

Examples

- If $a = 7$ and $b = 3$, then $q = 2$ and $r = 1$, since $7 = 3 \times 2 + 1$.

- If $a = 7$ and $b = -3$, then $q = -2$ and $r = 1$, since $7 = -3 \times (-2) + 1$.

- If $a = -7$ and $b = 3$, then $q = -3$ and $r = 2$, since $-7 = 3 \times (-3) + 2$.

- If $a = -7$ and $b = -3$, then $q = 3$ and $r = 2$, since $-7 = -3 \times 3 + 2$.

Proof

The proof consists of two parts — first, the proof of the existence of q and r, and second, the proof of the uniqueness of q and r.

Existence

Consider first the case $b < 0$. Setting $b' = -b$ and $q' = -q$, the equation $a = bq + r$ may be rewritten $a = b'q' + r$ and the inequality $0 \le r < |b|$ may be rewritten $0 \le r < |b'|$. This reduces the existence for the case $b < 0$ to that of the case $b > 0$.

Similarly, if $a < 0$ and $b > 0$, setting $a' = -a$, $q' = -q - 1$ and $r' = b - r$, the equation $a = bq + r$ may be rewritten $a' = bq' + r'$ and the inequality $0 \le r < b$ may be rewritten $0 \le r' < b$. Thus the proof of the existence is reduced to the case $a \ge 0$ and $b > 0$ and we consider only this case in the remainder of the proof.

Let q_1 and r_1, both nonnegative, such that $a = bq_1 + r_1$, for example $q_1 = 0$ and $r_1 = a$. If $r_1 < b$, we are done. Otherwise $q_2 = q_1 + 1$ and $r_2 = r_1 - b$ satisfy $a = bq_2 + r_2$ and $0 \le r_2 < r_1$. Repeating this process one gets eventually $q = q_k$ and $r = r_k$ such that $a = bq + r$ and $0 \le r < b$.

This proves the existence and also gives a simple division algorithm to compute the quotient and the remainder. However this algorithm needs q steps and is thus not efficient.

Uniqueness

Suppose there exists q, q', r, r' with $0 \le r, r' < |b|$ such that $a = bq + r$ and $a = bq' + r'$. Adding the two inequalities $0 \le r < |b|$ and $-|b| < -r' \le 0$ yields $-|b| < r - r' < |b|$, that is $|r - r'| < |b|$.

Subtracting the two equations yields: $b(q' - q) = (r - r')$. Thus $|b|$ divides $|r - r'|$. If $|r - r'| \ne 0$ this implies $|b| \le |r - r'|$, contradicting previous inequality. Thus, $r = r'$ and $b(q' - q) = 0$. As $b \ne 0$, this implies $q = q'$, proving uniqueness.

Other Proofs

Some proofs of the algorithm rely on the Well-ordering principle.

Effectiveness

Generally, an existence proof does not provide an algorithm to compute the existing object, but the above proof provides immediately an algorithm (Division algorithm#Division by repeated subtraction). However this is not a very efficient method, as it requires as many steps as the size of the quotient. This is related to the fact that it only uses addition, subtraction and comparison of the integers, without involving multiplication, nor any particular representation of the integers, such as decimal notation.

In terms of decimal notation, long division provides a much more efficient division algorithm. Its generalization to binary notation allows to use it in a computer. However, for large inputs, algorithms that reduce division to multiplication, like Newton–Raphson one, are usually preferred, because they need a time which is proportional to the time of the multiplication needed to verify the result, independently of the multiplication algorithm which is used.

Generalizations

In Other Domains

Euclidean domains (also known as Euclidean rings) are defined as integral domains which support the following generalization of Euclidean division:

> Given an element a and a non-zero element b in a Euclidean domain R equipped with a Euclidean function d (also known as a Euclidean valuation, or degree function), there exist q and r in R such that $a = bq + r$ and either $r = 0$ or $d(r) < d(b)$. Unlike in the integer case, q and r need not be unique.

Examples of Euclidean domains include fields, polynomial rings in one variable over a field, and the Gaussian integers. The Euclidean division of polynomials has been the object of specific developments. Polynomial long division, Polynomial greatest common divisor#Euclidean division and Polynomial greatest common divisor#Pseudo-remainder sequences.

References

- Schmid, Hermann (1974). Decimal Computation (1 ed.). Binghamton, New York, USA: John Wiley & Sons. ISBN 0-471-76180-X.

- Schmid, Hermann (1983) [1974]. Decimal Computation (1 (reprint) ed.). Malabar, Florida, USA: Robert E. Krieger Publishing Company. ISBN 0-89874-318-4.

- Boyer, Carl B. (revised by Merzbach, Uta C.) (1991). History of Mathematics. John Wiley and Sons, Inc. ISBN 0-471-54397-7.

- Derbyshire, John (2004). Prime Obsession: Bernhard Riemann and the Greatest Unsolved Problem in Mathematics. New York City: Penguin Books. ISBN 978-0452285255

- Fraleigh, John B. (1993), A First Course in Abstract Algebra (5th ed.), Addison-Wesley, ISBN 978-0-201-53467-2

- Rotman, Joseph J. (2006), A First Course in Abstract Algebra with Applications (3rd ed.), Prentice-Hall, ISBN 978-0-13-186267-8

6

Division and Multiplication Algorithm

Division algorithm calculates the quotient of two given integers N and D. Division algorithm falls under two major categories which are slow division and fast division. Multiplication algorithm, Euclidean algorithm, greatest common divisor, least common multiple and fundamental theorem of arithmetic are some of the aspects elucidated in the section. The chapter discusses the methods of division and multiplication algorithm in a critical manner providing key analysis to the subject matter.

Division Algorithm

A division algorithm is an algorithm which, given two integers N and D, computes their quotient and/or remainder, the result of division. Some are applied by hand, while others are employed by digital circuit designs and software.

Division algorithms fall into two main categories: slow division and fast division. Slow division algorithms produce one digit of the final quotient per iteration. Examples of slow division include restoring, non-performing restoring, non-restoring, and SRT division. Fast division methods start with a close approximation to the final quotient and produce twice as many digits of the final quotient on each iteration. Newton–Raphson and Goldschmidt fall into this category.

Discussion will refer to the form $N / D = (Q, R)$, where

- N = Numerator (dividend)
- D = Denominator (divisor)

is the input, and

- Q = Quotient
- R = Remainder

is the output.

Division by Repeated Subtraction

The simplest division algorithm, historically incorporated into a greatest common divisor algorithm presented in Euclid's *Elements*, Book VII, Proposition 1, finds the remainder given two positive integers using only subtractions and comparisons:

```
while  N ≥ D do

  N := N - D
```

```
end

return N
```

The proof that the quotient and remainder exist and are unique, ascertained by Euclidean division, gives rise to a complete division algorithm using additions, subtractions, and comparisons:

```
function divide(N, D)

  if D = 0 then error(DivisionByZero) end

  if D < 0 then (Q,R) := divide(N, -D); return (-Q, R) end

  if N < 0 then

    (Q,R) := divide(-N, D)

    if R = 0 then return (-Q, 0)

    else return (-Q-1, D-R) end

  end

  -- At this point, N ≥ 0 and D > 0

  Q := 0; R := N

  while R ≥ D do

    Q := Q + 1

    R := R - D

  end

  return (Q, R)

end
```

This procedure always produces $R \geq 0$. Although very simple, it takes $\Omega(Q)$ steps, and so is exponentially slower than even slow division algorithms like long division. It is useful if Q is known to be small (being an output-sensitive algorithm), and can serve as an executable specification.

Long Division

Long division is the standard algorithm used for pen-and-paper division of multidigit numbers expressed in decimal notation. It shifts gradually from the left to the right end of the dividend, subtracting the largest possible multiple of the divisor at each stage; the multiples become the digits of the quotient, and the final difference is the remainder. When used with a binary radix, it forms the basis for the integer division (unsigned) with remainder algorithm below. Short division is an abbreviated form of long division suitable for one-digit divisors. Chunking (also known as the partial quotients method or the hangman method) is a less-efficient form of long division which may be easier to understand.

Integer Division (Unsigned) With Remainder

The following algorithm, the binary version of the famous long division, will divide N by D, placing the quotient in Q and the remainder in R. All values are treated as unsigned integers.

```
if D == 0 then error(DivisionByZeroException) end

Q := 0                    -- initialize quotient and remainder to zero

R := 0

for i = n-1...0 do        -- where n is number of bits in N

  R := R << 1             -- left-shift R by 1 bit

  R(0) := N(i)            -- set the least-significant bit of R equal to bit
i of the numerator

  if R >= D then

    R := R - D

    Q(i) := 1

  end

end
```

Example

If we take N=1100$_2$ (12$_{10}$) and D=100$_2$ (4$_{10}$)

Step 1: Set R=0 and Q=0
Step 2: Take i=3 (one less than the number of bits in N)
Step 3: R=00 (left shifted by 1)
Step 4: R=01 (setting R(0) to N(i))
Step 5: R<D, so skip statement

Step 2: Set i=2
Step 3: R=010
Step 4: R=011
Step 5: R<D, statement skipped

Step 2: Set i=1
Step 3: R=0110
Step 4: R=0110
Step 5: R>=D, statement entered
Step 5b: R=10 (R−D)

Step 5c: Q=10 (setting Q(i) to 1)

Step 2: Set i=0
Step 3: R=100
Step 4: R=100
Step 5: R>=D, statement entered
Step 5b: R=0 (R−D)
Step 5c: Q=11 (setting Q(i) to 1)

end
Q=11_2 (3_{10}) and R=0.

Slow Division Methods

Slow division methods are all based on a standard recurrence equation

$$P_{j+1} = R \times P_j - q_{n-(j+1)} \times D \,,$$

where:

- P_j is *j*-th the partial remainder of the division

- R is the radix

- $q_{n-(j+1)}$ is the digit of the quotient in position *n–(j+1)*, where the digit positions are numbered from least-significant 0 to most significant *n−1*

- *n* is number of digits in the quotient

- *D* is the divisor

Restoring Division

Restoring division operates on fixed-point fractional numbers and depends on the following assumptions:

- $D < N$

- $0 < N, D < 1.$

The quotient digits *q* are formed from the digit set {0,1}.

The basic algorithm for binary (radix 2) restoring division is:

```
P := N

D := D << n              -- P and D need twice the word width of N and Q

for i = n-1..0 do        -- for example 31..0 for 32 bits

    P := 2P - D          -- trial subtraction from shifted value
```

```
if P >= 0 then

  q(i) := 1                  -- result-bit 1

else

  q(i) := 0                  -- result-bit 0

    P := P + D               -- new partial remainder is (restored) shifted
value

  end

end
```

-- Where: N = Numerator, D = Denominator, n = #bits, P = Partial remainder, q(i) = bit #i of quotient

The above restoring division algorithm can avoid the restoring step by saving the shifted value $2P$ before the subtraction in an additional register T (i.e., $T = P << 1$) and copying register T to P when the result of the subtraction $2P - D$ is negative.

Non-performing restoring division is similar to restoring division except that the value of $2*P[i]$ is saved, so D does not need to be added back in for the case of TP[i] ≤ 0.

Non-restoring Division

Non-restoring division uses the digit set {−1,1} for the quotient digits instead of {0,1}. The basic algorithm for binary (radix 2) non-restoring division of non-negative numbers is:

```
P := N

D := D << n                * P and D need twice the word width of N and Q

for i = n-1..0 do          * for example 31..0 for 32 bits

  if P >= 0 then

    q[i] := +1

    P := 2*P - D

  else

    q[i] := -1

    P := 2*P + D

  end if

end
```

* Note: N=Numerator, D=Denominator, n=#bits, P=Partial remainder, q(i)=bit #i of quotient.

Following this algorithm, the quotient is in a non-standard form consisting of digits of −1 and +1. This form needs to be converted to binary to form the final quotient. Example:

Convert the following quotient to the digit set {0,1}:	$Q = 111\overline{1}1\overline{1}1\overline{1}$
Steps:	
1. Form the positive term:	$P = 11101010$
2. Mask the negative term*:	$N = 00010101$
3. Subtract: P minus N :	$Q = 11010101$
*.(Signed binary notation with One's complement without Two's Complement)	

$Q := Q - (Q(-1))$ * Appropriate if −1 Digits in Q are Represented as zeros as is common.

Finally, quotients computed by this algorithm are always odd: 5 / 2 = 3 R −1. To correct this add the following after Q is converted from non-standard form to standard form:

```
if P < 0 then

    Q := Q - 1

    P := P + D            * Needed only if the Remainder is of interest.

end if
```

The actual remainder is P >> n.

SRT Division

Named for its creators (Sweeney, Robertson, and Tocher), SRT division is a popular method for division in many microprocessor implementations. SRT division is similar to non-restoring division, but it uses a lookup table based on the dividend and the divisor to determine each quotient digit. The Intel Pentium processor's infamous floating-point division bug was caused by an incorrectly coded lookup table. Five of the 1066 entries had been mistakenly omitted.

Pseudocode

The following computes the quotient of N and D with a precision of P binary places:

```
Express D as M × 2ᵉ where 1 ≤ M < 2 (standard floating point representa-
tion)

D' := D / 2ᵉ⁺¹    // scale between 0.5 and 1, can be performed with bit
shift / exponent subtraction

N' := N / 2ᵉ⁺¹

X := 48/17 - 32/17 × D'    // precompute constants with same precision as D
```

$$\text{repeat } \left\lceil \log_2 \frac{P+1}{\log_2 17} \right\rceil \text{ times} \quad \textit{// can be precomputed based on fixed P}$$

```
    X := X + X × (1 - D' × X)
end

return N' × X
```

For example, for a double-precision floating-point division, this method uses 10 multiplies, 9 adds, and 2 shifts.

Goldschmidt Division

Goldschmidt (after Robert Elliott Goldschmidt) division uses an iterative process of repeatedly multiplying both the dividend and divisor by a common factor F_i, chosen such that the divisor converges to 1. This causes the dividend to converge to the sought quotient Q:

$$Q = \frac{N}{D} \frac{F_1}{F_1} \frac{F_2}{F_2} \frac{F_{...}}{F_{...}}.$$

The steps for Goldschmidt division are:

1. Generate an estimate for the multiplication factor F_i.

2. Multiply the dividend and divisor by F_i.

3. If the divisor is sufficiently close to 1, return the dividend, otherwise, loop to step 1.

Assuming N/D has been scaled so that $0 < D < 1$, each F_i is based on D:

$$F_{i+1} = 2 - D_i.$$

Multiplying the dividend and divisor by the factor yields:

$$\frac{N_{i+1}}{D_{i+1}} = \frac{N_i}{D_i} \frac{F_{i+1}}{F_{i+1}}.$$

After a sufficient number k of iterations $Q = N_k$.

The Goldschmidt method is used in AMD Athlon CPUs and later models.

Large-integer Methods

Methods designed for hardware implementation generally do not scale to integers with thousands or millions of decimal digits; these frequently occur, for example, in modular reductions in cryptography. For these large integers, more efficient division algorithms transform the problem to use a small number of multiplications, which can then be done using an asymptotically efficient multiplication algorithm such as the Karatsuba algorithm, Toom–Cook multiplication or the Schönhage–Strassen algorithm. It results that the computational complexity of the division is of the same order (up to a multiplicative constant) as that of the multiplication. Examples include reduction to multiplication by Newton's method as described above, as well as the slightly faster Barrett reduction algorithm. Newton's method is particularly efficient in scenarios where one must divide by the same divisor many times, since after the initial Newton inversion only one (truncated) multiplication is needed for each division.

Division by A Constant

The division by a constant D is equivalent to the multiplication by its reciprocal. Since the denominator is constant, so is its reciprocal $(1/D)$. Thus it is possible to compute the value of $(1/D)$ once at compile time, and at run time perform the multiplication $N \cdot (1/D)$ rather than the division N/D. In floating-point arithmetic the use of $(1/D)$ presents little problem, but in integer arithmetic the reciprocal will always evaluate to zero (assuming $|D| > 1$).

It is not necessary to use specifically $(1/D)$; any value (X/Y) that reduces to $(1/D)$ may be used. For example, for division by 3, the factors $1/3$, $2/6$, $3/9$, or $194/582$ could be used. Consequently, if Y were a power of two so the division step reduces to a fast right bit shift. The effect of calculating N/D as $(N \cdot X)/Y$ replaces a division with a multiply and a shift. Note that the parentheses are important, as $N \cdot (X/Y)$ will evaluate to zero.

However, unless D itself is a power of two, there is no X and Y that satisfies the conditions above. Fortunately, $(N \cdot X)/Y$ gives exactly the same result as N/D in integer arithmetic even when (X/Y) is not exactly equal to $1/D$, but "close enough" that the error introduced by the approximation is in the bits that are discarded by the shift operation.

As a concrete example, for 32-bit unsigned integers, division by 3 can be replaced with a multiply by $2863311531/2^{33}$, a multiplication by 2863311531 (hexadecimal 0xAAAAAAAB) followed by a 33 right bit shift. The value of 2863311531 is calculated as $2^{33}/3$, then rounded up.

Likewise, division by 10 can be expressed as a multiplication by 3435973837 (0xCCCCCCCD) followed by division by 2^{35}.

In some cases, division by a constant can be accomplished in even less time by converting the "multiply by a constant" into a series of shifts and adds or subtracts. Of particular interest is division by 10, for which the exact quotient is obtained, with remainder if required.

Rounding Error

Round-off error can be introduced by division operations due to limited precision.

Multiplication Algorithm

A multiplication algorithm is an algorithm (or method) to multiply two numbers. Depending on the size of the numbers, different algorithms are in use. Efficient multiplication algorithms have existed since the advent of the decimal system.

Grid Method

The grid method (or box method) is an introductory method for multiple-digit multiplication that is often taught to pupils at primary school or elementary school level. It has been a standard part of the national primary-school mathematics curriculum in England and Wales since the late 1990s.

Both factors are broken up ("partitioned") into their hundreds, tens and units parts, and the products of the parts are then calculated explicitly in a relatively simple multiplication-only stage, before these contributions are then totalled to give the final answer in a separate addition stage.

The calculation 34×13, for example, could be computed using the grid:

$$
\begin{array}{r}
300 \\
40 \\
90 \\
+\ 12 \\
\hline
442
\end{array}
$$

×	**30**	**4**
10	300	40
3	90	12

followed by addition to obtain 442, either in a single sum (see right), or through forming the row-by-row totals $(300 + 40) + (90 + 12) = 340 + 102 = 442$.

This calculation approach (though not necessarily with the explicit grid arrangement) is also known as the partial products algorithm. Its essence is the calculation of the simple multiplications separately, with all addition being left to the final gathering-up stage.

The grid method can in principle be applied to factors of any size, although the number of sub-products becomes cumbersome as the number of digits increases. Nevertheless, it is seen as a usefully explicit method to introduce the idea of multiple-digit multiplications; and, in an age when most multiplication calculations are done using a calculator or a spreadsheet, it may in practice be the only multiplication algorithm that some students will ever need.

Long Multiplication

If a positional numeral system is used, a natural way of multiplying numbers is taught in schools as long multiplication, sometimes called grade-school multiplication, sometimes called Standard Algorithm: multiply the multiplicand by each digit of the multiplier and then add up all the properly shifted results. It requires memorization of the multiplication table for single digits.

This is the usual algorithm for multiplying larger numbers by hand in base 10. Computers initially used a very similar shift and add algorithm in base 2, but modern processors have optimized circuitry for fast multiplications using more efficient algorithms, at the price of a more complex hardware realization. A person doing long multiplication on paper will write down all the products and then add them together; an abacus-user will sum the products as soon as each one is computed.

Example

This example uses *long multiplication* to multiply 23,958,233 (multiplicand) by 5,830 (multiplier) and arrives at 139,676,498,390 for the result (product).

```
         23958233
    ×        5830
         00000000  ( =       23,958,233  ×        0)
         71874699  ( =       23,958,233  ×       30)
        191665864  ( =       23,958,233  ×      800)
    +   119791165  ( =       23,958,233  ×    5,000)
      139676498390 ( = 139,676,498,390              )
```

Below pseudocode describes the process of above multiplication. It keeps only one row to maintain sum which finally becomes the result. Note that the '+=' operator is used to denote sum to existing value and store operation (akin to languages such as Java and C) for compactness.

```
multiply(a[1..p], b[1..q], base)                        // Operands
containing rightmost digits at index 1

  product = [1..p+q]                                     //Allocate
space for result

  for b_i = 1 to q                                       // for all
digits in b

    carry = 0

    for a_i = 1 to p                                     //for all
digits in a

      product[a_i + b_i - 1] += carry + a[a_i] * b[b_i]

      carry = product[a_i + b_i - 1] / base

      product[a_i + b_i - 1] = product[a_i + b_i - 1] mod base

    product[b_i + p] += carry                            // last digit
comes from final carry

  return product
```

Space Complexity

Let n be the total number of digits in the two input numbers in base D. If the result must be kept in memory then the space complexity is trivially $\Theta(n)$. However, in certain applications, the entire result need not be kept in memory and instead the digits of the result can be streamed out as they are computed (for example, to system console or file). In these scenarios, long multiplication has the advantage that it can easily be formulated as a log space algorithm; that is, an algorithm that only needs working space proportional to the logarithm of the number of digits in the input

($\Theta(\log n)$). This is the *double* logarithm of the numbers being multiplied themselves ($\log \log N$). Note that operands themselves still need to be kept in memory and their $\Theta(n)$ space is not considered in this analysis.

The method is based on the observation that each digit of the result can be computed from right to left with only knowing the carry from the previous step. Let a_i and b_i be the i-th digit of the operand, r_i be the i-th digit of the result and c_i be the carry generated for r_i (i=1 is the right most digit) then

$$r_i = \left(c_{i-1} + \sum_{j+k=i} a_j b_k \right) \bmod D \quad c_i = \left\lfloor (c_{i-1} + \sum_{j+k=i} a_j b_k)/D \right\rfloor \quad c_0 = 0$$

A simple inductive argument shows that the carry can never exceed n and the total sum for r_i can never exceed $D * n$: the carry into the first column is zero, and for all other columns, there are at most n digits in the column, and a carry of at most n from the previous column (by the induction hypothesis). The sum is at most $D * n$, and the carry to the next column is at most $D * n / D$, or n. Thus both these values can be stored in $O(\log n)$ digits.

In pseudocode, the log-space algorithm is:

```
multiply(a[1..p], b[1..q], base)                          // Operands containing
rightmost digits at index 1

    tot = 0

    for ri = 1 to p + q - 1                               //For each digit of
result

        for bi = MAX(1, ri - p + 1) to MIN(ri, q) //Digits from b that
need to be considered

            ai = ri - bi + 1                             //Digits from a follow
"symmetry"

            tot = tot + (a[ai] * b[bi])

        product[ri] = tot mod base

        tot = floor(tot / base)

    product[p+q] = tot mod base                          //Last digit of the
result comes from last carry

    return product
```

Electronic Usage

Some chips implement this algorithm for various integer and floating-point sizes in computer hardware or in microcode. In arbitrary-precision arithmetic, it's common to use long multiplication with the base set to 2^w, where w is the number of bits in a word, for multiplying relatively small numbers.

To multiply two numbers with n digits using this method, one needs about n^2 operations. More formally: using a natural size metric of number of digits, the time complexity of multiplying two n-digit numbers using long multiplication is $\Theta(n^2)$.

When implemented in software, long multiplication algorithms have to deal with overflow during additions, which can be expensive. For this reason, a typical approach is to represent the number in a small base b such that, for example, $8b$ is a representable machine integer (for example Richard Brent used this approach in his Fortran package MP); we can then perform several additions before having to deal with overflow. When the number becomes too large, we add part of it to the result or carry and map the remaining part back to a number less than b; this process is called *normalization*.

Lattice Multiplication

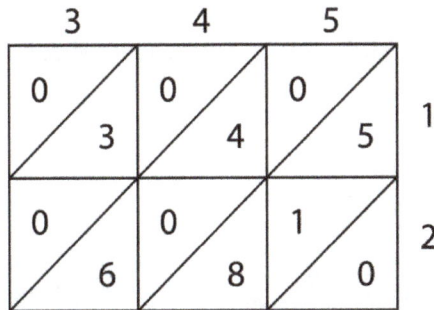

First, set up the grid by marking its rows and columns with the numbers to be multiplied. Then, fill in the boxes with tens digits in the top triangles and units digits on the bottom.

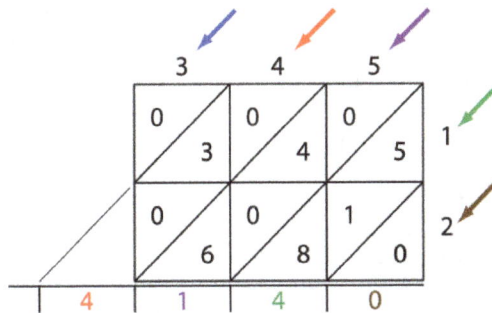

Finally, sum along the diagonal tracts and carry as needed to get the answer

Lattice, or sieve, multiplication is algorithmically equivalent to long multiplication. It requires the preparation of a lattice (a grid drawn on paper) which guides the calculation and separates all the multiplications from the additions. It was introduced to Europe in 1202 in Fibonacci's Liber Abaci. Leonardo described the operation as mental, using his right and left hands to carry the intermediate calculations. Matrakçı Nasuh presented 6 different variants of this method in this 16th-century book, Umdet-ul Hisab. It was widely used in Enderun schools across the Ottoman Empire. Napier's bones, or Napier's rods also used this method, as published by Napier in 1617, the year of his death.

As shown in the example, the multiplicand and multiplier are written above and to the right of a lattice, or a sieve. It is found in Muhammad ibn Musa al-Khwarizmi's "Arithmetic", one of Leonardo's sources mentioned by Sigler, author of "Fibonacci's Liber Abaci", 2002.

- During the multiplication phase, the lattice is filled in with two-digit products of the corresponding digits labeling each row and column: the tens digit goes in the top-left corner.

- During the addition phase, the lattice is summed on the diagonals.

- Finally, if a carry phase is necessary, the answer as shown along the left and bottom sides of the lattice is converted to normal form by carrying ten's digits as in long addition or multiplication.

Example

The pictures on the right show how to calculate 345×12 using lattice multiplication. As a more complicated example, consider the picture below displaying the computation of 23,958,233 multiplied by 5,830 (multiplier); the result is 139,676,498,390. Notice 23,958,233 is along the top of the lattice and 5,830 is along the right side. The products fill the lattice and the sum of those products (on the diagonal) are along the left and bottom sides. Then those sums are totaled as shown.

```
   2  3  9  5  8  2  3  3           01
  +---+---+---+---+---+---+---+---+-   002
  |1 /|1 /|4 /|2 /|4 /|1 /|1 /|1 /|    0017
  |/ |/ |/ |/ |/ |/ |/ |/ |5           00024
  01|/ 0|/ 5|/ 5|/ 5|/ 0|/ 0|/ 5|/ 5|  000026
  +---+---+---+---+---+---+---+---+-    0000015
  |1 /|2 /|7 /|4 /|6 /|1 /|2 /|2 /|     00000013
  |/ |/ |/ |/ |/ |/ |/ |/ |8            000000018
  02|/ 6|/ 4|/ 2|/ 0|/ 4|/ 6|/ 4|/ 4|   0000000017
  +---+---+---+---+---+---+---+---+-     00000000013
  |0 /|0 /|2 /|1 /|2 /|0 /|0 /|0 /|      000000000009
  |/ |/ |/ |/ |/ |/ |/ |/ |3             0000000000000
  17|/ 6|/ 9|/ 7|/ 5|/ 4|/ 6|/ 9|/ 9|   ------------
  +---+---+---+---+---+---+---+---+-     --
  |0 /|0 /|0 /|0 /|0 /|0 /|0 /|0 /|       139676498390
  |/ |/ |/ |/ |/ |/ |/ |/ |0
  24|/ 0|/ 0|/ 0|/ 0|/ 0|/ 0|/ 0|/ 0|   = 139,676,498,390
  +---+---+---+---+---+---+---+---+-
   26  15  13  18  17  13  09  00
```

Peasant or Binary Multiplication

In base 2, long multiplication reduces to a nearly trivial operation. For each '1' bit in the multiplier, shift the multiplicand an appropriate amount and then sum the shifted values. Depending on computer processor architecture and choice of multiplier, it may be faster to code this algorithm using hardware bit shifts and adds rather than depend on multiplication instructions, when the multiplier is fixed and the number of adds required is small.

This algorithm is also known as Peasant multiplication, because it has been widely used among those who are unschooled and thus have not memorized the multiplication tables required by long multiplication. The algorithm was also in use in ancient Egypt.

On paper, write down in one column the numbers you get when you repeatedly halve the multiplier, ignoring the remainder; in a column beside it repeatedly double the multiplicand. Cross out each row in which the last digit of the first number is even, and add the remaining numbers in the second column to obtain the product.

The main advantages of this method are that it can be taught quickly, no memorization is required, and it can be performed using tokens such as poker chips if paper and pencil are not available. It does however take more steps than long multiplication so it can be unwieldy when large numbers are involved.

Examples

This example uses peasant multiplication to multiply 11 by 3 to arrive at a result of 33.

```
Decimal:       Binary:

11   3         1011   11

5    6         101    110

2    12        10     1100

1    24        1      11000

     __               _____

     33               100001
```

Describing the steps explicitly:

- 11 and 3 are written at the top

- 11 is halved (5.5) and 3 is doubled (6). The fractional portion is discarded (5.5 becomes 5).

- 5 is halved (2.5) and 6 is doubled (12). The fractional portion is discarded (2.5 becomes 2). The figure in the left column (2) is even, so the figure in the right column (12) is discarded.

- 2 is halved (1) and 12 is doubled (24).

- All not-scratched-out values are summed: 3 + 6 + 24 = 33.

The method works because multiplication is distributive, so:

$$3 \times 11 = 3 \times (1 \times 2^0 + 1 \times 2^1 + 0 \times 2^2 + 1 \times 2^3) = 3 \times (1 + 2 + 8) = 3 + 6 + 24 = 33.$$

A more complicated example, using the figures from the earlier examples (23,958,233 and 5,830):

```
Decimal:              Binary:

5830   23958233       1011011000110   1011011011001001011011001

2915   47916466       101101100011    101101101100100101101100010

1457   95832932       10110110001     1011011011001001011011001100100
```

728	~~191665864~~	1011011000	~~1011011011001001011011001000~~
364	~~383331728~~	101101100	~~1011011011001001011011001000~~
182	~~766663456~~	10110110	~~1011011011001001011011001000~~
91	1533326912	1011011	1011011011001001011011001000000
45	3066653824	101101	1011011011001001011011001000000
22	~~6133307648~~	10110	~~1011011011001001011011001000000000~~
11	12266615296	1011	1011011011001001011011001000000000
5	24533230592	101	1011011011001001011011001000000000
2	~~49066461184~~	10	~~1011011011001001011011001000000000~~
1	98132922368	1	1011011011001001011011001000000000

1022143253354344244353353243222210110 (before carry)

139676498390   100000100001010101111000111001110110110

Shift and Add

Historically, computers used a "shift and add" algorithm to multiply small integers. Both base 2 long multiplication and base 2 peasant multiplication reduce to this same algorithm. In base 2, multiplying by the single digit of the multiplier reduces to a simple series of logical AND operations. Each partial product is added to a running sum as soon as each partial product is computed. Most currently available microprocessors implement this or other similar algorithms (such as Booth encoding) for various integer and floating-point sizes in hardware multipliers or in microcode.

On currently available processors, a bit-wise shift instruction is faster than a multiply instruction and can be used to multiply (shift left) and divide (shift right) by powers of two. Multiplication by a constant and division by a constant can be implemented using a sequence of shifts and adds or subtracts. For example, there are several ways to multiply by 10 using only bit-shift and addition.

```
((x << 2) + x) << 1 # Here 10*x is computed as (x*2^2 + x)*2

(x << 3) + (x << 1) # Here 10*x is computed as x*2^3 + x*2
```

In some cases such sequences of shifts and adds or subtracts will outperform hardware multipliers and especially dividers. A division by a number of the form 2^n or $2^n \pm 1$ often can be converted to such a short sequence.

These types of sequences have to always be used for computers that do not have a "multiply" instruction, and can also be used by extension to floating point numbers if one replaces the shifts with computation of $2*x$ as $x+x$, as these are logically equivalent.

Quarter Square Multiplication

Two quantities can be multiplied using quarter squares by employing the following identity involving the floor function that some sources attribute to Babylonian mathematics (2000–1600 BC).

$$\left\lfloor \frac{(x+y)^2}{4} \right\rfloor - \left\lfloor \frac{(x-y)^2}{4} \right\rfloor = \frac{1}{4}\left(\left(x^2+2xy+y^2\right)-\left(x^2-2xy+y^2\right)\right) = \frac{1}{4}\left(4xy\right) = xy.$$

If one of $x+y$ and $x-y$ is odd, the other is odd too; this means that the fractions, if any, will cancel out, and discarding the remainders does not introduce any error. Below is a lookup table of quarter squares with the remainder discarded for the digits 0 through 18; this allows for the multiplication of numbers up to 9×9.

n	0	1	2	3	4	5	6	7	8	9	10	11	12	13	14	15	16	17	18
$\lfloor n^2/4 \rfloor$	0	0	1	2	4	6	9	12	16	20	25	30	36	42	49	56	64	72	81

If, for example, you wanted to multiply 9 by 3, you observe that the sum and difference are 12 and 6 respectively. Looking both those values up on the table yields 36 and 9, the difference of which is 27, which is the product of 9 and 3.

Antoine Voisin published a table of quarter squares from 1 to 1000 in 1817 as an aid in multiplication. A larger table of quarter squares from 1 to 100000 was published by Samuel Laundy in 1856, and a table from 1 to 200000 by Joseph Blater in 1888.

Quarter square multipliers were used in analog computers to form an analog signal that was the product of two analog input signals. In this application, the sum and difference of two input voltages are formed using operational amplifiers. The square of each of these is approximated using piecewise linear circuits. Finally the difference of the two squares is formed and scaled by a factor of one fourth using yet another operational amplifier.

In 1980, Everett L. Johnson proposed using the quarter square method in a digital multiplier. To form the product of two 8-bit integers, for example, the digital device forms the sum and difference, looks both quantities up in a table of squares, takes the difference of the results, and divides by four by shifting two bits to the right. For 8-bit integers the table of quarter squares will have $2^9-1=511$ entries (one entry for the full range 0..510 of possible sums, the differences using only the first 256 entries in range 0..255) or $2^9-1=511$ entries (using for negative differences the technique of 2-complements and 9-bit masking, which avoids testing the sign of differences), each entry being 16-bit wide (the entry values are from $(0^2/4)=0$ to $(510^2/4)=65025$).

The Quarter square multiplier technique has also benefitted 8-bit systems that do not have any support for a hardware multiplier. Steven Judd implemented this for the 6502.

Fast Multiplication Algorithms for Large Inputs

Gauss's complex multiplication algorithm

Complex multiplication normally involves four multiplications and two additions.

$$(a+bi)(c+di) = (ac-bd)+(bc+ad)i. \text{ Or}$$

×	a	bi
c	ac	bci
di	adi	$-bd$

By 1805 Gauss had discovered a way of reducing the number of multiplications to three.

The product $(a + bi) \cdot (c + di)$ can be calculated in the following way.

$$k_1 = c \cdot (a + b)$$

$$k_2 = a \cdot (d - c)$$

$$k_3 = b \cdot (c + d)$$

$$\text{Real part} = k_1 - k_3$$

$$\text{Imaginary part} = k_1 + k_2.$$

This algorithm uses only three multiplications, rather than four, and five additions or subtractions rather than two. If a multiply is more expensive than three adds or subtracts, as when calculating by hand, then there is a gain in speed. On modern computers a multiply and an add can take about the same time so there may be no speed gain. There is a trade-off in that there may be some loss of precision when using floating point.

For fast Fourier transforms (FFTs) (or any linear transformation) the complex multiplies are by constant coefficients $c + di$ (called twiddle factors in FFTs), in which case two of the additions ($d-c$ and $c+d$) can be precomputed. Hence, only three multiplies and three adds are required. However, trading off a multiplication for an addition in this way may no longer be beneficial with modern floating-point units.

Karatsuba Multiplication

For systems that need to multiply numbers in the range of several thousand digits, such as computer algebra systems and bignum libraries, long multiplication is too slow. These systems may employ Karatsuba multiplication, which was discovered in 1960 (published in 1962). The heart of Karatsuba's method lies in the observation that two-digit multiplication can be done with only three rather than the four multiplications classically required. This is an example of what is now called a *divide and conquer algorithm*. Suppose we want to multiply two 2-digit numbers: $x_1x_2 \cdot y_1y_2$:

1. compute $x_1 \cdot y_1$, call the result A

2. compute $x_2 \cdot y_2$, call the result B

3. compute $(x_1 + x_2) \cdot (y_1 + y_2)$, call the result C

4. compute $C - A - B$, call the result K; this number is equal to $x_1 \cdot y_2 + x_2 \cdot y_1$

5. compute $A \cdot 100 + K \cdot 10 + B$.

Bigger numbers $x_1 x_2$ can be split into two parts x_1 and x_2. Then the method works analogously. To compute these three products of m-digit numbers, we can employ the same trick again, effectively using recursion. Once the numbers are computed, we need to add them together (step 5.), which takes about n operations.

Karatsuba multiplication has a time complexity of $O(n^{\log_2 3}) \approx O(n^{1.585})$, making this method significantly faster than long multiplication. Because of the overhead of recursion, Karatsuba's multiplication is slower than long multiplication for small values of n; typical implementations therefore switch to long multiplication if n is below some threshold.

Karatsuba's algorithm is the first known algorithm for multiplication that is asymptotically faster than long multiplication, and can thus be viewed as the starting point for the theory of fast multiplications.

Toom–Cook

Another method of multiplication is called Toom–Cook or Toom-3. The Toom–Cook method splits each number to be multiplied into multiple parts. The Toom–Cook method is one of the generalizations of the Karatsuba method. A three-way Toom–Cook can do a size-$3N$ multiplication for the cost of five size-N multiplications, improvement by a factor of $9/5$ compared to the Karatsuba method's improvement by a factor of $4/3$.

Although using more and more parts can reduce the time spent on recursive multiplications further, the overhead from additions and digit management also grows. For this reason, the method of Fourier transforms is typically faster for numbers with several thousand digits, and asymptotically faster for even larger numbers.

Fourier Transform Methods

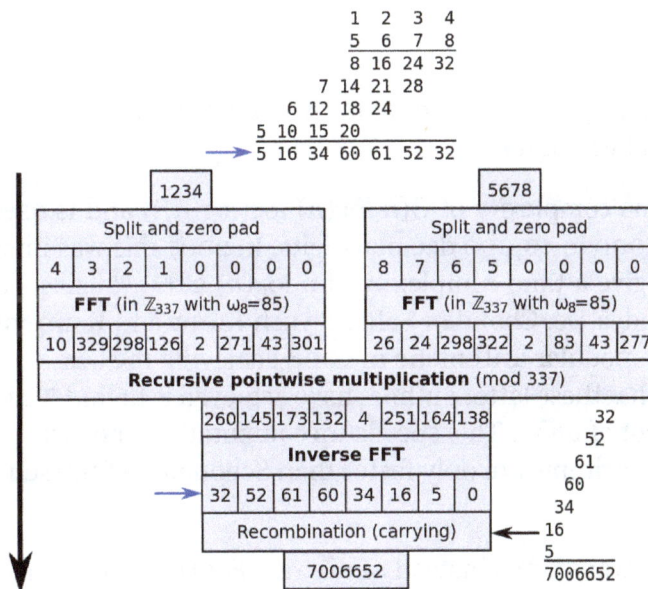

Demonstration of multiplying $1234 \times 5678 = 7006652$ using fast Fourier transforms (FFTs). Number-theoretic transforms in the integers modulo 337 are used, selecting 85 as an 8th root of unity. Base 10 is used in place of base 2^w for illustrative purposes.

The basic idea due to Strassen (1968) is to use fast polynomial multiplication to perform fast integer multiplication. The algorithm was made practical and theoretical guarantees were provided in 1971 by Schönhage and Strassen resulting in the Schönhage–Strassen algorithm. The details are the following: We choose the largest integer w that will not cause overflow during the process outlined below. Then we split the two numbers into m groups of w bits as follows

$$a = \sum_{i=0}^{m-1} a_i 2^{wi} \text{ and } b = \sum_{j=0}^{m-1} b_j 2^{wj}.$$

We look at these numbers as polynomials in x, where $x = 2^w$, to get,

$$a = \sum_{i=0}^{m-1} a_i x^i \text{ and } b = \sum_{j=0}^{m-1} b_j x^j.c$$

Then we can then say that,

$$ab = \sum_{i=0}^{m-1}\sum_{j=0}^{m-1} a_i b_j x^{(i+j)} = \sum_{k=0}^{2m-2} c_k x^k$$

Clearly the above setting is realized by polynomial multiplication, of two polynomials a and b. The crucial step now is to use Fast Fourier multiplication of polynomials to realize the multiplications above faster than in naive $O(m^2)$ time.

To remain in the modular setting of Fourier transforms, we look for a ring with a $2m^{th}$ root of unity. hence we do multiplication modulo N (and thus in the Z/NZ ring). Further, N must be chosen so that there is no 'wrap around', essentially, no reductions modulo N occur. Thus, the choice of N is crucial. For example, it could be done as,

$$N = 2^{3w} + 1$$

The ring Z/NZ would thus have a $2m^{th}$ root of unity, namely 8. Also, it can be checked that $c_k < N$, and thus no wrap around will occur.

The algorithm has a time complexity of $\Theta(n \log(n) \log(\log(n)))$ and is used in practice for numbers with more than 10,000 to 40,000 decimal digits. In 2007 this was improved by Martin Fürer (Fürer's algorithm) to give a time complexity of $n \log(n) 2^{\Theta(\log^*(n))}$ using Fourier transforms over complex numbers. Anindya De, Chandan Saha, Piyush Kurur and Ramprasad Saptharishi gave a similar algorithm using modular arithmetic in 2008 achieving the same running time. In context of the above material, what these latter authors have achieved is to find N much less than $2^{3k} + 1$, so that Z/NZ has a $2m^{th}$ root of unity. This speeds up computation and reduces the time complexity. However, these latter algorithms are only faster than Schönhage–Strassen for impractically large inputs.

Using number-theoretic transforms instead of discrete Fourier transforms avoids rounding error problems by using modular arithmetic instead of floating-point arithmetic. In order to apply the factoring which enables the FFT to work, the length of the transform must be factorable to small primes and must be a factor of N-1, where N is the field size. In particular, calculation using a Galois Field GF(k^2), where k is a Mersenne Prime, allows the use of a transform sized to a power of 2; e.g. $k = 2^{31}$-1 supports transform sizes up to 2^{32}.

Lower Bounds

There is a trivial lower bound of $\Omega(n)$ for multiplying two n-bit numbers on a single processor; no matching algorithm (on conventional Turing machines) nor any better lower bound is known. Multiplication lies outside of AC°[p] for any prime p, meaning there is no family of constant-depth, polynomial (or even subexponential) size circuits using AND, OR, NOT, and MOD$_p$ gates that can compute a product. This follows from a constant-depth reduction of MOD$_q$ to multiplication. Lower bounds for multiplication are also known for some classes of branching programs.

Polynomial Multiplication

All the above multiplication algorithms can also be expanded to multiply polynomials. For instance the Strassen algorithm may be used for polynomial multiplication Alternatively the Kronecker substitution technique may be used to convert the problem of multiplying polynomials into a single binary multiplication.

Euclidean Algorithm

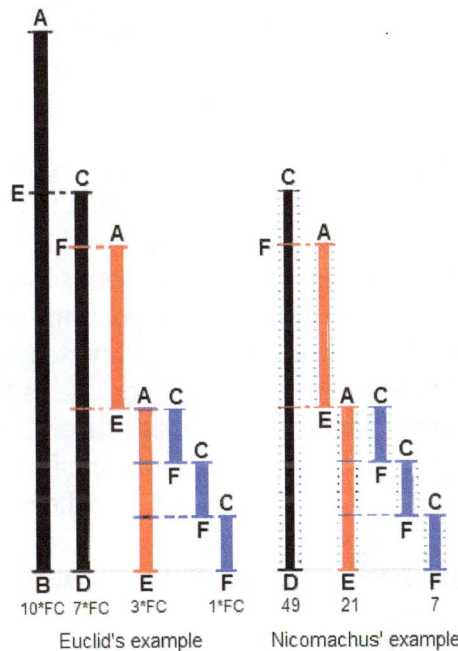

Euclid's method for finding the greatest common divisor (GCD) of two starting lengths BA and DC, both defined to be multiples of a common "unit" length. The length DC being shorter, it is used to "measure" BA, but only once because remainder EA is less than DC. EA now measures (twice) the shorter length DC, with remainder FC shorter than EA. Then FC measures (three times) length EA. Because there is no remainder, the process ends with FC being the GCD. On the right Nicomachus' example with numbers 49 and 21 resulting in their GCD of 7 (derived from Heath 1908:300).

In mathematics, the Euclidean algorithm, or Euclid's algorithm, is an efficient method for computing the greatest common divisor (GCD) of two numbers, the largest number that divides both of them without leaving a remainder. It is named after the ancient Greek mathematician Euclid, who first described it in Euclid's *Elements* (c. 300 BC). It is an example of an *algorithm*, a step-by-step procedure for performing a calculation according to well-defined rules, and is one of the oldest algorithms in common use. It can be used to reduce fractions to their simplest form, and is a part of many other number-theoretic and cryptographic calculations.

The Euclidean algorithm is based on the principle that the greatest common divisor of two numbers does not change if the larger number is replaced by its difference with the smaller number. For example, 21 is the GCD of 252 and 105 ($252 = 21 \times 12$ and $105 = 21 \times 5$), and the same number 21 is also the GCD of 105 and $147 = 252 - 105$. Since this replacement reduces the larger of the two numbers, repeating this process gives successively smaller pairs of numbers until the two numbers become equal. When that occurs, they are the GCD of the original two numbers. By reversing the steps, the GCD can be expressed as a sum of the two original numbers each multiplied by a positive or negative integer, e.g., $21 = 5 \times 105 + (-2) \times 252$. The fact that the GCD can always be expressed in this way is known as Bézout's identity.

The version of the Euclidean algorithm described above (and by Euclid) can take many subtraction steps to find the GCD when one of the given numbers is much bigger than the other. A more efficient version of the algorithm shortcuts these steps, instead replacing the larger of the two numbers by its remainder when divided by the smaller of the two (with this version, the algorithm stops when reaching a zero remainder). With this improvement, the algorithm never requires more steps than five times the number of digits (base 10) of the smaller integer. This was proven by Gabriel Lamé in 1844, and marks the beginning of computational complexity theory. Additional methods for improving the algorithm's efficiency were developed in the 20th century.

The Euclidean algorithm has many theoretical and practical applications. It is used for reducing fractions to their simplest form and for performing division in modular arithmetic. Computations using this algorithm form part of the cryptographic protocols that are used to secure internet communications, and in methods for breaking these cryptosystems by factoring large composite numbers. The Euclidean algorithm may be used to solve Diophantine equations, such as finding numbers that satisfy multiple congruences according to the Chinese remainder theorem, to construct continued fractions, and to find accurate rational approximations to real numbers. Finally, it can be used as a basic tool for proving theorems in number theory such as Lagrange's four-square theorem and the uniqueness of prime factorizations. The original algorithm was described only for natural numbers and geometric lengths (real numbers), but the algorithm was generalized in the 19th century to other types of numbers, such as Gaussian integers and polynomials of one variable. This led to modern abstract algebraic notions such as Euclidean domains.

Background: Greatest Common Divisor

The Euclidean algorithm calculates the greatest common divisor (GCD) of two natural numbers a and b. The greatest common divisor g is the largest natural number that divides both a and b without leaving a remainder. Synonyms for the GCD include the *greatest common factor* (GCF), the *highest common factor* (HCF), the *highest common divisor* (HCD), and the *greatest common measure* (GCM). The greatest common divisor is often written as $\gcd(a, b)$ or, more simply, as

(a, b), although the latter notation is also used for other mathematical concepts, such as two-dimensional vectors.

If $\gcd(a, b) = 1$, then a and b are said to be coprime (or relatively prime). This property does not imply that a or b are themselves prime numbers. For example, neither 6 nor 35 is a prime number, since they both have two prime factors: $6 = 2 \times 3$ and $35 = 5 \times 7$. Nevertheless, 6 and 35 are coprime. No natural number other than 1 divides both 6 and 35, since they have no prime factors in common.

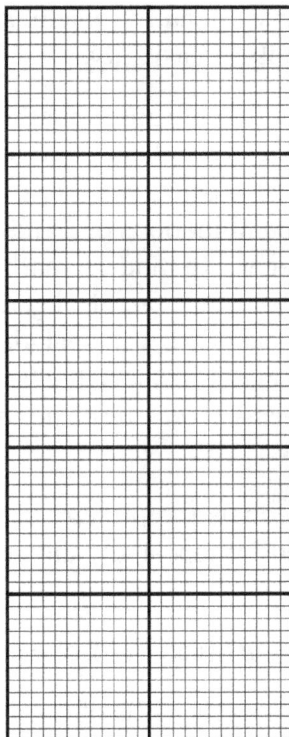

A 24-by-60 rectangle is covered with ten 12-by-12 square tiles, where 12 is the GCD of 24 and 60. More generally, an a-by-b rectangle can be covered with square tiles of side-length c only if c is a common divisor of a and b.

Let $g = \gcd(a, b)$. Since a and b are both multiples of g, they can be written $a = mg$ and $b = ng$, and there is no larger number $G > g$ for which this is true. The natural numbers m and n must be coprime, since any common factor could be factored out of m and n to make g greater. Thus, any other number c that divides both a and b must also divide g. The greatest common divisor g of a and b is the unique (positive) common divisor of a and b that is divisible by any other common divisor c.

The GCD can be visualized as follows. Consider a rectangular area a by b, and any common divisor c that divides both a and b exactly. The sides of the rectangle can be divided into segments of length c, which divides the rectangle into a grid of squares of side length c. The greatest common divisor g is the largest value of c for which this is possible. For illustration, a 24-by-60 rectangular area can be divided into a grid of: 1-by-1 squares, 2-by-2 squares, 3-by-3 squares, 4-by-4 squares, 6-by-6 squares or 12-by-12 squares. Therefore, 12 is the greatest common divisor of 24 and 60. A 24-by-60 rectangular area can be divided into a grid of 12-by-12 squares, with two squares along one edge (24/12 = 2) and five squares along the other (60/12 = 5).

The GCD of two numbers a and b is the product of the prime factors shared by the two numbers, where a same prime factor can be used multiple times, but only as long as the product of these factors divides both a and b. For example, since 1386 can be factored into $2 \times 3 \times 3 \times 7 \times 11$, and 3213 can be factored into $3 \times 3 \times 3 \times 7 \times 17$, the greatest common divisor of 1386 and 3213 equals $63 = 3 \times 3 \times 7$, the product of their shared prime factors. If two numbers have no prime factors in common, their greatest common divisor is 1 (obtained here as an instance of the empty product), in other words they are coprime. A key advantage of the Euclidean algorithm is that it can find the GCD efficiently without having to compute the prime factors. Factorization of large integers is believed to be a computationally very difficult problem, and the security of many widely used cryptographic protocols is based upon its infeasibility.

Another definition of the GCD is helpful in advanced mathematics, particularly ring theory. The greatest common divisor g of two nonzero numbers a and b is also their smallest positive integral linear combination, that is, the smallest positive number of the form $ua + vb$ where u and v are integers. The set of all integral linear combinations of a and b is actually the same as the set of all multiples of g (mg, where m is an integer). In modern mathematical language, the ideal generated by a and b is the ideal generated by g alone (an ideal generated by a single element is called a principal ideal, and all ideals of the integers are principal ideals). Some properties of the GCD are in fact easier to see with this description, for instance the fact that any common divisor of a and b also divides the GCD (it divides both terms of $ua + vb$). The equivalence of this GCD definition with the other definitions is described below.

The GCD of three or more numbers equals the product of the prime factors common to all the numbers, but it can also be calculated by repeatedly taking the GCDs of pairs of numbers. For example,

$$\gcd(a, b, c) = \gcd(a, \gcd(b, c)) = \gcd(\gcd(a, b), c) = \gcd(\gcd(a, c), b).$$

Thus, Euclid's algorithm, which computes the GCD of two integers, suffices to calculate the GCD of arbitrarily many integers.

Description

Procedure

The Euclidean algorithm proceeds in a series of steps such that the output of each step is used as an input for the next one. Let k be an integer that counts the steps of the algorithm, starting with zero. Thus, the initial step corresponds to $k = 0$, the next step corresponds to $k = 1$, and so on.

Each step begins with two nonnegative remainders r_{k-1} and r_{k-2}. Since the algorithm ensures that the remainders decrease steadily with every step, r_{k-1} is less than its predecessor r_{k-2}. The goal of the kth step is to find a quotient q_k and remainder r_k that satisfy the equation

$$r_{k-2} = q_k r_{k-1} + r_k$$

and that have $r_k < r_{k-1}$. In other words, multiples of the smaller number r_{k-1} are subtracted from the larger number r_{k-2} until the remainder r_k is smaller than r_{k-1}.

In the initial step ($k = 0$), the remainders r_{-2} and r_{-1} equal a and b, the numbers for which the GCD is sought. In the next step ($k = 1$), the remainders equal b and the remainder r_0 of the initial step, and so on. Thus, the algorithm can be written as a sequence of equations

$$a = q_0 b + r_0 b = q_1 r_0 + r_1 r_0 = q_2 r_1 + r_2 r_1 = q_3 r_2 + r_3 \cdots$$

If a is smaller than b, the first step of the algorithm swaps the numbers. For example, if $a < b$, the initial quotient q_0 equals zero, and the remainder r_0 is a. Thus, r_k is smaller than its predecessor r_{k-1} for all $k \geq 0$.

Since the remainders decrease with every step but can never be negative, a remainder r_N must eventually equal zero, at which point the algorithm stops. The final nonzero remainder r_{N-1} is the greatest common divisor of a and b. The number N cannot be infinite because there are only a finite number of nonnegative integers between the initial remainder r_0 and zero.

Proof of Validity

The validity of the Euclidean algorithm can be proven by a two-step argument. In the first step, the final nonzero remainder r_{N-1} is shown to divide both a and b. Since it is a common divisor, it must be less than or equal to the greatest common divisor g. In the second step, it is shown that any common divisor of a and b, including g, must divide r_{N-1}; therefore, g must be less than or equal to r_{N-1}. These two conclusions are inconsistent unless $r_{N-1} = g$.

To demonstrate that r_{N-1} divides both a and b (the first step), r_{N-1} divides its predecessor r_{N-2}

$$r_{N-2} = q_N r_{N-1}$$

since the final remainder r_N is zero. r_{N-1} also divides its next predecessor r_{N-3}

$$r_{N-3} = q_{N-1} r_{N-2} + r_{N-1}$$

because it divides both terms on the right-hand side of the equation. Iterating the same argument, r_{N-1} divides all the preceding remainders, including a and b. None of the preceding remainders r_{N-2}, r_{N-3}, etc. divide a and b, since they leave a remainder. Since r_{N-1} is a common divisor of a and b, $r_{N-1} \leq g$.

In the second step, any natural number c that divides both a and b (in other words, any common divisor of a and b) divides the remainders r_k. By definition, a and b can be written as multiples of c: $a = mc$ and $b = nc$, where m and n are natural numbers. Therefore, c divides the initial remainder r_0, since $r_0 = a - q_0 b = mc - q_0 nc = (m - q_0 n)c$. An analogous argument shows that c also divides the subsequent remainders r_1, r_2, etc. Therefore, the greatest common divisor g must divide r_{N-1}, which implies that $g \leq r_{N-1}$. Since the first part of the argument showed the reverse ($r_{N-1} \leq g$), it follows that $g = r_{N-1}$. Thus, g is the greatest common divisor of all the succeeding pairs:

$$g = \gcd(a, b) = \gcd(b, r_0) = \gcd(r_0, r_1) = \ldots = \gcd(r_{N-2}, r_{N-1}) = r_{N-1}.$$

Worked Example

Subtraction-based animation of the Euclidean algorithm. The initial rectangle has dimensions $a = 1071$ and $b = 462$. Squares of size 462×462 are placed within it leaving a 462×147 rectangle. This rectangle is tiled with 147×147 squares until a 21×147 rectangle is left, which in turn is tiled with 21×21 squares, leaving no uncovered area. The smallest square size, 21, is the GCD of 1071 and 462.

For illustration, the Euclidean algorithm can be used to find the greatest common divisor of $a = 1071$ and $b = 462$. To begin, multiples of 462 are subtracted from 1071 until the remainder is less than 462. Two such multiples can be subtracted ($q_0 = 2$), leaving a remainder of 147:

$$1071 = 2 \times 462 + 147.$$

Then multiples of 147 are subtracted from 462 until the remainder is less than 147. Three multiples can be subtracted ($q_1 = 3$), leaving a remainder of 21:

$$462 = 3 \times 147 + 21.$$

Then multiples of 21 are subtracted from 147 until the remainder is less than 21. Seven multiples can be subtracted ($q_2 = 7$), leaving no remainder:

$$147 = 7 \times 21 + 0.$$

Since the last remainder is zero, the algorithm ends with 21 as the greatest common divisor of 1071 and 462. This agrees with the gcd(1071, 462) found by prime factorization above. In tabular form, the steps are

Step k	Equation	Quotient and remainder
0	$1071 = q_0\, 462 + r_0$	$q_0 = 2$ and $r_0 = 147$
1	$462 = q_1\, 147 + r_1$	$q_1 = 3$ and $r_1 = 21$
2	$147 = q_2\, 21 + r_2$	$q_2 = 7$ and $r_2 = 0$; algorithm ends

Visualization

The Euclidean algorithm can be visualized in terms of the tiling analogy given above for the greatest common divisor. Assume that we wish to cover an a-by-b rectangle with square tiles exactly, where a is the larger of the two numbers. We first attempt to tile the rectangle using b-by-b square tiles; however, this leaves an r_0-by-b residual rectangle untiled, where $r_0 < b$. We then attempt to tile the residual rectangle with r_0-by-r_0 square tiles. This leaves a second residual rectangle r_1-by-r_0, which we attempt to tile using r_1-by-r_1 square tiles, and so on. The sequence ends when there is no residual rectangle, i.e., when the square tiles cover the previous residual rectangle exactly. The length of the sides of the smallest square tile is the GCD of the dimensions of the original rectangle. For example, the smallest square tile in the adjacent figure is 21-by-21 (shown in red), and 21 is the GCD of 1071 and 462, the dimensions of the original rectangle (shown in green).

Euclidean Division

At every step k, the Euclidean algorithm computes a quotient q_k and remainder r_k from two numbers r_{k-1} and r_{k-2}

$$r_{k-2} = q_k \, r_{k-1} + r_k$$

where the magnitude of r_k is strictly less than that of r_{k-1}. The theorem which underlies the definition of the Euclidean division ensures that such a quotient and remainder always exist and are unique.

In Euclid's original version of the algorithm, the quotient and remainder are found by repeated subtraction; that is, r_{k-1} is subtracted from r_{k-2} repeatedly until the remainder r_k is smaller than r_{k-1}. After that r_k and r_{k-1} are exchanged and the process is iterated. Euclidean division reduces all the steps between two exchanges into a single step, which is thus more efficient. Moreover, the quotients are not needed, thus one may replace Euclidean division by the modulo operation, which gives only the remainder. Thus the iteration of the Euclidean algorithm becomes simply

$$r_k = r_{k-2} \bmod r_{k-1}.$$

Implementations

Implementations of the algorithm may be expressed in pseudocode. For example, the division-based version may be programmed as

```
function gcd(a, b)
    while b ≠ 0
        t := b;
        b := a mod b;
        a := t;
    return a;
```

At the beginning of the kth iteration, the variable b holds the latest remainder r_{k-1}, whereas the variable a holds its predecessor, r_{k-2}. The step $b := a \bmod b$ is equivalent to the above recursion formula $r_k \equiv r_{k-2} \bmod r_{k-1}$. The temporary variable t holds the value of r_{k-1} while the next remainder r_k is being calculated. At the end of the loop iteration, the variable b holds the remainder r_k, whereas the variable a holds its predecessor, r_{k-1}.

In the subtraction-based version which was Euclid's original version, the remainder calculation ($b = a \bmod b$) is replaced by repeated subtraction. Contrary to the division-based version, which works with arbitrary integers as input, the subtraction-based version supposes that the input consists of positive integers and stops when $a = b$:

```
function gcd(a, b)

    while a ≠ b

        if a > b

            a := a - b;

        else

            b := b - a;

    return a;
```

The variables a and b alternate holding the previous remainders r_{k-1} and r_{k-2}. Assume that a is larger than b at the beginning of an iteration; then a equals r_{k-2}, since $r_{k-2} > r_{k-1}$. During the loop iteration, a is reduced by multiples of the previous remainder b until a is smaller than b. Then a is the next remainder r_k. Then b is reduced by multiples of a until it is again smaller than a, giving the next remainder r_{k+1}, and so on.

The recursive version is based on the equality of the GCDs of successive remainders and the stopping condition $\gcd(r_{N-1}, 0) = r_{N-1}$.

```
function gcd(a, b)

    if b = 0

        return a;

    else

        return gcd(b, a mod b);
```

For illustration, the $\gcd(1071, 462)$ is calculated from the equivalent $\gcd(462, 1071 \bmod 462) = \gcd(462, 147)$. The latter GCD is calculated from the $\gcd(147, 462 \bmod 147) = \gcd(147, 21)$, which in turn is calculated from the $\gcd(21, 147 \bmod 21) = \gcd(21, 0) = 21$.

Method of Least Absolute Remainders

In another version of Euclid's algorithm, the quotient at each step is increased by one if the resulting negative remainder is smaller in magnitude than the typical positive remainder. Previously, the equation

$$r_{k-2} = q_k\, r_{k-1} + r_k$$

assumed that $|r_{k-1}| > r_k > 0$. However, an alternative negative remainder e_k can be computed:

$$r_{k-2} = (q_k + 1)\, r_{k-1} + e_k$$

if $r_{k-1} > 0$ or

$$r_{k-2} = (q_k - 1)\, r_{k-1} + e_k$$

if $r_{k-1} < 0$.

If r_k is replaced by e_k. when $|e_k| < |r_k|$, then one gets a variant of Euclidean algorithm such that

$$|r_k| \leq |r_{k-1}| / 2$$

at each step.

Leopold Kronecker has shown that this version requires the least number of steps of any version of Euclid's algorithm. More generally, it has been proven that, for every input numbers a and b,

the number of steps is minimal if and only if q_k is chosen in order that $\left|\dfrac{r_{k+1}}{r_k}\right| < \dfrac{1}{\varphi} \sim 0.618,$ where φ is the golden ratio.

Historical Development

The Euclidean algorithm was probably invented centuries before Euclid, shown here holding a compass.

The Euclidean algorithm is one of the oldest algorithms in common use. It appears in Euclid's *Elements* (c. 300 BC), specifically in Book 7 (Propositions 1–2) and Book 10 (Propositions 2–3). In Book 7, the algorithm is formulated for integers, whereas in Book 10, it is formulated for lengths of line segments. (In modern usage, one would say it was formulated there for real numbers. But lengths, areas, and volumes, represented as real numbers in modern usage, are not measured in the same units and there is no natural unit of length, area, or volume; the concept of real numbers was unknown at that time.) The latter algorithm is geometrical. The GCD of two lengths a and b corresponds to the greatest length g that measures a and b evenly; in other words, the lengths a and b are both integer multiples of the length g.

The algorithm was probably not discovered by Euclid, who compiled results from earlier mathematicians in his *Elements*. The mathematician and historian B. L. van der Waerden suggests that Book VII derives from a textbook on number theory written by mathematicians in the school of Pythagoras. The algorithm was probably known by Eudoxus of Cnidus (about 375 BC). The algorithm may even pre-date Eudoxus, judging from the use of the technical term ἀνθυφαίρεσις (*anthyphairesis*, reciprocal subtraction) in works by Euclid and Aristotle.

Centuries later, Euclid's algorithm was discovered independently both in India and in China, primarily to solve Diophantine equations that arose in astronomy and making accurate calendars. In the late 5th century, the Indian mathematician and astronomer Aryabhata described the algorithm as the "pulverizer", perhaps because of its effectiveness in solving Diophantine equations. Although a special case of the Chinese remainder theorem had already been described by Chinese mathematician and astronomer Sun Tzu, the general solution was published by Qin Jiushao in his 1247 book *Shushu Jiuzhang* (數書九章 *Mathematical Treatise in Nine Sections*). The Euclidean algorithm was first described in Europe in the second edition of Bachet's *Problèmes plaisants et délectables* (*Pleasant and enjoyable problems*, 1624). In Europe, it was likewise used to solve Diophantine equations and in developing continued fractions. The extended Euclidean algorithm was published by the English mathematician Nicholas Saunderson, who attributed it to Roger Cotes as a method for computing continued fractions efficiently.

In the 19th century, the Euclidean algorithm led to the development of new number systems, such as Gaussian integers and Eisenstein integers. In 1815, Carl Gauss used the Euclidean algorithm to demonstrate unique factorization of Gaussian integers, although his work was first published in 1832. Gauss mentioned the algorithm in his *Disquisitiones Arithmeticae* (published 1801), but only as a method for continued fractions. Peter Gustav Lejeune Dirichlet seems to have been the first to describe the Euclidean algorithm as the basis for much of number theory. Lejeune Dirichlet noted that many results of number theory, such as unique factorization, would hold true for any other system of numbers to which the Euclidean algorithm could be applied. Lejeune Dirichlet's lectures on number theory were edited and extended by Richard Dedekind, who used Euclid's algorithm to study algebraic integers, a new general type of number. For example, Dedekind was the first to prove Fermat's two-square theorem using the unique factorization of Gaussian integers. Dedekind also defined the concept of a Euclidean domain, a number system in which a generalized version of the Euclidean algorithm can be defined (as described below). In the closing decades of the 19th century, the Euclidean algorithm gradually became eclipsed by Dedekind's more general theory of ideals.

"[The Euclidean algorithm] is the granddaddy of all algorithms, because it is the oldest nontrivial algorithm that has survived to the present day."

Donald Knuth, *The Art of Computer Programming, Vol. 2: Seminumerical Algorithms*, 2nd edition (1981), p. 318.

Other applications of Euclid's algorithm were developed in the 19th century. In 1829, Charles Sturm showed that the algorithm was useful in the Sturm chain method for counting the real roots of polynomials in any given interval.

The Euclidean algorithm was the first integer relation algorithm, which is a method for finding integer relations between commensurate real numbers. Several novel integer relation algorithms have been developed, such as the algorithm of Helaman Ferguson and R.W. Forcade (1979) and the LLL algorithm.

In 1969, Cole and Davie developed a two-player game based on the Euclidean algorithm, called *The Game of Euclid*, which has an optimal strategy. The players begin with two piles of a and b stones. The players take turns removing m multiples of the smaller pile from the larger. Thus, if the two piles consist of x and y stones, where x is larger than y, the next player can reduce the larger pile from x stones to $x - my$ stones, as long as the latter is a nonnegative integer. The winner is the first player to reduce one pile to zero stones.

Mathematical Applications

Bézout's Identity

Bézout's identity states that the greatest common divisor g of two integers a and b can be represented as a linear sum of the original two numbers a and b. In other words, it is always possible to find integers s and t such that $g = sa + tb$.

The integers s and t can be calculated from the quotients q_0, q_1, etc. by reversing the order of equations in Euclid's algorithm. Beginning with the next-to-last equation, g can be expressed in terms of the quotient q_{N-1} and the two preceding remainders, r_{N-2} and r_{N-3}:

$$g = r_{N-1} = r_{N-3} - q_{N-1} r_{N-2}.$$

Those two remainders can be likewise expressed in terms of their quotients and preceding remainders,

$$r_{N-2} = r_{N-4} - q_{N-2} r_{N-3} \text{ and}$$

$$r_{N-3} = r_{N-5} - q_{N-3} r_{N-4}.$$

Substituting these formulae for r_{N-2} and r_{N-3} into the first equation yields g as a linear sum of the remainders r_{N-4} and r_{N-5}. The process of substituting remainders by formulae involving their predecessors can be continued until the original numbers a and b are reached:

$$r_2 = r_0 - q_2 r_1$$

$$r_1 = b - q_1 r_0$$

$$r_0 = a - q_0 b.$$

After all the remainders r_0, r_1, etc. have been substituted, the final equation expresses g as a linear sum of a and b: $g = sa + tb$. Bézout's identity, and therefore the previous algorithm, can both be generalized to the context of Euclidean domains.

Principal Ideals and Related Problems

Bézout's identity provides yet another definition of the greatest common divisor g of two numbers a and b. Consider the set of all numbers $ua + vb$, where u and v are any two integers. Since a and b are both divisible by g, every number in the set is divisible by g. In other words, every number of the set is an integer multiple of g. This is true for every common divisor of a and b. However, unlike other common divisors, the greatest common divisor is a member of the set; by Bézout's identity, choosing $u = s$ and $v = t$ gives g. A smaller common divisor cannot be a member of the set, since every member of the set must be divisible by g. Conversely, any multiple m of g can be obtained by choosing $u = ms$ and $v = mt$, where s and t are the integers of Bézout's identity. This may be seen by multiplying Bézout's identity by m,

$$mg = msa + mtb.$$

Therefore, the set of all numbers $ua + vb$ is equivalent to the set of multiples m of g. In other words, the set of all possible sums of integer multiples of two numbers (a and b) is equivalent to the set of multiples of $\gcd(a, b)$. The GCD is said to be the generator of the ideal of a and b. This GCD definition led to the modern abstract algebraic concepts of a principal ideal (an ideal generated by a single element) and a principal ideal domain (a domain in which every ideal is a principal ideal).

Certain problems can be solved using this result. For example, consider two measuring cups of volume a and b. By adding/subtracting u multiples of the first cup and v multiples of the second cup, any volume $ua + vb$ can be measured out. These volumes are all multiples of $g = \gcd(a, b)$.

Extended Euclidean Algorithm

The integers s and t of Bézout's identity can be computed efficiently using the extended Euclidean algorithm. This extension adds two recursive equations to Euclid's algorithm

$$s_k = s_{k-2} - q_k s_{k-1}$$

$$t_k = t_{k-2} - q_k t_{k-1}$$

with the starting values

$$s_{-2} = 1,\ t_{-2} = 0$$

$$s_{-1} = 0,\ t_{-1} = 1.$$

Using this recursion, Bézout's integers s and t are given by $s = s_N$ and $t = t_N$, where $N+1$ is the step on which the algorithm terminates with $r_{N+1} = 0$.

The validity of this approach can be shown by induction. Assume that the recursion formula is correct up to step $k - 1$ of the algorithm; in other words, assume that

$$r_j = s_j\, a + t_j\, b$$

for all j less than k. The kth step of the algorithm gives the equation

$$r_k = r_{k-2} - q_k r_{k-1}.$$

Since the recursion formula has been assumed to be correct for r_{k-2} and r_{k-1}, they may be expressed in terms of the corresponding s and t variables

$$r_k = (s_{k-2} \, a + t_{k-2} \, b) - q_k(s_{k-1} \, a + t_{k-1} \, b).$$

Rearranging this equation yields the recursion formula for step k, as required

$$r_k = s_k \, a + t_k \, b = (s_{k-2} - q_k s_{k-1}) \, a + (t_{k-2} - q_k t_{k-1}) \, b.$$

Matrix Method

The integers s and t can also be found using an equivalent matrix method. The sequence of equations of Euclid's algorithm

$$a = q_0 \, b + r_0$$
$$b = q_1 \, r_0 + r_1$$
$$\cdots$$
$$r_{N-2} = q_N \, r_{N-1} + 0$$

can be written as a product of 2-by-2 quotient matrices multiplying a two-dimensional remainder vector

$$\begin{pmatrix} a \\ b \end{pmatrix} = \begin{pmatrix} q_0 & 1 \\ 1 & 0 \end{pmatrix} \begin{pmatrix} b \\ r_0 \end{pmatrix} = \begin{pmatrix} q_0 & 1 \\ 1 & 0 \end{pmatrix} \begin{pmatrix} q_1 & 1 \\ 1 & 0 \end{pmatrix} \begin{pmatrix} r_0 \\ r_1 \end{pmatrix} = \cdots = \prod_{i=0}^{N} \begin{pmatrix} q_i & 1 \\ 1 & 0 \end{pmatrix} \begin{pmatrix} r_{N-1} \\ 0 \end{pmatrix}.$$

Let M represent the product of all the quotient matrices

$$\mathbf{M} = \begin{pmatrix} m_{11} & m_{12} \\ m_{21} & m_{22} \end{pmatrix} = \prod_{i=0}^{N} \begin{pmatrix} q_i & 1 \\ 1 & 0 \end{pmatrix} = \begin{pmatrix} q_0 & 1 \\ 1 & 0 \end{pmatrix} \begin{pmatrix} q_1 & 1 \\ 1 & 0 \end{pmatrix} \cdots \begin{pmatrix} q_N & 1 \\ 1 & 0 \end{pmatrix}.$$

This simplifies the Euclidean algorithm to the form

$$\begin{pmatrix} a \\ b \end{pmatrix} = \mathbf{M} \begin{pmatrix} r_{N-1} \\ 0 \end{pmatrix} = \mathbf{M} \begin{pmatrix} g \\ 0 \end{pmatrix}.$$

To express g as a linear sum of a and b, both sides of this equation can be multiplied by the inverse of the matrix M. The determinant of M equals $(-1)^{N+1}$, since it equals the product of the determinants of the quotient matrices, each of which is negative one. Since the determinant of M is never zero, the vector of the final remainders can be solved using the inverse of M

$$\begin{pmatrix} g \\ 0 \end{pmatrix} = \mathbf{M}^{-1} \begin{pmatrix} a \\ b \end{pmatrix} = (-1)^{N+1} \begin{pmatrix} m_{22} & -m_{12} \\ -m_{21} & m_{11} \end{pmatrix} \begin{pmatrix} a \\ b \end{pmatrix}.$$

Since the top equation gives

$$g = (-1)^{N+1} \left(m_{22} \, a - m_{12} \, b \right),$$

the two integers of Bézout's identity are $s = (-1)^{N+1} m_{22}$ and $t = (-1)^N m_{12}$. The matrix method is as efficient as the equivalent recursion, with two multiplications and two additions per step of the Euclidean algorithm.

Euclid's Lemma and Unique Factorization

Bézout's identity is essential to many applications of Euclid's algorithm, such as demonstrating the unique factorization of numbers into prime factors. To illustrate this, suppose that a number L can be written as a product of two factors u and v, that is, $L = uv$. If another number w also divides L but is coprime with u, then w must divide v, by the following argument: If the greatest common divisor of u and w is 1, then integers s and t can be found such that

$1 = su + tw$.

by Bézout's identity. Multiplying both sides by v gives the relation

$v = suv + twv = sL + twv$.

Since w divides both terms on the right-hand side, it must also divide the left-hand side, v. This result is known as Euclid's lemma. Specifically, if a prime number divides L, then it must divide at least one factor of L. Conversely, if a number w is coprime to each of a series of numbers $a_1, a_2, ...,$ a_n, then w is also coprime to their product, $a_1 \times a_2 \times ... \times a_n$.

Euclid's lemma suffices to prove that every number has a unique factorization into prime numbers. To see this, assume the contrary, that there are two independent factorizations of L into m and n prime factors, respectively

$$L = p_1 p_2 ... p_m = q_1 q_2 ... q_n \ .$$

Since each prime p divides L by assumption, it must also divide one of the q factors; since each q is prime as well, it must be that $p = q$. Iteratively dividing by the p factors shows that each p has an equal counterpart q; the two prime factorizations are identical except for their order. The unique factorization of numbers into primes has many applications in mathematical proofs, as shown below.

Linear Diophantine Equations

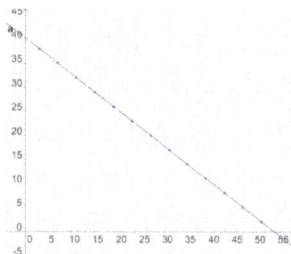

Plot of a linear Diophantine equation, $9x + 12y = 483$. The solutions are shown as blue circles.

Diophantine equations are equations in which the solutions are restricted to integers; they are named after the 3rd-century Alexandrian mathematician Diophantus. A typical *linear* Diophantine equation seeks integers x and y such that

$$ax + by = c$$

where a, b and c are given integers. This can be written as an equation for x in modular arithmetic:

$$ax \equiv c \bmod b.$$

Let g be the greatest common divisor of a and b. Both terms in $ax + by$ are divisible by g; therefore, c must also be divisible by g, or the equation has no solutions. By dividing both sides by c/g, the equation can be reduced to Bezout's identity

$$sa + tb = g$$

where s and t can be found by the extended Euclidean algorithm. This provides one solution to the Diophantine equation, $x_1 = s\,(c/g)$ and $y_1 = t\,(c/g)$.

In general, a linear Diophantine equation has no solutions, or an infinite number of solutions. To find the latter, consider two solutions, (x_1, y_1) and (x_2, y_2), where

$$ax_1 + by_1 = c = ax_2 + by_2$$

or equivalently

$$a(x_1 - x_2) = b(y_2 - y_1).$$

Therefore, the smallest difference between two x solutions is b/g, whereas the smallest difference between two y solutions is a/g. Thus, the solutions may be expressed as

$$x = x_1 - bu/g$$

$$y = y_1 + au/g.$$

By allowing u to vary over all possible integers, an infinite family of solutions can be generated from a single solution (x_1, y_1). If the solutions are required to be *positive* integers ($x > 0$, $y > 0$), only a finite number of solutions may be possible. This restriction on the acceptable solutions allows some systems of Diophantine equations with more unknowns than equations to have a finite number of solutions; this is impossible for a system of linear equations when the solutions can be any real number (see Underdetermined system).

Multiplicative Inverses and The RSA Algorithm

A finite field is a set of numbers with four generalized operations. The operations are called addition, subtraction, multiplication and division and have their usual properties, such as commutativity, associativity and distributivity. An example of a finite field is the set of 13 numbers {0, 1, 2, ..., 12} using modular arithmetic. In this field, the results of any mathematical operation (addition, subtraction, multiplication, or division) is reduced modulo 13; that is, multiples of 13 are added or subtracted until the result is brought within the range 0–12. For example, the result of $5 \times 7 = 35 \bmod 13 = 9$. Such finite fields can be defined for any prime p; using more sophisticated definitions, they can also be defined for any power m of a prime p^m. Finite fields are often called Galois fields, and are abbreviated as $GF(p)$ or $GF(p^m)$.

In such a field with m numbers, every nonzero element a has a unique modular multiplicative inverse, a^{-1} such that $aa^{-1} = a^{-1}a \equiv 1 \bmod m$. This inverse can be found by solving the congruence equation $ax \equiv 1 \bmod m$, or the equivalent linear Diophantine equation

$$ax + my = 1.$$

This equation can be solved by the Euclidean algorithm, as described above. Finding multiplicative inverses is an essential step in the RSA algorithm, which is widely used in electronic commerce; specifically, the equation determines the integer used to decrypt the message. Note that although the RSA algorithm uses rings rather than fields, the Euclidean algorithm can still be used to find a multiplicative inverse where one exists. The Euclidean algorithm also has other applications in error-correcting codes; for example, it can be used as an alternative to the Berlekamp–Massey algorithm for decoding BCH and Reed–Solomon codes, which are based on Galois fields.

Chinese Remainder Theorem

Euclid's algorithm can also be used to solve multiple linear Diophantine equations. Such equations arise in the Chinese remainder theorem, which describes a novel method to represent an integer x. Instead of representing an integer by its digits, it may be represented by its remainders x_i modulo a set of N coprime numbers m_i:

$$x_1 \equiv x \bmod m_1, x_2 \equiv x \bmod m_2 \vdots x_N \equiv x \bmod m_N.$$

The goal is to determine x from its N remainders x_i. The solution is to combine the multiple equations into a single linear Diophantine equation with a much larger modulus M that is the product of all the individual moduli m_i, and define M_i as

$$M_i = \frac{M}{m_i}.$$

Thus, each M_i is the product of all the moduli *except* m_i. The solution depends on finding N new numbers h_i such that

$$M_i h_i \equiv 1 \bmod m_i.$$

With these numbers h_i, any integer x can be reconstructed from its remainders x_i by the equation

$$x \equiv (x_1 M_1 h_1 + x_2 M_2 h_2 + \cdots + x_N M_N h_N) \bmod M.$$

Since these numbers h_i are the multiplicative inverses of the M_i, they may be found using Euclid's algorithm as described in the previous subsection.

Stern–Brocot Tree

The Euclidean algorithm can be used to arrange the set of all positive rational numbers into an infinite binary search tree, called the Stern–Brocot tree. The number 1 (expressed as a fraction 1/1) is placed at the root of the tree, and the location of any other number a/b can be found by computing $\gcd(a,b)$ using the original form of the Euclidean algorithm, in which each step replaces the larger of the two given numbers by its difference with the smaller number (not its remainder), stopping when two equal numbers are reached. A step of the Euclidean algorithm that replaces the first of the two numbers corresponds to a step in the tree from a node to its right child, and a step that replaces the second of the two numbers corresponds to a step in the tree from a node to its left child. The sequence of steps constructed in this way does not depend on whether a/b is given in lowest terms, and forms a path from the root to a node containing the number a/b. This fact can be used to prove that each positive rational number appears exactly once in this tree.

For example, 3/4 can be found by starting at the root, going to the left once, then to the right twice:

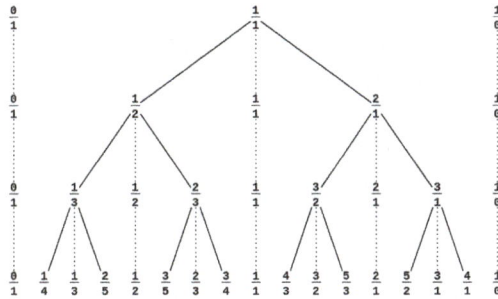

The Stern–Brocot tree, and the Stern–Brocot sequences of order i for i = 1, 2, 3, 4

$$\gcd(3,4) \longleftarrow= \gcd(3,1) \longrightarrow= \gcd(2,1) \longrightarrow= \gcd(1,1).$$

The Euclidean algorithm has almost the same relationship to another binary tree on the rational numbers called the Calkin–Wilf tree. The difference is that the path is reversed: instead of producing a path from the root of the tree to a target, it produces a path from the target to the root.

Continued Fractions

The Euclidean algorithm has a close relationship with continued fractions. The sequence of equations can be written in the form

$$\frac{a}{b} = q_0 + \frac{r_0}{b}\frac{b}{r_0} = q_1 + \frac{r_1}{r_0}\frac{r_0}{r_1} = q_2 + \frac{r_2}{r_1}:\frac{r_{k-2}}{r_{k-1}} = q_k + \frac{r_k}{r_{k-1}}:\frac{r_{N-2}}{r_{N-1}} = q_N.$$

The last term on the right-hand side always equals the inverse of the left-hand side of the next equation. Thus, the first two equations may be combined to form

$$\frac{a}{b} = q_0 + \cfrac{1}{q_1 + \cfrac{r_1}{r_0}}.$$

The third equation may be used to substitute the denominator term r_1/r_0, yielding

$$\frac{a}{b} = q_0 + \cfrac{1}{q_1 + \cfrac{1}{q_2 + \cfrac{r_2}{r_1}}}.$$

The final ratio of remainders r_k/r_{k-1} can always be replaced using the next equation in the series, up to the final equation. The result is a continued fraction

$$\frac{a}{b} = q_0 + \cfrac{1}{q_1 + \cfrac{1}{q_2 + \cfrac{1}{\ddots + \cfrac{1}{q_N}}}} = [q_0; q_1, q_2, \ldots, q_N].$$

In the worked example above, the gcd(1071, 462) was calculated, and the quotients q_k were 2, 3 and 7, respectively. Therefore, the fraction 1071/462 may be written

$$\frac{1071}{462} = 2 + \cfrac{1}{3 + \cfrac{1}{7}} = [2; 3, 7]$$

as can be confirmed by calculation.

Factorization Algorithms

Calculating a greatest common divisor is an essential step in several integer factorization algorithms, such as Pollard's rho algorithm, Shor's algorithm, Dixon's factorization method and the Lenstra elliptic curve factorization. The Euclidean algorithm may be used to find this GCD efficiently. Continued fraction factorization uses continued fractions, which are determined using Euclid's algorithm.

Algorithmic Efficiency

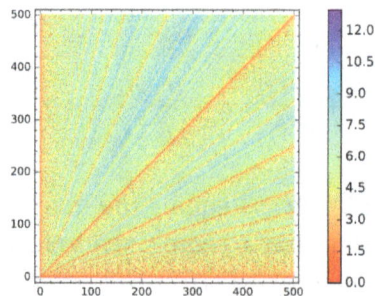

Number of steps in the Euclidean algorithm for gcd(x,y). Lighter (red and yellow) points indicate relatively few steps, whereas darker (violet and blue) points indicate more steps. The largest dark area follows the line $y = \Phi x$, where Φ represents the Golden ratio.

The computational efficiency of Euclid's algorithm has been studied thoroughly. This efficiency can be described by the number of division steps the algorithm requires, multiplied by the computational expense of each step. The first known analysis of Euclid's algorithm is due to A.-A.-L. Reynaud in 1811, who showed that the number of division steps on input (u, v) is bounded by v; later he improved this to $v/2 + 2$. Later, in 1841, P.-J.-E. Finck showed that the number of division steps is at most $2 \log_2 v + 1$, and hence Euclid's algorithm runs in time polynomial in the size of the input. Émile Léger, in 1837, studied the worst case, which is when the inputs are consecutive Fibonacci numbers. Finck's analysis was refined by Gabriel Lamé in 1844, who showed that the number of steps required for completion is never more than five times the number h of base-10 digits of the smaller number b.

In the uniform cost model (suitable for analyzing the complexity of gcd calculation on numbers that fit into a single machine word), each step of the algorithm takes constant time, and Lamé's analysis implies that the total running time is also $O(h)$. However, in a model of computation suitable for computation with larger numbers, the computational expense of a single remainder computation in the algorithm can be as large as $O(h^2)$. In this case the total time for all of the steps of the algorithm can be analyzed using a telescoping series, showing that it is also $O(h^2)$. Modern algorithmic techniques based on the Schönhage–Strassen algorithm for fast integer multiplication can be used to speed this up, leading to quasilinear algorithms for the GCD.

Number of Steps

The number of steps to calculate the GCD of two natural numbers, a and b, may be denoted by $T(a, b)$. If g is the GCD of a and b, then $a = mg$ and $b = ng$ for two coprime numbers m and n. Then

$$T(a, b) = T(m, n)$$

as may be seen by dividing all the steps in the Euclidean algorithm by g. By the same argument, the number of steps remains the same if a and b are multiplied by a common factor w: $T(a, b) = T(wa, wb)$. Therefore, the number of steps T may vary dramatically between neighboring pairs of numbers, such as $T(a, b)$ and $T(a, b + 1)$, depending on the size of the two GCDs.

The recursive nature of the Euclidean algorithm gives another equation

$$T(a, b) = 1 + T(b, r_0) = 2 + T(r_0, r_1) = \ldots = N + T(r_{N-2}, r_{N-1}) = N + 1$$

where $T(x, 0) = 0$ by assumption.

Worst-case

If the Euclidean algorithm requires N steps for a pair of natural numbers $a > b > 0$, the smallest values of a and b for which this is true are the Fibonacci numbers F_{N+2} and F_{N+1}, respectively. This can be shown by induction. If $N = 1$, b divides a with no remainder; the smallest natural numbers for which this is true is $b = 1$ and $a = 2$, which are F_2 and F_3, respectively. Now assume that the result holds for all values of N up to $M - 1$. The first step of the M-step algorithm is $a = q_0 b + r_0$, and the second step is $b = q_1 r_0 + r_1$. Since the algorithm is recursive, it required $M - 1$ steps to find gcd(b, r_0) and their smallest values are F_{M+1} and F_M. The smallest value of a is therefore when $q_0 = 1$, which gives $a = b + r_0 = F_{M+1} + F_M = F_{M+2}$. This proof, published by Gabriel Lamé in 1844, represents the beginning of computational complexity theory, and also the first practical application of the Fibonacci numbers.

This result suffices to show that the number of steps in Euclid's algorithm can never be more than five times the number of its digits (base 10). For if the algorithm requires N steps, then b is greater than or equal to F_{N+1} which in turn is greater than or equal to φ^{N-1}, where φ is the golden ratio. Since $b \geq \varphi^{N-1}$, then $N - 1 \leq \log_\varphi b$. Since $\log_{10} \varphi > 1/5$, $(N - 1)/5 < \log_{10} \varphi \log_\varphi b = \log_{10} b$. Thus, $N \leq 5 \log_{10} b$. Thus, the Euclidean algorithm always needs less than $O(h)$ divisions, where h is the number of digits in the smaller number b.

Average

The average number of steps taken by the Euclidean algorithm has been defined in three different ways. The first definition is the average time $T(a)$ required to calculate the GCD of a given number a and a smaller natural number b chosen with equal probability from the integers 0 to $a - 1$

$$T(a) = \frac{1}{a} \sum_{0 \leq b < a} T(a,b).$$

However, since $T(a, b)$ fluctuates dramatically with the GCD of the two numbers, the averaged function $T(a)$ is likewise "noisy".

To reduce this noise, a second average $\tau(a)$ is taken over all numbers coprime with a

$$\tau(a) = \frac{1}{\varphi(a)} \sum_{\substack{0 \leq b < a \\ \gcd(a,b)=1}} T(a,b).$$

There are $\varphi(a)$ coprime integers less than a, where φ is Euler's totient function. This tau average grows smoothly with a

$$\tau(a) = \frac{12}{\pi^2} \ln 2 \ln a + C + O(a^{-1/6-\epsilon})$$

with the residual error being of order $a^{-(1/6) + \epsilon}$, where ε is infinitesimal. The constant C (*Porter's Constant*) in this formula equals

$$C = -\frac{1}{2} + \frac{6 \ln 2}{\pi^2} (4\gamma - 24\pi^2 \zeta'(2) + 3 \ln 2 - 2) \approx 1.467$$

where γ is the Euler–Mascheroni constant and ζ is the derivative of the Riemann zeta function. The leading coefficient $(12/\pi^2) \ln 2$ was determined by two independent methods.

Since the first average can be calculated from the tau average by summing over the divisors d of a

$$T(a) = \frac{1}{a} \sum_{d|a} \varphi(d)\tau(d)$$

it can be approximated by the formula

$$T(a) \approx C + \frac{12}{\pi^2} \ln 2 \left(\ln a - \sum_{d|a} \frac{\Lambda(d)}{d} \right)$$

where $\Lambda(d)$ is the Mangoldt function.

A third average $Y(n)$ is defined as the mean number of steps required when both a and b are chosen randomly (with uniform distribution) from 1 to n

$$Y(n) = \frac{1}{n^2} \sum_{a=1}^{n} \sum_{b=1}^{n} T(a,b) = \frac{1}{n} \sum_{a=1}^{n} T(a).$$

Substituting the approximate formula for $T(a)$ into this equation yields an estimate for $Y(n)$

$$Y(n) \approx \frac{12}{\pi^2} \ln 2 \ln n + 0.06.$$

Computational Expense Per Step

In each step k of the Euclidean algorithm, the quotient q_k and remainder r_k are computed for a given pair of integers r_{k-2} and r_{k-1}

$$r_{k-2} = q_k r_{k-1} + r_k.$$

The computational expense per step is associated chiefly with finding q_k, since the remainder r_k can be calculated quickly from r_{k-2}, r_{k-1}, and q_k

$$r_k = r_{k-2} - q_k r_{k-1}.$$

The computational expense of dividing h-bit numbers scales as $O(h(\ell+1))$, where ℓ is the length of the quotient.

For comparison, Euclid's original subtraction-based algorithm can be much slower. A single integer division is equivalent to the quotient q number of subtractions. If the ratio of a and b is very large, the quotient is large and many subtractions will be required. On the other hand, it has been shown that the quotients are very likely to be small integers. The probability of a given quotient q is approximately $\ln|u/(u-1)|$ where $u = (q+1)^2$. For illustration, the probability of a quotient of 1, 2, 3, or 4 is roughly 41.5%, 17.0%, 9.3%, and 5.9%, respectively. Since the operation of subtraction is faster than division, particularly for large numbers, the subtraction-based Euclid's algorithm is competitive with the division-based version. This is exploited in the binary version of Euclid's algorithm.

Combining the estimated number of steps with the estimated computational expense per step shows that the Euclid's algorithm grows quadratically (h^2) with the average number of digits h in the initial two numbers a and b. Let $h_0, h_1, ..., h_{N-1}$ represent the number of digits in the successive remainders $r_0, r_1, ..., r_{N-1}$. Since the number of steps N grows linearly with h, the running time is bounded by

$$O\left(\sum_{i<N} h_i(h_i - h_{i+1} + 2)\right) \subseteq O\left(h\sum_{i<N}(h_i - h_{i+1} + 2)\right) \subseteq O(h(h_0 + 2N)) \subseteq O(h^2).$$

Alternative Methods

Euclid's algorithm is widely used in practice, especially for small numbers, due to its simplicity. For comparison, the efficiency of alternatives to Euclid's algorithm may be determined.

One inefficient approach to finding the GCD of two natural numbers a and b is to calculate all their common divisors; the GCD is then the largest common divisor. The common divisors can be found by dividing both numbers by successive integers from 2 to the smaller number b. The number of steps of this approach grows linearly with b, or exponentially in the number of digits. Another inefficient approach is to find the prime factors of one or both numbers. As noted above, the GCD equals the product of the prime factors shared by the two numbers a and b. Present methods for prime factorization are also inefficient; many modern cryptography systems even rely on that inefficiency.

The binary GCD algorithm is an efficient alternative that substitutes division with faster operations by exploiting the binary representation used by computers. However, this alternative also scales like $O(h^2)$. It is generally faster than the Euclidean algorithm on real computers, even though it scales in the same way. Additional efficiency can be gleaned by examining only the leading digits of the two numbers a and b. The binary algorithm can be extended to other bases (k-ary algorithms), with up to fivefold increases in speed. Lehmer's GCD algorithm uses the same general principle as the binary algorithm to speed up GCD computations in arbitrary bases.

A recursive approach for very large integers (with more than 25,000 digits) leads to quasilinear integer GCD algorithms, such as those of Schönhage, and Stehlé and Zimmermann. These algorithms exploit the 2×2 matrix form of the Euclidean algorithm given above. These quasilinear methods generally scale as $O(h (\log h)^2 (\log \log h))$.

Generalizations

Although the Euclidean algorithm is used to find the greatest common divisor of two natural numbers (positive integers), it may be generalized to the real numbers, and to other mathematical objects, such as polynomials, quadratic integers and Hurwitz quaternions. In the latter cases, the Euclidean algorithm is used to demonstrate the crucial property of unique factorization, i.e., that such numbers can be factored uniquely into irreducible elements, the counterparts of prime numbers. Unique factorization is essential to many proofs of number theory.

Rational and Real Numbers

Euclid's algorithm can be applied to real numbers, as described by Euclid in Book 10 of his *Elements*. The goal of the algorithm is to identify a real number g such that two given real numbers, a and b, are integer multiples of it: $a = mg$ and $b = ng$, where m and n are integers. This identification is equivalent to finding an integer relation among the real numbers a and b; that is, it determines integers s and t such that $sa + tb = 0$. Euclid uses this algorithm to treat the question of incommensurable lengths.

The real-number Euclidean algorithm differs from its integer counterpart in two respects. First, the remainders r_k are real numbers, although the quotients q_k are integers as before. Second, the algorithm is not guaranteed to end in a finite number N of steps. If it does, the fraction a/b is a rational number, i.e., the ratio of two integers

$$a/b = mg/ng = m/n$$

and can be written as a finite continued fraction $[q_0; q_1, q_2, ..., q_N]$. If the algorithm does not stop, the fraction a/b is an irrational number and can be described by an infinite continued fraction $[q_0; q_1, q_2, ...]$. Examples of infinite continued fractions are the golden ratio $\varphi = [1; 1, 1, ...]$ and the square root of two, $\sqrt{2} = [1; 2, 2, ...]$. The algorithm is unlikely to stop, since almost all ratios a/b of two real numbers are irrational.

An infinite continued fraction may be truncated at a step k $[q_0; q_1, q_2, ..., q_k]$ to yield an approximation to a/b that improves as k is increased. The approximation is described by convergents m_k/n_k; the numerator and denominators are coprime and obey the recurrence relation

$$m_k = q_k \, m_{k-1} + m_{k-2}$$

$$n_k = q_k \, n_{k-1} + n_{k-2}$$

where $m_{-1} = n_{-2} = 1$ and $m_{-2} = n_{-1} = 0$ are the initial values of the recursion. The convergent m_k/n_k is the best rational number approximation to a/b with denominator n_k:

$$\left| \frac{a}{b} - \frac{m_k}{n_k} \right| < \frac{1}{n_k^2}.$$

Polynomials

Polynomials in a single variable x can be added, multiplied and factored into irreducible polynomials, which are the analogs of the prime numbers for integers. The greatest common divisor polynomial $g(x)$ of two polynomials $a(x)$ and $b(x)$ is defined as the product of their shared irreducible polynomials, which can be identified using the Euclidean algorithm. The basic procedure is similar to integers. At each step k, a quotient polynomial $q_k(x)$ and a remainder polynomial $r_k(x)$ are identified to satisfy the recursive equation

$$r_{k-2}(x) = q_k(x) \, r_{k-1}(x) + r_k(x)$$

where $r_{-2}(x) = a(x)$ and $r_{-1}(x) = b(x)$. The quotient polynomial is chosen so that the leading term of $q_k(x) \, r_{k-1}(x)$ equals the leading term of $r_{k-2}(x)$; this ensures that the degree of each remainder is smaller than the degree of its predecessor $\deg[r_k(x)] < \deg[r_{k-1}(x)]$. Since the degree is a nonnegative integer, and since it decreases with every step, the Euclidean algorithm concludes in a finite number of steps. The final nonzero remainder is the greatest common divisor of the original two polynomials, $a(x)$ and $b(x)$.

For example, consider the following two quartic polynomials, which each factor into two quadratic polynomials

$$a(x) = x^4 - 4x^3 + 4x^2 - 3x + 14 = (x^2 - 5x + 7)(x^2 + x + 2)$$

and

$$b(x) = x^4 + 8x^3 + 12x^2 + 17x + 6 = (x^2 + 7x + 3)(x^2 + x + 2).$$

Dividing $a(x)$ by $b(x)$ yields a remainder $r_o(x) = x^3 + (2/3)x^2 + (5/3)x - (2/3)$. In the next step, $b(x)$ is divided by $r_o(x)$ yielding a remainder $r_1(x) = x^2 + x + 2$. Finally, dividing $r_o(x)$ by $r_1(x)$ yields a zero remainder, indicating that $r_1(x)$ is the greatest common divisor polynomial of $a(x)$ and $b(x)$, consistent with their factorization.

Many of the applications described above for integers carry over to polynomials. The Euclidean algorithm can be used to solve linear Diophantine equations and Chinese remainder problems for polynomials; continued fractions of polynomials can also be defined.

The polynomial Euclidean algorithm has other applications, such as Sturm chains, a method for counting the zeros of a polynomial that lie inside a given real interval. This in turn has applications in several areas, such as the Routh–Hurwitz stability criterion in control theory.

Finally, the coefficients of the polynomials need not be drawn from integers, real numbers or even the complex numbers. For example, the coefficients may be drawn from a general field, such as the finite fields GF(p) described above. The corresponding conclusions about the Euclidean algorithm and its applications hold even for such polynomials.

Gaussian Integers

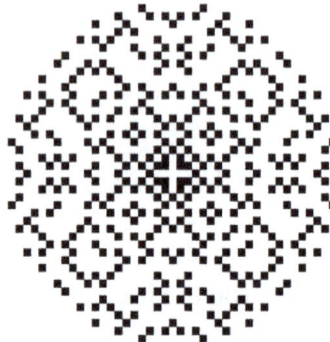

Distribution of Gaussian primes $u + vi$ in the complex plane, with norms $u^2 + v^2$ less than 500

The Gaussian integers are complex numbers of the form $\alpha = u + vi$, where u and v are ordinary integers and i is the square root of negative one. By defining an analog of the Euclidean algorithm, Gaussian integers can be shown to be uniquely factorizable, by the argument above. This unique factorization is helpful in many applications, such as deriving all Pythagorean triples or proving Fermat's theorem on sums of two squares. In general, the Euclidean algorithm is convenient in such applications, but not essential; for example, the theorems can often be proven by other arguments.

The Euclidean algorithm developed for two Gaussian integers α and β is nearly the same as that for normal integers, but differs in two respects. As before, the task at each step k is to identify a quotient q_k and a remainder r_k such that

$$r_k = r_{k-2} - q_k r_{k-1}$$

where $r_{k-2} = \alpha$, $r_{k-1} = \beta$, and every remainder is strictly smaller than its predecessor, $|r_k| < |r_{k-1}|$. The first difference is that the quotients and remainders are themselves Gaussian integers, and thus are complex numbers. The quotients q_k are generally found by rounding the real and complex parts of the exact ratio (such as the complex number α/β) to the nearest integers. The second difference lies in the necessity of defining how one complex remainder can be "smaller" than another. To do this, a norm function $f(u + vi) = u^2 + v^2$ is defined, which converts every Gaussian integer $u + vi$ into a normal integer. After each step k of the Euclidean algorithm, the norm of the remainder $f(r_k)$ is smaller than the norm of the preceding remainder, $f(r_{k-1})$. Since the norm is a nonnegative integer and decreases with every step, the Euclidean algorithm for Gaussian integers ends in a finite number of steps. The final nonzero remainder is the gcd(α,β), the Gaussian integer of largest norm that divides both α and β; it is unique up to multiplication by a unit, ± 1 or $\pm i$.

Many of the other applications of the Euclidean algorithm carry over to Gaussian integers. For example, it can be used to solve linear Diophantine equations and Chinese remainder problems for Gaussian integers; continued fractions of Gaussian integers can also be defined.

Euclidean Domains

A set of elements under two binary operations, + and −, is called a Euclidean domain if it forms a commutative ring R and, roughly speaking, if a generalized Euclidean algorithm can be performed on them. The two operations of such a ring need not be the addition and multiplication of ordinary arithmetic; rather, they can be more general, such as the operations of a mathematical group or monoid. Nevertheless, these general operations should respect many of the laws governing ordinary arithmetic, such as commutativity, associativity and distributivity.

The generalized Euclidean algorithm requires a *Euclidean function*, i.e., a mapping f from R into the set of nonnegative integers such that, for any two nonzero elements a and b in R, there exist q and r in R such that $a = qb + r$ and $f(r) < f(b)$. An example of this mapping is the norm function used to order the Gaussian integers above. The function f can be the magnitude of the number, or the degree of a polynomial. The basic principle is that each step of the algorithm reduces f inexorably; hence, if f can be reduced only a finite number of times, the algorithm must stop in a finite number of steps. This principle relies heavily on the natural well-ordering of the non-negative integers; roughly speaking, this requires that every non-empty set of non-negative integers has a smallest member.

The fundamental theorem of arithmetic applies to any Euclidean domain: Any number from a Euclidean domain can be factored uniquely into irreducible elements. Any Euclidean domain is a unique factorization domain (UFD), although the converse is not true. The Euclidean domains and the UFD's are subclasses of the GCD domains, domains in which a greatest common divisor of two numbers always exists. In other words, a greatest common divisor may exist (for all pairs of elements in a domain), although it may not be possible to find it using a Euclidean algorithm. A Euclidean domain is always a principal ideal domain (PID), an integral domain in which every ideal is a principal ideal. Again, the converse is not true: not every PID is a Euclidean domain.

The unique factorization of Euclidean domains is useful in many applications. For example, the unique factorization of the Gaussian integers is convenient in deriving formulae for all Pythagorean triples and in proving Fermat's theorem on sums of two squares. Unique factorization was also a key element in an attempted proof of Fermat's Last Theorem published in 1847 by Gabriel Lamé, the same mathematician who analyzed the efficiency of Euclid's algorithm, based on a suggestion of Joseph Liouville. Lamé's approach required the unique factorization of numbers of the form $x + \omega y$, where x and y are integers, and $\omega = e^{2i\pi/n}$ is an nth root of 1, that is, $\omega^n = 1$. Although this approach succeeds for some values of n (such as $n=3$, the Eisenstein integers), in general such numbers do *not* factor uniquely. This failure of unique factorization in some cyclotomic fields led Ernst Kummer to the concept of ideal numbers and, later, Richard Dedekind to ideals.

Unique Factorization of Quadratic Integers

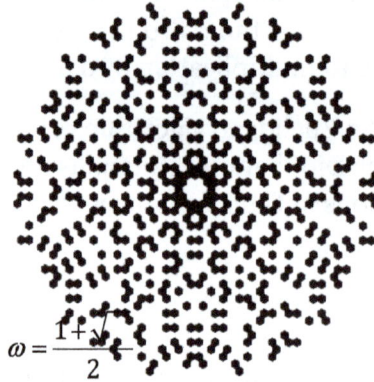

$$\omega = \frac{1+\sqrt{-3}}{2}$$

Distribution of Eisenstein primes $u + v\omega$ in the complex plane, with norms less than 500. The number ω equals the cube root of 1.

The quadratic integer rings are helpful to illustrate Euclidean domains. Quadratic integers are generalizations of the Gaussian integers in which the imaginary unit i is replaced by a number ω. Thus, they have the form $u + v\omega$, where u and v are integers and ω has one of two forms, depending on a parameter D. If D does not equal a multiple of four plus one, then

$$\omega = \sqrt{D}.$$

If, however, D does equal a multiple of four plus one, then

$$\omega = \frac{1+\sqrt{D}}{2}$$

If the function f corresponds to a norm function, such as that used to order the Gaussian integers above, then the domain is known as *norm-Euclidean*. The norm-Euclidean rings of quadratic integers are exactly those where $D = -11, -7, -3, -2, -1, 2, 3, 5, 6, 7, 11, 13, 17, 19, 21, 29, 33, 37, 41, 57$ or 73. The quadratic integers with $D = -1$ and -3 are known as the Gaussian integers and Eisenstein integers, respectively.

If f is allowed to be any Euclidean function, then the list of possible D values for which the domain is Euclidean is not yet known. The first example of a Euclidean domain that was not norm-Euclidean (with $D = 69$) was published in 1994. In 1973, Weinberger proved that a quadratic integer ring with $D > 0$ is Euclidean if, and only if, it is a principal ideal domain, provided that the generalized Riemann hypothesis holds.

Noncommutative Rings

The Euclidean algorithm may be applied to noncommutative rings such as the set of Hurwitz quaternions. Let α and β represent two elements from such a ring. They have a common right divisor δ if $\alpha = \xi\delta$ and $\beta = \eta\delta$ for some choice of ξ and η in the ring. Similarly, they have a common left divisor if $\alpha = \delta\xi$ and $\beta = \delta\eta$ for some choice of ξ and η in the ring. Since multiplication is not commutative, there are two versions of the Euclidean algorithm, one for right divisors and one for left divisors. Choosing the right divisors, the first step in finding the gcd(α, β) by the Euclidean algorithm can be written

$$\rho_0 = \alpha - \psi_0\beta = (\xi - \psi_0\eta)\delta$$

where ψ_0 represents the quotient and ρ_0 the remainder. This equation shows that any common right divisor of α and β is likewise a common divisor of the remainder ρ_0. The analogous equation for the left divisors would be

$$\rho_0 = \alpha - \beta\psi_0 = \delta(\xi - \eta\psi_0) \, .$$

With either choice, the process is repeated as above until the greatest common right or left divisor is identified. As in the Euclidean domain, the "size" of the remainder ρ_0 must be strictly smaller than β, and there must be only a finite number of possible sizes for ρ_0, so that the algorithm is guaranteed to terminate.

Most of the results for the GCD carry over to noncommutative numbers. For example, Bézout's identity states that the right gcd(α, β) can be expressed as a linear combination of α and β. In other words, there are numbers σ and τ such that

$$\Gamma_{\text{right}} = \sigma\alpha + \tau\beta$$

The analogous identity for the left GCD is nearly the same:

$$\Gamma_{\text{left}} = \alpha\sigma + \beta\tau \, .$$

Bézout's identity can be used to solve Diophantine equations. For instance, one of the standard proofs of Lagrange's four-square theorem, that every positive integer can be represented as a sum of four squares, is based on quaternion GCDs in this way.

Greatest Common Divisor

In mathematics, the greatest common divisor (gcd) of two or more integers, when at least one of them is not zero, is the largest positive integer that divides the numbers without a remainder. For example, the GCD of 8 and 12 is 4.

The greatest common divisor is also known as the greatest common factor (gcf), highest common factor (hcf), greatest common measure (gcm), or highest common divisor.

This notion can be extended to polynomials and other commutative rings (see below).

Overview

Notation

In this article we will denote the greatest common divisor of two integers a and b as $\gcd(a,b)$. Some textbooks use (a,b).

Reducing Fractions

The greatest common divisor is useful for reducing fractions to be in lowest terms. For example, $\gcd(42, 56) = 14$, therefore,

$$\frac{42}{56} = \frac{3 \cdot 14}{4 \cdot 14} = \frac{3}{4}.$$

Coprime Numbers

Two numbers are called *relatively prime*, or *coprime*, if their greatest common divisor equals 1. For example, 9 and 28 are relatively prime.

A Geometric View

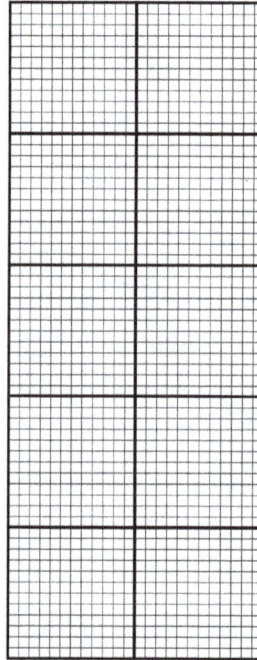

A 24-by-60 rectangle is covered with ten 12-by-12 square tiles, where 12 is the GCD of 24 and 60. More generally, an a-by-b rectangle can be covered with square tiles of side length c only if c is a common divisor of a and b.

For example, a 24-by-60 rectangular area can be divided into a grid of: 1-by-1 squares, 2-by-2 squares, 3-by-3 squares, 4-by-4 squares, 6-by-6 squares or 12-by-12 squares. Therefore, 12 is the greatest common divisor of 24 and 60. A 24-by-60 rectangular area can be divided into a grid of 12-by-12 squares, with two squares along one edge (24/12 = 2) and five squares along the other (60/12 = 5).

Calculation

Using Prime Factorizations

Greatest common divisors can in principle be computed by determining the prime factorizations of the two numbers and comparing factors, as in the following example: to compute gcd(18, 84), we find the prime factorizations $18 = 2 \cdot 3^2$ and $84 = 2^2 \cdot 3 \cdot 7$ and notice that the "overlap" of the two expressions is $2 \cdot 3$; so gcd(18, 84) = 6. In practice, this method is only feasible for small numbers; computing prime factorizations in general takes far too long.

Here is another concrete example, illustrated by a Venn diagram. Suppose it is desired to find the greatest common divisor of 48 and 180. First, find the prime factorizations of the two numbers:

$$48 = 2 \times 2 \times 2 \times 2 \times 3,$$

$$180 = 2 \times 2 \times 3 \times 3 \times 5.$$

What they share in common is two "2"s and a "3":

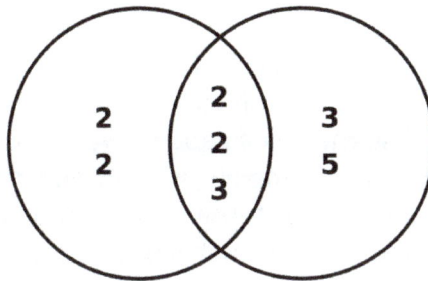

Least common multiple = $2 \times 2 \times (2 \times 2 \times 3) \times 3 \times 5 = 720$
Greatest common divisor = $2 \times 2 \times 3 = 12$.

Using Euclid's Algorithm

A much more efficient method is the Euclidean algorithm, which uses a division algorithm such as long division in combination with the observation that the gcd of two numbers also divides their difference. To compute gcd(48,18), divide 48 by 18 to get a quotient of 2 and a remainder of 12. Then divide 18 by 12 to get a quotient of 1 and a remainder of 6. Then divide 12 by 6 to get a remainder of 0, which means that 6 is the gcd. Note that we ignored the quotient in each step except to notice when the remainder reached 0, signalling that we had arrived at the answer. Formally the algorithm can be described as:

$$\gcd(a,0) = a$$

$$\gcd(a,b) = \gcd(b, a \bmod b),,$$

where

$$a \bmod b = a - b \left\lfloor \frac{a}{b} \right\rfloor..$$

If the arguments are both greater than zero then the algorithm can be written in more elementary terms as follows:

$$\gcd(a,a)=a,,$$

$$\gcd(a,b)=\gcd(a-b,b) \quad \text{, if } a>b$$

$$\gcd(a,b)=\gcd(a,b-a) \quad \text{, if } b>a$$

Animation showing an application of the Euclidean Algorithm to find the Great Common Divisor of 62 and 36 which is 2.

Complexity of Euclidean Method

The existence of the Euclidean algorithm places (the decision problem version of) the greatest common divisor problem in P, the class of problems solvable in polynomial time. The GCD problem is not known to be in NC, and so there is no known way to parallelize its computation across many processors; nor is it known to be P-complete, which would imply that it is unlikely to be possible to parallelize GCD computation. In this sense the GCD problem is analogous to e.g. the integer factorization problem, which has no known polynomial-time algorithm, but is not known to be NP-complete. Shallcross et al. showed that a related problem (EUGCD, determining the remainder sequence arising during the Euclidean algorithm) is NC-equivalent to the problem of integer linear programming with two variables; if either problem is in NC or is P-complete, the other is as well. Since NC contains NL, it is also unknown whether a space-efficient algorithm for computing the GCD exists, even for nondeterministic Turing machines.

Although the problem is not known to be in NC, parallel algorithms asymptotically faster than the Euclidean algorithm exist; the best known deterministic algorithm is by Chor and Goldreich, which (in the CRCW-PRAM model) can solve the problem in $O(n/\log n)$ time with $n^{1+\varepsilon}$ processors. Randomized algorithms can solve the problem in $O((\log n)^2)$ time on $\exp\left[O\left(\sqrt{n \log n}\right)\right]$ processors (note this is superpolynomial).

Binary Method

An alternative method of computing the gcd is the binary gcd method which uses only subtraction and division by 2. In outline the method is as follows: Let a and b be the two non negative integers. Also set the integer d to 0. There are five possibilities:

- $a = b$.

As $\gcd(a, a) = a$, the desired gcd is $a \times 2^d$ (as a and b are changed in the other cases, and d records the number of times that a and b have been both divided by 2 in the next step, the gcd of the initial pair is the product of a by 2^d).

- Both a and b are even.

In this case 2 is a common divisor. Divide both a and b by 2, increment d by 1 to record the number of times 2 is a common divisor and continue.

- a is even and b is odd.

In this case 2 is not a common divisor. Divide a by 2 and continue.

- a is odd and b is even.

As in the previous case 2 is not a common divisor. Divide b by 2 and continue.

- Both a and b are odd.

As $\gcd(a,b) = \gcd(b,a)$ and we have already considered the case $a = b$, we may assume that $a > b$. The number $c = a - b$ is smaller than a yet still positive. Any number that divides a and b must also divide c so every common divisor of a and b is also a common divisor of b and c. Similarly, $a = b + c$ and every common divisor of b and c is also a common divisor of a and b. So the two pairs (a, b) and (b, c) have the same common divisors, and thus $\gcd(a,b) = \gcd(b,c)$. Moreover, as a and b are both odd, c is even, and one may replace c by $c/2$ without changing the gcd. Thus the process can be continued with the pair (a, b) replaced by the smaller numbers $(c/2, b)$.

Each of the above steps reduces at least one of a and b towards 0 and so can only be repeated a finite number of times. Thus one must eventually reach the case $a = b$, which is the only stopping case. Then, as quoted above, the gcd is $a \times 2^d$.

This algorithm may easily programmed as follows:

```
Input: a, b positive integers

Output: g and d such that g is odd and gcd(a, b) = g×2^d

    d := 0

    while a and b are both even do

        a := a/2

        b := b/2

        d := d + 1

    while a ≠ b do

        if a is even then a := a/2

        else if b is even then b := b/2

        else if a > b then a := (a - b)/2

        else b := (b - a)/2
```

```
g := a

output g, d
```

Example: $(a, b, d) = (48, 18, 0) \rightarrow (24, 9, 1) \rightarrow (12, 9, 1) \rightarrow (6, 9, 1) \rightarrow (3, 9, 1) \rightarrow (3, 6, 1) \rightarrow (3, 3, 1)$; the original gcd is thus $2^d = 2^1$ times $a = b = 3$, that is 6.

The Binary GCD algorithm is particularly easy to implement on binary computers. The test for whether a number is divisible by two can be performed by testing the lowest bit in the number. Division by two can be achieved by shifting the input number by one bit. Each step of the algorithm makes at least one such shift. Subtracting two numbers smaller than a and b costs $O(\log a + \log b)$ bit operations. Each step makes at most one such subtraction. The total number of steps is at most the sum of the numbers of bits of a and b, hence the computational complexity is

$$O((\log a + \log b)^2)..$$

Properties

- Every common divisor of a and b is a divisor of $\gcd(a, b)$.

- $\gcd(a, b)$, where a and b are not both zero, may be defined alternatively and equivalently as the smallest positive integer d which can be written in the form $d = a \cdot p + b \cdot q$, where p and q are integers. This expression is called Bézout's identity. Numbers p and q like this can be computed with the extended Euclidean algorithm.

- $\gcd(a, 0) = |a|$, for $a \neq 0$, since any number is a divisor of 0, and the greatest divisor of a is $|a|$. This is usually used as the base case in the Euclidean algorithm.

- If a divides the product $b \cdot c$, and $\gcd(a, b) = d$, then a/d divides c.

- If m is a non-negative integer, then $\gcd(m \cdot a, m \cdot b) = m \cdot \gcd(a, b)$.

- If m is any integer, then $\gcd(a + m \cdot b, b) = \gcd(a, b)$.

- If m is a nonzero common divisor of a and b, then $\gcd(a/m, b/m) = \gcd(a, b)/m$.

- The gcd is a multiplicative function in the following sense: if a_1 and a_2 are relatively prime, then $\gcd(a_1 \cdot a_2, b) = \gcd(a_1, b) \cdot \gcd(a_2, b)$. In particular, recalling that gcd is a positive integer valued function (i.e., gets natural values only) we obtain that $\gcd(a, b \cdot c) = 1$ if and only if $\gcd(a, b) = 1$ and $\gcd(a, c) = 1$.

- The gcd is a commutative function: $\gcd(a, b) = \gcd(b, a)$.

- The gcd is an associative function: $\gcd(a, \gcd(b, c)) = \gcd(\gcd(a, b), c)$.

- The gcd of three numbers can be computed as $\gcd(a, b, c) = \gcd(\gcd(a, b), c)$, or in some different way by applying commutativity and associativity. This can be extended to any number of numbers.

- $\gcd(a, b)$ is closely related to the least common multiple $\text{lcm}(a, b)$: we have

$$\gcd(a, b) \cdot \text{lcm}(a, b) = a \cdot b.$$

This formula is often used to compute least common multiples: one first computes the gcd with Euclid's algorithm and then divides the product of the given numbers by their gcd.

- The following versions of distributivity hold true:

  gcd(a, lcm(b, c)) = lcm(gcd(a, b), gcd(a, c))

  lcm(a, gcd(b, c)) = gcd(lcm(a, b), lcm(a, c)).

- If we have the unique prime factorizations of $a = p_1^{e_1} p_2^{e_2} ... p_m^{e_m}$ and $b = p_1^{f_1} p_2^{f_2} ... p_m^{f_m}$ where $e_i \geq 0$ and $f_i \geq 0$, then the gcd of a and b is

  gcd(a,b) = $p_1^{\min(e_1,f_1)} p_2^{\min(e_2,f_2)} ... p_m^{\min(e_m,f_m)}$

- It is sometimes useful to define gcd(0, 0) = 0 and lcm(0, 0) = 0 because then the natural numbers become a complete distributive lattice with gcd as meet and lcm as join operation. This extension of the definition is also compatible with the generalization for commutative rings given below.

- In a Cartesian coordinate system, gcd(a, b) can be interpreted as the number of segments between points with integral coordinates on the straight line segment joining the points (0, 0) and (a, b).

Probabilities and Expected Value

In 1972, James E. Nymann showed that k integers, chosen independently and uniformly from $\{1,...,n\}$, are coprime with probability $1/\zeta(k)$ as n goes to infinity, where ζ refers to the Riemann zeta function. This result was extended in 1987 to show that the probability that k random integers have greatest common divisor d is $d^{-k}/\zeta(k)$.

Using this information, the expected value of the greatest common divisor function can be seen (informally) to not exist when $k = 2$. In this case the probability that the gcd equals d is $d^{-2}/\zeta(2)$, and since $\zeta(2) = \pi^2/6$ we have

$$E(2) = \sum_{d=1}^{\infty} d \frac{6}{\pi^2 d^2} = \frac{6}{\pi^2} \sum_{d=1}^{\infty} \frac{1}{d}.$$

This last summation is the harmonic series, which diverges. However, when $k \geq 3$, the expected value is well-defined, and by the above argument, it is

$$E(k) = \sum_{d=1}^{\infty} d^{1-k} \zeta(k)^{-1} = \frac{\zeta(k-1)}{\zeta(k)}.$$

For $k = 3$, this is approximately equal to 1.3684. For $k = 4$, it is approximately 1.1106.

The Gcd In Commutative Rings

The notion of greatest common divisor can more generally be defined for elements of an arbitrary commutative ring, although in general there need not exist one for every pair of elements.

If R is a commutative ring, and a and b are in R, then an element d of R is called a *common divisor* of a and b if it divides both a and b (that is, if there are elements x and y in R such that $d \cdot x = a$ and $d \cdot y = b$). If d is a common divisor of a and b, and every common divisor of a and b divides d, then d is called a *greatest common divisor* of a and b.

Note that with this definition, two elements a and b may very well have several greatest common divisors, or none at all. If R is an integral domain then any two gcd's of a and b must be associate elements, since by definition either one must divide the other; indeed if a gcd exists, any one of its associates is a gcd as well. Existence of a gcd is not assured in arbitrary integral domains. However if R is a unique factorization domain, then any two elements have a gcd, and more generally this is true in gcd domains. If R is a Euclidean domain in which euclidean division is given algorithmically (as is the case for instance when $R = F[X]$ where F is a field, or when R is the ring of Gaussian integers), then greatest common divisors can be computed using a form of the Euclidean algorithm based on the division procedure.

The following is an example of an integral domain with two elements that do not have a gcd:

$$R = \mathbb{Z}\left[\sqrt{-3}\right], \quad a = 4 = 2 \cdot 2 = \left(1 + \sqrt{-3}\right)\left(1 - \sqrt{-3}\right), \quad b = \left(1 + \sqrt{-3}\right) \cdot 2.$$

The elements 2 and $1 + \sqrt{-3}$ are two "maximal common divisors" (i.e. any common divisor which is a multiple of 2 is associated to 2, the same holds for $1 + \sqrt{-3}$, but they are not associated, so there is no greatest common divisor of a and b.

Corresponding to the Bézout property we may, in any commutative ring, consider the collection of elements of the form $pa + qb$, where p and q range over the ring. This is the ideal generated by a and b, and is denoted simply (a, b). In a ring all of whose ideals are principal (a principal ideal domain or PID), this ideal will be identical with the set of multiples of some ring element d; then this d is a greatest common divisor of a and b. But the ideal (a, b) can be useful even when there is no greatest common divisor of a and b. (Indeed, Ernst Kummer used this ideal as a replacement for a gcd in his treatment of Fermat's Last Theorem, although he envisioned it as the set of multiples of some hypothetical, or *ideal*, ring element d, whence the ring-theoretic term.)

Least Common Multiple

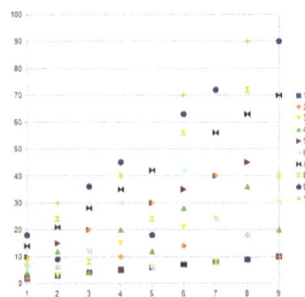

LCMs of numbers 1 through 10 with numbers 2 through 10. Line labels = first number. X axis = second number minus 1. Y axis = LCM of the two numbers.

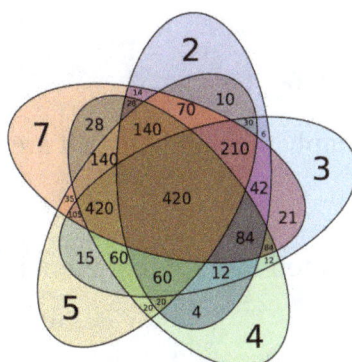

A Venn Diagram showing the least common multiples of combinations of 2, 3, 4, 5 and 7 (6 is skipped as it is 2 × 3, both of which are already represented).
For example, a card game which requires its cards to be divided equally among up to 5 players requires at least 60 cards, the number at the intersection of the 2, 3, 4 and 5 sets, but not the 7 set.

In arithmetic and number theory, the least common multiple, lowest common multiple, or smallest common multiple of two integers a and b, usually denoted by LCM(a, b), is the smallest positive integer that is divisible by both a and b. Since division of integers by zero is undefined, this definition has meaning only if a and b are both different from zero. However, some authors define lcm(a,0) as 0 for all a, which is the result of taking the lcm to be the least upper bound in the lattice of divisibility.

The LCM is familiar from elementary-school arithmetic as the "lowest common denominator" (LCD) that must be determined before fractions can be added, subtracted or compared. The LCM of more than two integers is also well-defined: it is the smallest positive integer that is divisible by each of them.

Overview

A multiple of a number is the product of that number and an integer. For example, 10 is a multiple of 5 because 5 × 2 = 10, so 10 is divisible by 5 and 2. Because 10 is the smallest positive integer that is divisible by both 5 and 2, it is the least common multiple of 5 and 2. By the same principle, 10 is the least common multiple of −5 and −2 as well.

Notation

In this article we will denote the least common multiple of two integers a and b as lcm(a, b).

Some older textbooks use $[a, b]$.

The programming language J uses a*.b

Example

What is the LCM of 4 and 6?

Multiples of 4 are:

4, 8, 12, 16, 20, 24, 28, 32, 36, 40, 44, 48, 52, 56, 60, 64, 68, 72, 76, ...

and the multiples of 6 are:

6, 12, 18, 24, 30, 36, 42, 48, 54, 60, 66, 72, ...

Common multiples of 4 and 6 are simply the numbers that are in both lists:

12, 24, 36, 48, 60, 72, ...

So, from this list of the first few common multiples of the numbers 4 and 6, their least common multiple is 12.

Applications

When adding, subtracting, or comparing vulgar fractions, it is useful to find the least common multiple of the denominators, often called the lowest common denominator, because each of the fractions can be expressed as a fraction with this denominator. For instance,

$$\frac{2}{21} + \frac{1}{6} = \frac{4}{42} + \frac{7}{42} = \frac{11}{42}$$

where the denominator 42 was used because it is the least common multiple of 21 and 6.

Computing The Least Common Multiple

Reduction by The Greatest Common Divisor

The following formula reduces the problem of computing the least common multiple to the problem of computing the greatest common divisor (GCD), also known as the greatest common factor:

$$\text{lcm}(a,b) = \frac{|a \cdot b|}{\gcd(a,b)}.$$

This formula is also valid when exactly one of a and b is 0, since $\gcd(a, 0) = |a|$. However, if both a and b are 0, this formula would cause division by zero; $\text{lcm}(0, 0) = 0$ is a special case.

There are fast algorithms for computing the GCD that do not require the numbers to be factored, such as the Euclidean algorithm. To return to the example above,

$$\text{lcm}(21, 6) = \frac{21 \cdot 6}{\gcd(21,6)} = \frac{21 \cdot 6}{\gcd(3,6)} = \frac{21 \cdot 6}{3} = \frac{126}{3} = 42.$$

Because $\gcd(a, b)$ is a divisor of both a and b, it is more efficient to compute the LCM by dividing *before* multiplying:

$$\text{lcm}(a,b) = \left(\frac{|a|}{\gcd(a,b)} \right) \cdot |b| = \left(\frac{|b|}{\gcd(a,b)} \right) \cdot |a|.$$

This reduces the size of one input for both the division and the multiplication, and reduces the required storage needed for intermediate results (overflow in the $a \times b$ computation). Because gcd(a,

b) is a divisor of both a and b, the division is guaranteed to yield an integer, so the intermediate result can be stored in an integer. Done this way, the previous example becomes:

$$\text{lcm}(21,6) = \frac{21}{\gcd(21,6)} \cdot 6 = \frac{21}{\gcd(3,6)} \cdot 6 = \frac{21}{3} \cdot 6 = 7 \cdot 6 = 42.$$

Finding Least Common Multiples by Prime Factorization

The unique factorization theorem says that every positive integer greater than 1 can be written in only one way as a product of prime numbers. The prime numbers can be considered as the atomic elements which, when combined together, make up a composite number.

For example:

$$90 = 2^1 \cdot 3^2 \cdot 5^1 = 2 \cdot 3 \cdot 3 \cdot 5.$$

Here we have the composite number 90 made up of one atom of the prime number 2, two atoms of the prime number 3 and one atom of the prime number 5.

This knowledge can be used to find the LCM of a set of numbers.

Example: Find the value of lcm(8,9,21).

First, factor each number and express it as a product of prime number powers.

$$8 = 2^3 \quad 9 = 3^2 \quad 21 = 3^1 \cdot 7^1$$

The lcm will be the product of multiplying the highest power of each prime number together. The highest power of the three prime numbers 2, 3, and 7 is 2^3, 3^2, and 7^1, respectively. Thus,

$$\text{lcm}(8,9,21) = 2^3 \cdot 3^2 \cdot 7^1 = 8 \cdot 9 \cdot 7 = 504.$$

This method is not as efficient as reducing to the greatest common divisor, since there is no known general efficient algorithm for integer factorization, but is useful for illustrating concepts.

This method can be illustrated using a Venn diagram as follows. Find the prime factorization of each of the two numbers. Put the prime factors into a Venn diagram with one circle for each of the two numbers, and *all* factors they share in common in the intersection. To find the LCM, just multiply all of the prime numbers in the diagram.

Here Is An Example:

48 = 2 × 2 × 2 × 2 × 3,

180 = 2 × 2 × 3 × 3 × 5,

and what they share in common is two "2"s and a "3":

Greatest common divisor = 2 × 2 × 3 = 12

This also works for the greatest common divisor (GCD), except that instead of multiplying all of the numbers in the Venn diagram, one multiplies only the prime factors that are in the intersection. Thus the GCD of 48 and 180 is $2 \times 2 \times 3 = 12$.

A Simple Algorithm

This method works as easily for finding the LCM of several integers.

Let there be a finite sequence of positive integers $X = (x_1, x_2, ..., x_n)$, $n > 1$. The algorithm proceeds in steps as follows: on each step m it examines and updates the sequence $X^{(m)} = (x_1^{(m)}, x_2^{(m)}, ..., x_n^{(m)})$, $X^{(1)} = X$, where $X^{(m)}$ is the mth iteration of X, i.e. X at step m of the algorithm, etc. The purpose of the examination is to pick the least (perhaps, one of many) element of the sequence $X^{(m)}$. Assuming $x_{k_0}^{(m)}$ is the selected element, the sequence $X^{(m+1)}$ is defined as

$$x_k^{(m+1)} = x_k^{(m)}, k \neq k_0$$
$$x_{k0}^{(m+1)} = x_{k0}^{(m)} + x_{k0}^{(1)}.$$

In other words, the least element is increased by the corresponding x whereas the rest of the elements pass from $X^{(m)}$ to $X^{(m+1)}$ unchanged.

The algorithm stops when all elements in sequence $X^{(m)}$ are equal. Their common value L is exactly LCM(X).

A Method Using A Table

This method works for any number of factors. One begins by listing all of the numbers vertically in a table (in this example 4, 7, 12, 21, and 42):

4

7

12

21

42

The process begins by dividing all of the factors by 2. If any of them divides evenly, write 2 at the top of the table and the result of division by 2 of each factor in the space to the right of each factor and below the 2. If a number does not divide evenly, just rewrite the number again. If 2 does not divide evenly into any of the numbers, try 3.

x	2
4	2
7	7
12	6
21	21
42	21

Now, check if 2 divides again:

x	2	2
4	2	1
7	7	7
12	6	3
21	21	21
42	21	21

Once 2 no longer divides, divide by 3. If 3 no longer divides, try 5 and 7. Keep going until all of the numbers have been reduced to 1.

x	2	2	3	7
4	2	1	1	1
7	7	7	7	1
12	6	3	1	1
21	21	21	7	1
42	21	21	7	1

Now, multiply the numbers on the top and you have the LCM. In this case, it is $2 \times 2 \times 3 \times 7 = 84$. You will get to the LCM the quickest if you use prime numbers and start from the lowest prime, 2.

As a general computational algorithm, the above is quite inefficient. One would never want to implement it in software: it takes too many steps, and requires too much storage space. A far more efficient numerical algorithm can be obtained simply by using Euclid's algorithm to compute the gcd first, and then obtaining the lcm by division.

The LCM in Commutative Rings

The least common multiple can be defined generally over commutative rings as follows: Let a and b be elements of a commutative ring R. A common multiple of a and b is an element m of R such that both a and b divide m (i.e. there exist elements x and y of R such that $ax = m$ and $by = m$). A least common multiple of a and b is a common multiple that is minimal in the sense that for any other common multiple n of a and b, m divides n.

In general, two elements in a commutative ring can have no least common multiple or more than one. However, any two least common multiples of the same pair of elements are associates. In a unique factorization domain, any two elements have a least common multiple. In a principal ideal domain, the least common multiple of a and b can be characterised as a generator of the intersection of the ideals generated by a and b (the intersection of a collection of ideals is always an ideal).

Fundamental Theorem of Arithmetic

In number theory, the fundamental theorem of arithmetic, also called the unique factorization theorem or the unique-prime-factorization theorem, states that every integer greater than 1 either is prime itself or is the product of prime numbers, and that this product is unique, up to the order of the factors. For example,

$1200 = 2^4 \times 3^1 \times 5^2 = 3 \times 2 \times 2 \times 2 \times 2 \times 5 \times 5 = 5 \times 2 \times 3 \times 2 \times 5 \times 2 \times 2 =$ etc.

The theorem is stating two things: first, that 1200 *can* be represented as a product of primes, and second, no matter how this is done, there will always be four 2s, one 3, two 5s, and no other primes in the product.

The requirement that the factors be prime is necessary: factorizations containing composite numbers may not be unique (e.g. $12 = 2 \times 6 = 3 \times 4$).

This theorem is one of the main reasons why 1 is not considered a prime number: if 1 were prime, the factorization would not be unique, as, for example, $2 = 2 \times 1 = 2 \times 1 \times 1 = \dots$

History

Book VII, propositions 30, 31 and 32, and Book IX, proposition 14 of Euclid's Elements are essentially the statement and proof of the fundamental theorem.

If two numbers by multiplying one another make some number, and any prime number measure the product, it will also measure one of the original numbers.

—Euclid, Elements Book VII, Proposition 30

Proposition 30 is referred to as Euclid's lemma. And it is the key in the proof of the fundamental theorem of arithmetic.

Any composite number is measured by some prime number.

—Euclid, Elements Book VII, Proposition 31

Proposition 31 is proved directly by infinite descent.

Any number either is prime or is measured by some prime number.

—Euclid, Elements Book VII, Proposition 32

Proposition 32 is derived from proposition 31, and prove that the decomposition is possible.

If a number be the least that is measured by prime numbers, it will not be measured by any other prime number except those originally measuring it.

—Euclid, Elements Book IX, Proposition 14

Book IX, proposition 14 is derived from Book VII, proposition 30, and prove partially that the decomposition is unique – a point critically noted by André Weil. Indeed, in this proposition the exponents are all equal to one, so nothing is said for the general case.

Article 16 of Gauss' *Disquisitiones Arithmeticae* is an early modern statement and proof employing modular arithmetic.

Applications

Canonical Representation of A Positive Integer

Every positive integer $n > 1$ can be represented in exactly one way as a product of prime powers:

$$n = p_1^{\alpha_1} p_2^{\alpha_2} \cdots p_k^{\alpha_k} = \prod_{i=1}^{k} p_i^{\alpha_i}$$

where $p_1 < p_2 < \ldots < p_k$ are primes and the α_i are positive integers. This representation is commonly extended to all positive integers, including one, by the convention that the empty product is equal to 1 (the empty product corresponds to $k = 0$).

This representation is called the canonical representation of n, or the standard form of n.

For example $999 = 3^3 \times 37$, $1000 = 2^3 \times 5^3$, $1001 = 7 \times 11 \times 13$

Note that factors $p^0 = 1$ may be inserted without changing the value of n (e.g. $1000 = 2^3 \times 3^0 \times 5^3$). In fact, any positive integer can be uniquely represented as an infinite product taken over all the positive prime numbers,

$$n = 2^{n_1} 3^{n_2} 5^{n_3} 7^{n_4} \cdots = \prod p_i^{n_i},$$

where a finite number of the n_i are positive integers, and the rest are zero. Allowing negative exponents provides a canonical form for positive rational numbers.

Arithmetic Operations

The canonical representation, when it is known, is convenient for easily computing products, gcd, and lcm:

$$a \cdot b = 2^{a_2 + b_2} \, 3^{a_3 + b_3} \, 5^{a_5 + b_5} \, 7^{a_7 + b_7} \cdots = \prod p_i^{a_{p_i} + b_{p_i}},$$

$$\gcd(a, b) = 2^{\min(a_2, b_2)} \, 3^{\min(a_3, b_3)} \, 5^{\min(a_5, b_5)} \, 7^{\min(a_7, b_7)} \cdots = \prod p_i^{\min(a_{p_i}, b_{p_i})},$$

$$\mathrm{lcm}(a, b) = 2^{\max(a_2, b_2)} \, 3^{\max(a_3, b_3)} \, 5^{\max(a_5, b_5)} \, 7^{\max(a_7, b_7)} \cdots = \prod p_i^{\max(a_{p_i}, b_{p_i})}.$$

However, as Integer factorization of large integers is much harder than computing their product, gcd or lcm, these formulas have, in practice, a limited usage.

Arithmetical Functions

Many arithmetical functions are defined using the canonical representation. In particular, the values of additive and multiplicative functions are determined by their values on the powers of prime numbers.

Proof

The proof uses Euclid's lemma (*Elements* VII, 30): if a prime p divides the product of two natural numbers a and b, then p divides a or p divides b.

Existence

We need to show that every integer greater than 1 is either prime or a product of primes. For the base case, note that 2 is prime. By induction: assume true for all numbers between 1 and n. If n is prime, there is nothing more to prove. Otherwise, there are integers a and b, where $n = ab$ and $1 < a \le b < n$. By the induction hypothesis, $a = p_1p_2...p_j$ and $b = q_1q_2...q_k$ are products of primes. But then $n = ab = p_1p_2...p_jq_1q_2...q_k$ is a product of primes.

Uniqueness

Assume that $s > 1$ is the product of prime numbers in two different ways:

$$s = p_1p_2 \cdots p_m = q_1q_2 \cdots q_n.$$

We must show $m = n$ and that the q_j are a rearrangement of the p_i.

By Euclid's lemma, p_1 must divide one of the q_j; relabeling the q_j if necessary, say that p_1 divides q_1. But q_1 is prime, so its only divisors are itself and 1. Therefore, $p_1 = q_1$, so that

$$\frac{s}{p_1} = p_2 \cdots p_m = q_2 \cdots q_n.$$

Reasoning the same way, p_2 must equal one of the remaining q_j. Relabeling again if necessary, say $p_2 = q_2$. Then

$$\frac{s}{p_1p_2} = p_3 \cdots p_m$$
$$= q_3 \cdots q_n.$$

This can be done for each of the m p_i's, showing that $m \le n$ and every p_i is a q_j. Applying the same argument with the p's and q's reversed shows $n \le m$ (hence $m = n$) and every q_j is a p_i.

Elementary Proof of Uniqueness

The fundamental theorem of arithmetic can also be proved without using Euclid's lemma, as follows:

Assume that $s > 1$ is the smallest positive integer which is the product of prime numbers in two different ways. If s were prime then it would factor uniquely as itself, so there must be at least two primes in each factorization of s:

$$s = p_1p_2 \cdots p_m = q_1q_2 \cdots q_n.$$

If any $p_i = q_j$ then, by cancellation, $s/p_i = s/q_j$ would be another positive integer, different from s, which is greater than 1 and also has two distinct factorizations. But s/p_i is smaller than s, meaning s would not actually be the smallest such integer. Therefore every p_i must be distinct from every q_j.

Without loss of generality, take $p_1 < q_1$ (if this is not already the case, switch the p and q designations.) Consider

$$t = (q_1 - p_1)(q_2 \cdots q_n),$$

and note that $1 < q_2 \leq t < s$. Therefore t must have a unique prime factorization. By rearrangement we see,

$$
\begin{aligned}
t \quad &= q_1(q_2 \cdots q_n) - p_1(q_2 \cdots q_n) \\
&= s - p_1(q_2 \cdots q_n) \\
&= p_1((p_2 \cdots p_m) - (q_2 \cdots q_n)).
\end{aligned}
$$

Here $u = ((p_2 \ldots p_m) - (q_2 \ldots q_n))$ is positive, for if it were negative or zero then so would be its product with p_1, but that product equals t which is positive. So u is either 1 or factors into primes. In either case, $t = p_1 u$ yields a prime factorization of t, which we know to be unique, so p_1 appears in the prime factorization of t.

If $(q_1 - p_1)$ equaled 1 then the prime factorization of t would be all q's, which would preclude p_1 from appearing. Thus $(q_1 - p_1)$ is not 1, but is positive, so it factors into primes: $(q_1 - p_1) = (r_1 \ldots r_h)$. This yields a prime factorization of

$$t = (r_1 \cdots r_h)(q_2 \cdots q_n),$$

which we know is unique. Now, p_1 appears in the prime factorization of t, and it is not equal to any q, so it must be one of the r's. That means p_1 is a factor of $(q_1 - p_1)$, so there exists a positive integer k such that $p_1 k = (q_1 - p_1)$, and therefore

$$p_1(k+1) = q_1.$$

But that means q_1 has a proper factorization, so it is not a prime number. This contradiction shows that s does not actually have two different prime factorizations. As a result, there is no smallest positive integer with multiple prime factorizations, hence all positive integers greater than 1 factor uniquely into primes.

References

- von zur Gathen, Joachim; Gerhard, Jürgen (1999), Modern Computer Algebra, Cambridge University Press, pp. 243–244, ISBN 978-0-521-64176-0.

- Lovász, L.; Pelikán, J.; Vesztergombi, K. (2003). Discrete Mathematics: Elementary and Beyond. New York: Springer-Verlag. pp. 100–101. ISBN 0-387-95584-4.

- Jones, A. (1994). "Greek mathematics to AD 300". Companion encyclopedia of the history and philosophy of the mathematical sciences. New York: Routledge. pp. 46–48. ISBN 0-415-09238-8.

- Fowler, D. H. (1987). The Mathematics of Plato's Academy: A New Reconstruction. Oxford: Oxford University Press. pp. 31–66. ISBN 0-19-853912-6.

- Roberts, J. (1977). Elementary Number Theory: A Problem Oriented Approach. Cambridge, MA: MIT Press. pp. 1–8. ISBN 0-262-68028-9.

- Koshy, T. (2002). Elementary Number Theory with Applications. Burlington, MA: Harcourt/Academic Press.

pp. 167–169. ISBN 0-12-421171-2.

- Moon, T. K. (2005). Error Correction Coding: Mathematical Methods and Algorithms. John Wiley and Sons. p. 266. ISBN 0-471-64800-0.

- Aho, A.; Hopcroft, J.; Ullman, J. (1974). The Design and Analysis of Computer Algorithms. New York: Addison–Wesley. pp. 300–310. ISBN 0-201-00029-6.

- Fuks, D. B.; Tabachnikov, Serge (2007). Mathematical Omnibus: Thirty Lectures on Classic Mathematics. American Mathematical Society. p. 13. ISBN 9780821843161.

- Darling, David (2004). "Khintchine's constant". The Universal Book of Mathematics: From Abracadabra to Zeno's Paradoxes. John Wiley & Sons. p. 175. ISBN 9780471667001.

- Dedekind, Richard (1996). Theory of Algebraic Integers. Cambridge Mathematical Library. Cambridge University Press. pp. 22–24. ISBN 9780521565189.

- Johnston, Bernard L.; Richman, Fred (1997). Numbers and Symmetry: An Introduction to Algebra. CRC Press. p. 44. ISBN 9780849303012.

- Lauritzen, Niels (2003). Concrete Abstract Algebra: From Numbers to Gröbner Bases. Cambridge University Press. p. 130. ISBN 9780521534109.

- LeVeque, W. J. (2002) [1956]. Topics in Number Theory, Volumes I and II. New York: Dover Publications. pp. II:57,81. ISBN 978-0-486-42539-9. Zbl 1009.11001.

- Hardy, G. H.; Wright, E. M. (1979), An Introduction to the Theory of Numbers (Fifth ed.), Oxford: Oxford University Press, ISBN 978-0-19-853171-5

- Gauss, Carl Friedrich; Clarke, Arthur A. (translator into English) (1986), Disquisitiones Arithemeticae (Second, corrected edition), New York: Springer, ISBN 978-0-387-96254-2

- Baker, Alan (1984), A Concise Introduction to the Theory of Numbers, Cambridge, UK: Cambridge University Press, ISBN 978-0-521-28654-1

- Riesel, Hans (1994), Prime Numbers and Computer Methods for Factorization (second edition), Boston: Birkhäuser, ISBN 0-8176-3743-5

- Weil, André (2007) [1984]. Number Theory: An Approach through History from Hammurapi to Legendre. Modern Birkhäuser Classics. Boston, MA: Birkhäuser. ISBN 978-0-817-64565-6.

Permissions

Index

www.ingramcontent.com/pod-product-compliance
Lightning Source LLC
Chambersburg PA
CBHW061317190326
41458CB00011B/3831